Proceedings in Adaptation, Learning and Optimization

Volume 5

About this Series

The role of adaptation, learning and optimization are becoming increasingly essential and intertwined. The capability of a system to adapt either through modification of its physiological structure or via some revalidation process of internal mechanisms that directly dictate the response or behavior is crucial in many real world applications. Optimization lies at the heart of most machine learning approaches while learning and optimization are two primary means to effect adaptation in various forms. They usually involve computational processes incorporated within the system that trigger parametric updating and knowledge or model enhancement, giving rise to progressive improvement. This book series serves as a channel to consolidate work related to topics linked to adaptation, learning and optimization in systems and structures. Topics covered under this series include:

- complex adaptive systems including evolutionary computation, memetic computing, swarm intelligence, neural networks, fuzzy systems, tabu search, simulated annealing, etc.
- machine learning, data mining & mathematical programming
- hybridization of techniques that span across artificial intelligence and computational intelligence for synergistic alliance of strategies for problem-solving
- aspects of adaptation in robotics
- agent-based computing
- autonomic/pervasive computing
- dynamic optimization/learning in noisy and uncertain environment
- systemic alliance of stochastic and conventional search techniques
- all aspects of adaptations in man-machine systems.

This book series bridges the dichotomy of modern and conventional mathematical and heuristic/meta-heuristics approaches to bring about effective adaptation, learning and optimization. It propels the maxim that the old and the new can come together and be combined synergistically to scale new heights in problem-solving. To reach such a level, numerous research issues will emerge and researchers will find the book series a convenient medium to track the progresses made.

More information about this series at http://www.springer.com/series/13543

Kittichai Lavangnananda
Somnuk Phon-Amnuaisuk
Worrawat Engchuan · Jonathan H. Chan
Editors

Intelligent and Evolutionary Systems

The 19th Asia Pacific Symposium, IES 2015, Bangkok, Thailand, November 2015, Proceedings

Editors
Kittichai Lavangnananda
School of Information Technology
King Mongkut's University of Technology Thonburi
Bangkok
Thailand

Worrawat Engchuan
School of Information Technology
King Mongkut's University of Technology Thonburi
Bangkok
Thailand

Somnuk Phon-Amnuaisuk
School of Computing and Informatics
Institut Teknologi Brunei
Gadong
Brunei Darussalam

Jonathan H. Chan
School of Information Technology
King Mongkut's University of Technology Thonburi
Bangkok
Thailand

ISSN 2363-6084 ISSN 2363-6092 (electronic)
Proceedings in Adaptation, Learning and Optimization
ISBN 978-3-319-26999-3 ISBN 978-3-319-27000-5 (eBook)
DOI 10.1007/978-3-319-27000-5

Library of Congress Control Number: 2015954990

Springer Cham Heidelberg New York Dordrecht London

Printed on acid-free paper

Springer International Publishing AG Switzerland is part of Springer Science+Business Media (www.springer.com)

Preface

On behalf of the organizing committee, it is an honor and a great pleasure to welcome you all to the 19th Asia Pacific Symposium on Intelligent and Evolutionary Systems (IES 2015).

IES conference is an annual event, initiated back in 1997 in Canberra, Australia, and it has been held annually in the Asia-Pacific region ever since. As mentioned, IES 2015 this year is the latest and the 19th in the series. It is held from 22nd to 25th November, 2015. It aims to bring together researchers from countries of the Asian Pacific Rim, in the fields of intelligent systems and evolutionary computation, to exchange ideas, present recent results and discuss possible collaborations. Researchers beyond Asian Pacific Rim countries are also welcome and encouraged to participate. The theme for IES 2015 is "*Transforming Big Data into Knowledge and Technological Breakthroughs*".

The host organization for IES2015 is the School of Information Technology (SIT), King Mongkut's University of Technology Thonburi (KMUTT), and it is technically sponsored by the International Neural Network Society (INNS). IES2015 is collocated with three other conferences; namely, The 6th International Conference on Computational Systems-Biology and Bioinformatics 2015 (CSBio 2015), The 7th International Conference on Advances in Information Technology 2015 (IAIT 2015) and The 10th International Conference on e-Business 2015 (iNCEB 2015) as a major part of series of events to celebrate the SIT 20th anniversary and the KMUTT 55th anniversary.

IES 2015 attracted 73 total submissions altogether from 19 countries. The reviewing process was as rigorous as could be, without a target for number of acceptance. The acceptance of each submission was decided solely on its technical merit. In short, quality took priority over quantity. Most submissions were subjected to at least 3 reviews. Some submissions with rather unique and difficult topics managed to be assessed by at least 2 reviewers. The reviewing committee comprises multifarious group of experts from 20 countries. Finally, 34 submissions were deemed suitable for presentation and publication.

The organizing committee believes that with four collocated conferences, it will be a great forum for all participants from various fields to meet, discuss and exchange knowledge and experiences in a friendly non-competitive atmosphere. The organizing committee wishes to express gratitude and appreciation to members of the reviewing committee for their time and invaluable effort, without them IES 2015 will not materialize. IES 2015 is technically and academically stimulated by several keynote and plenary speakers. Their anticipated contributions are much appreciated and acknowledged. Smooth collaboration with Springers is gratefully acknowledged here. It is hoped that IES 2015 will be both professionally and personally a rewarding experience to all.

Finally, on behalf of the organizing committee, we would also like to take this opportunity to thank you all participants for taking an interest in IES 2015 and wish you a pleasant stay in Thailand.

September 2015

Kittichai Lavangnananda
Somnuk Phon-Amnuaisuk
Worrawat Engchuan
Jonathan H. Chan

Organization

Organizer

School of Information Technology (SIT), King Mongkut's University of Technology Thonburi (KMUTT), Bangkok, Thailand

Technical Sponsors

International Neural Network Society (INNS)
INNS Thailand Region Chapter
VeriGuide^{TM}

Conference Committee

Steering Committee

Akira Namatame (Chair)	National Defense Academy, Japan
Sung-Bae Cho (Co-chair)	Yonsei University, Korea
Hussein A. Abbass	University of New South Wales, Australia
Shu-Heng Chen	National Chengchi University, Taiwan
Meng-Hiot Lim	Nanyang Technological University, Singapore
Kazuhiro Ohkura	Hiroshima University, Japan

International Advisory Board

Jun Wang	The Chinese University of Hong Kong, China
Jun Zhang	Sun Yat-Sen University, China
Laszlo T. Koczy	Budapest University of Technology and Economics, Hungary
Pablo Moscato	University of Newcastle, Australia
Tribeni Prasad Banerjee	Dr. B.C. Roy Engineering College, India

General Chair

Jonathan H. Chan	Thailand

Program Co-chairs

Kittichai Lavangnananda	Thailand
Somnuk Phon-Amnuaisuk	Brunei Darussalam

Secretary

Worrawat Engchuan	Thailand

Special Sessions Chair

Lakhmi C. Jain	Australia

Special Sessions Organizers

Special Session on Smart Workspace (SoSWS)

Somnuk Phon-Amnuaisuk	Institut Teknologi Brunei, Brunei
Saiful Omar	Institut Teknologi Brunei, Brunei
Au Thien Wan	Institut Teknologi Brunei, Brunei

Abdollah Dehzangi	Griffith University, Australia
Abdelrahman osman Elfaki	Tabuk University, Kingdom of Saudi Arabia
Adham Atyabi	Salford University, United Kingdom
Palaniappan Ramaswamy	University of Kent, United Kingdom
Wida Susanty Binti Hj Suhaili	Institut Teknologi Brunei, Brunei
Yun-Li Lee	Sunway University, Malaysia

Special Session on Nature Inspired Creative Computing (SoNIC)

Ahmad Rafi Mohamed Eshaq	Multimedia University, Malaysia
Somnuk Phon-Amnuaisuk	Institut Teknologi Brunei, Brunei
Voon Nyuk Hiong	Institut Teknologi Brunei, Brunei
Chuan-Kang Ting	National Chung Cheng University, Taiwan
Deenina Salleh	Institut Teknologi Brunei, Brunei
Eduardo R. Miranda	Plymouth University, United Kingdom
Penousal Machado	University of Coimbra, Portugal
Simon Colton	Imperial College, London, United Kingdom
Siti Noorfatimah Binti Hj Awg Safar @ Hj Sapar	Institut Teknologi Brunei, Brunei

Special Session on Data Mining and its Applications (SoDA)

Kok-Chin Khor	Multimedia University, Malaysia
Keng-Hoong Ng	Multimedia University, Malaysia
Ten Ying Wah	University of Malaya, Malaysia
Ang Tan Fong	University of Malaya, Malaysia
Foo Lee Kien	Multimedia University, Malaysia
Lunda Chua Sook Ling	Multimedia University, Malaysia
Goh Hui Ngo	Multimedia University, Malaysia

Special Session on Computer and Deep Vision

Vembasaran Vaitheeswanran	University of Malaya, Malaysia
Jonathan Chan	King Mongkut's University of Technology Thonburi, Thailand
Weng Kin Lai	Tunku Abdul Rahman University College, Malaysia
Chee Seng Chan	University of Malaya, Malaysia

Publicity Co-chairs

Chan Chee Seng	Malaysia
Praisan Pradungweang	Thailand
Aki-Hiro Sato	Japan
Kitsuchart Pasupa	Thailand

Local Organizing Chair

Vithida Chongsuphajaisiddhi	Thailand

International Program Committee

Giovanni Acampora, UK
Martin Allen, USA
Adham Atyabi, UK
Pascal Bouvry, Luxemburg
Erik Cambria, Singapore
Chee Seng Chan, Malaysia
Jonathan H. Chan, Thailand
Ying-Ping Chen, Taiwan
Sung-Bae Cho, Korea
Sook Ling Chua, Malaysia
Gregoire Danoy, Luxemburg
Kusum Deep, India
Abdollah Dehzangi, Australia
Bo Du, China
Abdelrahman Elfaki, Saudi Arabia
Saber Elsayed, Australia
Worrawat Engchuan, Thailand
Simon Fong, Macau
Lee Kien Foo, Malaysia
Hui-Ngo Goh, Malaysia
Wenyin Gong, China
Frédéric Guinand, France
Mohamed Saleem Haja Nazmudeen, Brunei Darussalam
Hisashi Handa, Japan
Alfredo G. Hernández-Díaz, Spain
Hisao Ishibuchi, Japan
Mohamad Izani, Malaysia
Saori Iwanaga, Japan
Yasushi Kambayashi, Japan
Yoshitaka Kameya, Japan
Hiroshi Kawakami, Japan
Nittaya Kerdprasop, Thailand
Kok Chin Khor, Malaysia
Kyung-Joong Kim, Korea
Laszlo T. Koczy, Hungary
Werasak Kurutach, Thailand
Kittichai Lavangnananda, Thailand
Yunli Lee, Malaysia
C P Lim, Malaysia
Stefan Menzel, Japan
Pornchai Mongkolnam, Thailand
Amir Nakib, France
Ponrudee Netisopakul, Thailand
Keng Hoong Ng, Malaysia
Praisan Padungweang, Thailand
Natapon Pantuwong, Thailand
Kitsuchart Pasupa, Thailand
Preecha Patumcharoenpol, Thailand
Somnuk Phon-Amnuaisuk, Brunei Darussalam
Kai Qin, Australia
Ahmad Rafi, Malaysia
Palaniappan Ramaswamy, UK

Chotirat Ann Ratanamahatana, Thailand
Tapabrata Ray, Australia
Athikom Roeksabutr, Thailand
Roberto Santana, Spain
Ruhul Sarker, Australia
Aki-Hiro Sato, Japan
Hiroshi Sato, Japan
John See, Malaysia
Udom Silparcha, USA
Thomas Stützle, Belgium
Simon Su, USA
Wallapak Tavanapong, USA
Vembarasan Vaitheeswaran, Malaysia
Vajirasak Vanijja, Thailand
Yuping Wang, China
Bunthit Watanapa, Thailand
Saowaluk Watanapa, Thailand
Kuntpong Woraratpanya, Thailand
Hong Qu, China

Contents

Part I
Agents and Complex Systems

An Experimental Analysis of a Robust Pheromone-Based Algorithm for the Patrolling Problem

Shigeo Doi

Abstract Recently, the necessity to resolve the patrolling problem has become pressing. This problem is modeled using an undirected graph structure in which one or more agents patrol the graph and regularly visit each node with the shortest time interval possible. Some central controlled algorithms have been proposed to solve this problem. However, the reliability of these algorithms, which depends on the central controller and communication between the controller and each agent, is considered insufficient. Thus, algorithms with a central controller are not applicable to critical environments. As an alternative approach, some autonomous and distributed algorithms have been proposed to achieve higher reliability and robustness. In a previous paper, we proposed an autonomous and distributed algorithm, called pheromone- and inverse-degree-based Probabilistic Vertex-Ant-Walk (pidPVAW). pidPVAW uses a pheromone model corresponding to fixed points for agent communication and cooperative patrolling as an extension of pheromone-based PVAW (pPVAW). In this paper, we introduce a new parameter k to control the effect of the degree of the neighbor nodes on the agent decision to move. When $k = 0$, pidPVAW behaves like pPVAW; therefore, pidPVAW includes pPVAW. The parameter k controls how easily nodes with lower connectivity can be visited. We ran some computer simulations for the parameter k on square grid graphs and scale-free graphs, and showed its effect on the system.

Keywords Patrolling problem · Ant colony system · Multi-agent systems

1 Introduction

When security officers patrol a monitored area at midnight, as part of their tasks, they usually visit certain fixed locations at regular intervals and check whether intruders

S. Doi(✉)
National Institute of Technology, Tomakomai College, 443 Aza-nishikioka, Tomakomai, Hokkaido 059-1275, Japan
e-mail: doi@jo.tomakomai-ct.ac.jp

K. Lavangnananda et al. (eds.), *Intelligent and Evolutionary Systems*,
Proceedings in Adaptation, Learning and Optimization 5,
DOI: 10.1007/978-3-319-27000-5_1

have entered the area. However, during their duties, the officers can be potentially ambushed by the intruders, resulting in serious incidents that may pose a risk on their safety.

Consequently, security-patrolling robots have been recently developed and partially used to substitute human security officers. However, as their purpose is to check for emergencies or security incidents by travelling along a fixed path, they are incapable to modify their behavior according to conditions. Ideally, the robots should adapt their actions to different situations. The ability to change behavior is thus of fundamental importance. These robots should be autonomous and able to modify their actions independently.

In addition, the introduction of a central control system to control the robots may lead to low reliability, as the robots would then have to communicate with each other through the central control system. Therefore, if the communication channels between the system and the robots were lost, they could not work properly and accomplish their tasks. Thus, the robots should be able to work effectively and communicate with each other to exchange information without a central controller.

In this paper, we design an autonomous, robust, and distributed algorithm to deal with environments that utilize security robots instead of security officers, using pheromone information corresponding to fixed points that the robots should visit at regular intervals in the monitored area.

In other words, this approach uses pheromone information instead of a direct communication channel between the robots. In addition, we also consider the ease of visiting such fixed points in the monitored area by examining the degree of the neighbor nodes. The monitored area is expressed as an undirected graph. A node corresponds to a fixed point that the robots should visit at regular intervals. We use the degree of each node to represent the ease of visiting that particular node. For example, a node with low connectivity, such as a corner node in the grid environment, cannot be easily visited by the robots. We developed a pheromone- and inverse-degree-based Probabilistic Vertex-Ant-Walk (pidPVAW) algorithm [1]. This feature is effective for graphs that have a biased degree distribution. A comparison of pidPVAW and pheromone-based PVAW (pPVAW) using computer simulations shows that the performance of pidPVAW is superior to that of pPVAW on scale-free graphs [2]. In this paper, we have also evaluated the grid network and focused on the nodes located at the corners of the grid.

2 Problem Definition and Related Work

First, let us define the patrolling problem. Let $G = (V, E)$ be an undirected graph, where V is a set of nodes and E is a set of links. The nodes represent the places that agents should visit at regular intervals. If $(i, j) \in E$, an agent is able to move on the link from i to j using one unit of simulation time. A patrolling problem is defined as the problem of enabling the agents to visit all the nodes in the shortest intervals possible during the simulation time. In general, as the number of agents increases, the visiting interval for all the nodes should decrease.

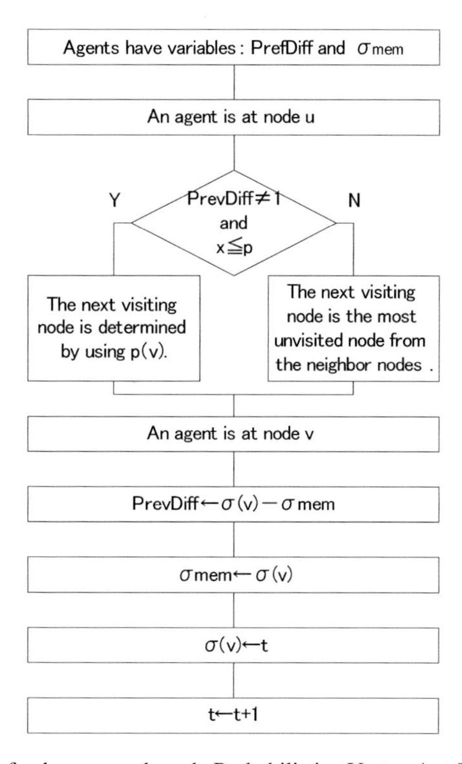

Fig. 1 Flowchart of pheromone-based Probabilistic Vertex-Ant-Walk (pPVAW) and pheromone- and inverse-degree-based PVAW (pidPVAW). The difference between pidPVAW and pPVAW resides only in the equation used to calculate $p(v)$.

Architectures for multi-agent patrolling were discussed in [3]. Spanning-tree based algorithms were proposed in [4] and [5]. Theoretical analyses were proposed in [6] and [7]. In addition, various algorithms, such as graph partitioning [8], reinforcement learning [9], and the ant colony system [10], were proposed under static environments. Various approaches for the patrolling problem were discussed in [11].

Elor et al. proposed Probabilistic Vertex-Ant-Walk (PVAW) [12][13]. In PVAW, when an agent determines the next neighbor node to visit, it primarily selects the least visited neighbor node with probability $1 - p$. Otherwise, with probability p, it selects a neighbor node randomly. We developed an improved version of PVAW, called pPVAW [14]. The difference between pPVAW and PVAW is that a robot using pPVAW selects its next node in proportion to the quantity of pheromones with probability p , whereas a robot using PVAW selects its next node randomly. In this paper, we describe a algorithm, pidPVAW, which is based on pPVAW. To visit nodes with lower connectivity, the selection of the next node by a robot using pidPVAW is based on the inverse proportionality to the degree of the neighbor nodes, in addition to the quantity of pheromone.

2.1 pidPVAW

pidPVAW is an improved version of the pPVAW algorithm[14]. In this algorithm, we focus on the assumption that the agents may find difficult visiting lower degree nodes. In addition, we consider the degree of each neighbor node so that the lower degree nodes can be more easily visited using pidPVAW than using pPVAW. In a scale-free graph, the frequency of visiting nodes by the pidPVAW algorithm is higher than that of pPVAW [1] because the improvement is considered effective for scale-free networks.

The difference between pidPVAW and pPVAW is only in the equation of $p(v)$. Now, we assume that $D(w)$ is the degree of $w \in V$, and the agent is at the node w. The probability of the agent moving to $v \in N(w)$ is given as

$$p(v) = \frac{\frac{1}{D(v)^k}\left(\max_{n \in N(w)} \sigma(n) - \sigma(v)\right)}{\sum_{m \in N(w)} \left(\frac{1}{D(m)^k}\left(\max_{n \in N(w)} \sigma(n) - \sigma(m)\right)\right)} \tag{1}$$

The equation of $p(v)$ indicates that $p(v)$ is inversely proportional to the degree of the neighbor nodes, and k is a parameter. If k is high, the lower degree nodes are more likely to be visited.

In a grid environment, the behavior of an agent using pidPVAW is often similar to that of an agent using pPVAW. In most cases, when an agent tries to determine the next node to visit among the neighbor nodes, the degree of all the neighbor nodes is equal and, therefore, $p(v)$ obtained by pidPVAW is equal to $p(v)$ obtained by pPVAW. With the exception described above, when an agent located at an edge node in a grid environment determines the next node to visit, an edge node of the grid is more likely to be selected than the other nodes. In pidPVAW, we assume that an agent can acquire additional information, e.g., the degree of the neighbor nodes, when determining the next node; on the other hand, pPVAW does not require this piece of information. In this paper, we introduce the parameter k to the original pidPVAW algorithm[1]. If k is high, nodes that have less connectivity can be visited more easily by the agents. By introducing this parameter, pidPVAW can work identically to pPVAW. If the parameter k is set to 0, Eq. 1 does not consider the degree of the neighbor nodes; therefore, when $k = 0$, pidPVAW is identical to pPVAW.

3 Simulation

3.1 Overview

In this paper, we present two scenarios for the evaluation of the parameter k in pidPVAW. The purpose of the first scenario is to evaluate the adaptiveness of the parameter k in relation to the visit of corner nodes in grid environments. The purpose of the second scenario is to evaluate the performance on scale-free graphs, which

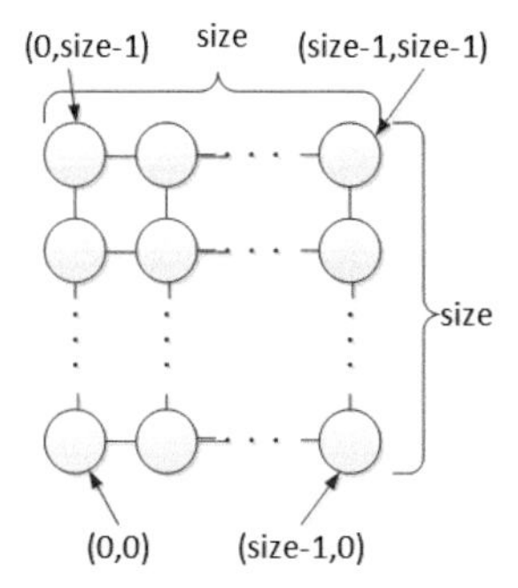

Fig. 2 Graph structure using the simulations in Scenario 1. The parameter $size$ was set to 10 or 100.

have a biased node degree of distribution. Some simulation parameter settings were identical for the first and the second scenarios; for example, the simulation time for each run was set to 1000000 units for both scenarios and the probability p was set to 0.1. Both Scenario 1 and Scenario 2 ran under a static environment. In other words, the graph structure did not change during the simulations.

3.2 Scenario 1: Static Square Grid Graph

The first scenario, Scenario 1, used the square grid graph shown in Fig. 2. Owing to their connectivity, four corner nodes of the grid are not easily visited by the agents, as shown in Fig. 2. In this scenario, we changed the parameter $size$, i.e., the length of edges of the square grid, to 10 or 100, and increased the number of agents from 2 to 10 with increments of 2.

3.3 Scenario 2: Static Scale-Free Graph

The second scenario, Scenario 2, used two scale-free graphs, whose characteristics are shown in Table 1. As this graph appears coarse, it was difficult for the agents in the graph to visit all the nodes. In this scenario, we used a number of agents 2 or 10.

3.4 Evaluation Criteria

We used the following values of s and m from Eqs.(2) and (3) as our evaluation criteria for the dynamic behavior for both scenarios. As t is the current time of each simulation, $t - \sigma(v)$ represents the elapsed time since the last visit by the agents.

In scenario 1, we also considered the number of times that the agents have visited the corner nodes in the square grids. If the number is high, it is easier for the agents to visit nodes with lower connectivity.

Table 1 Characteristics of two scale-free graphs

Graph	A	B
The Number of Nodes	10000	10000
Average Degree	2.000	2.000
Diameter	27	23
Radius	14	12
a of Approximate Eq. $y = ax^b$	3784.6	2508.8
b of Approximate Eq. $y = ax^b$	-2.158	-1.969
Correlation Coefficient R^2 of Approx. Eq.	0.900	0.876

Table 2 Average number of visits for each corner node obtained from Scenario 1. The number of agents was set to 2 and the parameter $size$ was set to 10.

$k\backslash$ coordinate	(0,0)	(9,0)	(0,9)	(9,9)	Average	Stdev.
0	19641.3	19652.7	19633.1	19644.7	19642.95	6.290786914
1	19291.0	19302.7	19289.0	19285.9	19292.15	5.685244058
2	19440.4	19435.9	19442.0	19438.0	19439.075	2.075933525
3	19430.6	19433.6	19436.5	19428.7	19432.35	2.652168924
4	19496.8	19494.8	19495.4	19497.7	19496.175	1.020539073

$$s = \sum_{v \in V} (t - \sigma(v)) \tag{2}$$

$$m = \max_{v \in V} (t - \sigma(v)) \tag{3}$$

Table 3 Average number of visits for each corner node obtained from Scenario 1. The number of agents was set to 10 and the parameter $size$ was set to 10.

$k\backslash$ coordinate	(0,0)	(9,0)	(0,9)	(9,9)	Average	Stdev.
0	83820.8	83731.7	83876.9	83777.8	83801.8	47.93958698
1	84412.0	84477.0	84393.2	84402.0	84421.05	29.49857624
2	84934.7	84913.3	85005.4	85000.4	84963.45	35.96300877
3	85492.4	85473.9	85501.6	85390.0	85464.475	39.48019124
4	85811.6	85818.1	85889.2	85873.5	85848.1	30.2212508

Table 4 Average number of visits for each corner node obtained from Scenario 1. The number of agents was set to 2 and the parameter $size$ was set to 100.

$k\backslash$ coordinate	(0,0)	(99,0)	(0,99)	(99,99)	Average	Stdev.
0	162.1	156.6	159.4	167.2	161.325	3.497070202
1	168.0	158.6	169.3	162.2	164.525	3.882975663
2	163.8	160.2	173.1	165.9	165.75	4.210700654
3	167.4	162.3	170.8	166.1	166.65	2.720661684
4	163.8	164.3	169.9	165.4	165.85	2.154530111

Table 5 Average number of visits for each corner node obtained from Scenario 1. The number of agents was set to 10 and the parameter $size$ was set to 100.

$k\backslash$ coordinate	(0,0)	(99,0)	(0,99)	(99,99)	Average	Stdev.
0	826.3	809.4	822.0	818.4	819.025	5.564305887
1	833.1	830.9	823.2	820.4	826.9	4.698510402
2	839.8	833.1	831.7	825.7	832.575	4.483469639
3	840.6	844.9	849.9	848.4	845.95	3.204059925
4	840.0	840.5	851.1	842.7	843.575	3.990676634

Table 6 Average number of visits for each corner node obtained from Scenario 1. The statistics are the average of the four corner nodes for each k and $agent$. The parameter $size$ was set to 10.

$k\backslash agents$	2	4	6	8	10
0	19643.0	33587.8	50269.7	66944.7	83801.8
1	19292.2	33892.3	50706.3	67496.4	84421.05
2	19439.1	34279.6	51045.5	67959.9	84963.45
3	19432.4	34340.0	51334.7	68360.2	85464.475
4	19496.2	34509.8	51582.4	68667.7	85848.1

4 Results

4.1 *Scenario 1: Static Square Grid Graph*

The average graphs obtained from 10 simulations for s and m are shown in Fig. 3 and Fig. 4. The statistical results obtained from 10 simulations are shown in Tables 2-7.

For Scenario 1, the average numbers of visits for each corner node in the square grid graphs obtained from 10 simulations are shown in Tables 2-7. Tables 2-5 show how many times the agents visited the four corner nodes in the square grid when the parameter k changed from 0 to 4 with an increment of 1. Table 6 and Table 7 show how many times the agents visited the four corner nodes when k changed from 0 to

Table 7 Average number of visits for each corner node obtained from Scenario 1. The statistics are the average of the four corner nodes for each k and $agent$. The parameter $size$ was set to 100.

$k \backslash agents$	2	4	6	8	10
0	161.3	324.8	490.7	651.5	819.0
1	164.5	331.9	493.3	660.2	826.9
2	165.8	332.1	496.1	667.2	832.6
3	166.7	334.8	499.5	672.0	846.0
4	165.9	336.8	503.1	679.0	843.6

4 with an increment of 1 and $agent$ changes from 2 to 10 with increments of 2. The average numbers of visits of all corner nodes in the square grid graphs obtained from 10 simulations are shown in Table 6 and Table 7. These results are the averages over 10 simulations for each setting.

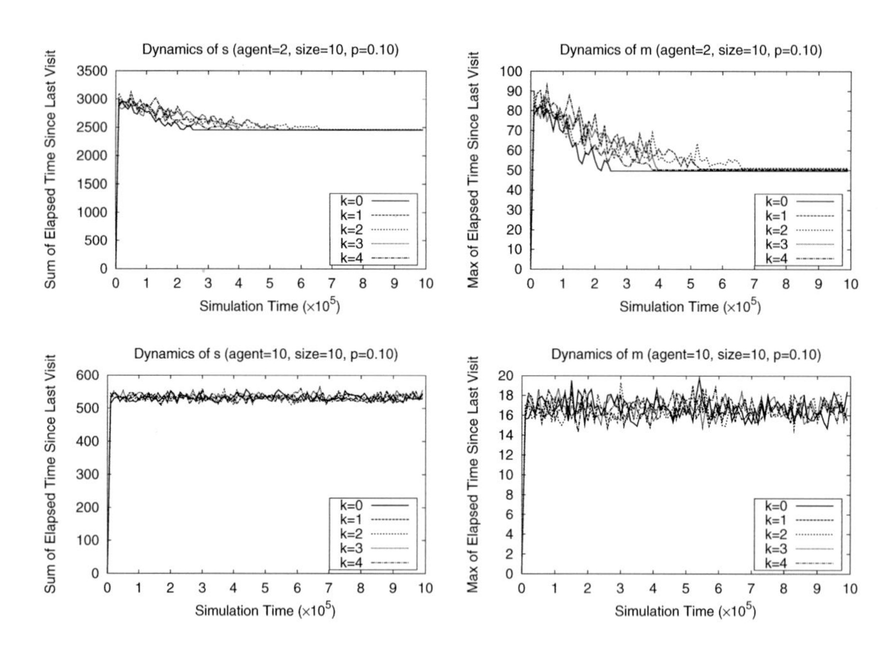

Fig. 3 Simulation result for Scenario 1 with the settings $size = 10$ and $p = 0.10$. In the two graphs in the top row, the setting $agent = 2$ was used, whereas, in the two graphs in the bottom row, the setting $agent = 10$ was used. For each row, the graph on the left shows s, the sum of the visiting intervals for all nodes, whereas the graph on the right shows m, the maximum visiting interval for all nodes.

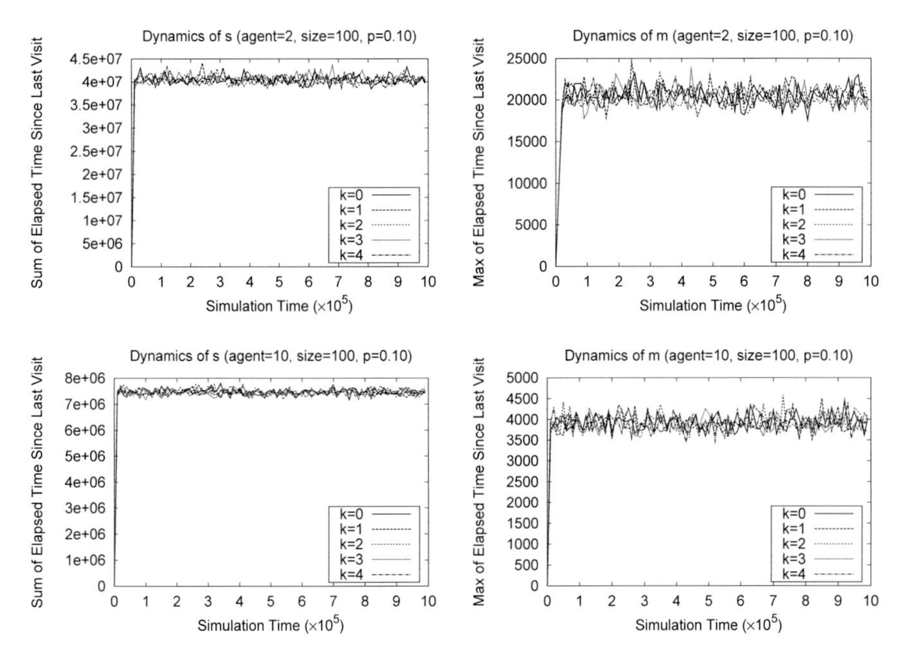

Fig. 4 Simulation result for Scenario 1 with the settings $size = 100$ and $p = 0.10$. In the two graphs in the top row, the setting $agent = 2$ was used, whereas, in the two graphs in the bottom row, the setting $agent = 10$ was used. For each row, the graph on the left shows s, the sum of the visiting intervals for all nodes, whereas the graph on the right shows m, the maximum visiting interval for all nodes.

As shown in Tables 2-7, if the number of agents is considered sufficient, an improvement in performance can be observed when the parameter k is high. Especially, there seems no significant variation in the case $agent = 2$ and $size = 10$ while k changes. When the number of agents is 10, the performance related to the visits of the corner nodes is superior for greater values of k. If the number of agents is higher, the difference will be larger.

The results s and m of pidPVAW obtained from Figs. 3 and 4 were similar for different values of the parameter k. The reason is that the distribution of degree of the square grid graphs used in Scenario 1 is almost constant. All the nodes, except those on the edges or corners, had four neighbors. Therefore, almost no difference could be observed between the results for different values of k . In the case $agent = 2$ and $size = 10$, Fig. 3 shows that m converged to a value of approximately 50 after that half of each simulation time had elapsed. In other words, the agents could walk on a fixed path while communicating to other agents their next node without global knowledge. This behavior may be effective for graph monitoring.

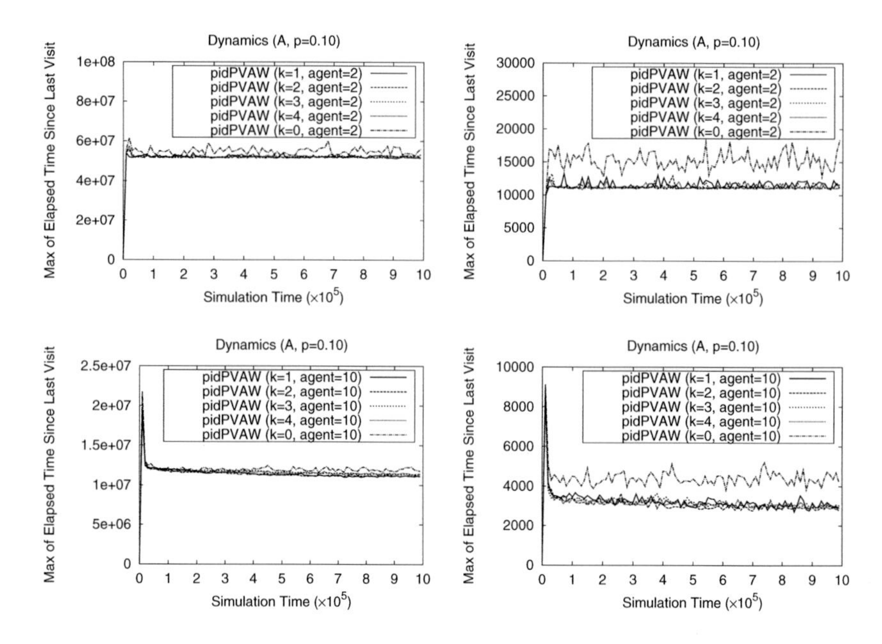

Fig. 5 Simulation result of graph A for Scenario 2. In the two graphs in the top row, the setting $agent = 2$ was used, whereas, in the two graphs in the bottom row, the setting $agent = 10$ was used. For each row, the graph on the left shows s, the sum of the visiting intervals for all nodes, and the graph on the right shows m, the maximum visiting interval for all nodes.

4.2 Scenario 2: Static Scale-Free Graph

In Scenario 2, the results in Fig. 5 show that the performance of pidPVAW, with $k > 0$ were superior to that of pidPVAW with $k = 0$. Furthermore there seems little difference among the results of $k > 0$. As shown in Fig. 6, the performance of pidPVAW with $k > 0$ was worse than that of pidPVAW with $k = 0$ when we used the setting of $agent = 10$ and graph B. If the number of agents in the environment is sufficient, the agents using pidPVAW with $k = 0$ can move to hub nodes of a scale-free graph, as opposed to the agents using pidPVAW with $k > 0$; therefore, the agents using pidPVAW with $k = 0$ can visit different lower connectivity more easily than the agents using pidPVAW with $k > 0$. In addition, in the case of $k = 0$, the measurement of m oscillated. However, for $k > 0$, the measurements of s and m mostly converged to a specific value quickly; therefore, the agents using pidPVAW with $k = 0$ can visit hub nodes more frequently than the agents using pidPVAW with $k > 0$. As shown in Figs. 5 and 6, little difference can be noticed between the algorithms with $k > 0$. Notably, the setting $k > 0$ showed similar performance to the others; therefore, $k = 1$ is considered to be the best setting because of computation costs. As a result, an algorithm considering the degree of the neighbor nodes can lead to superior performances in relation to the visit of square grid corner nodes.

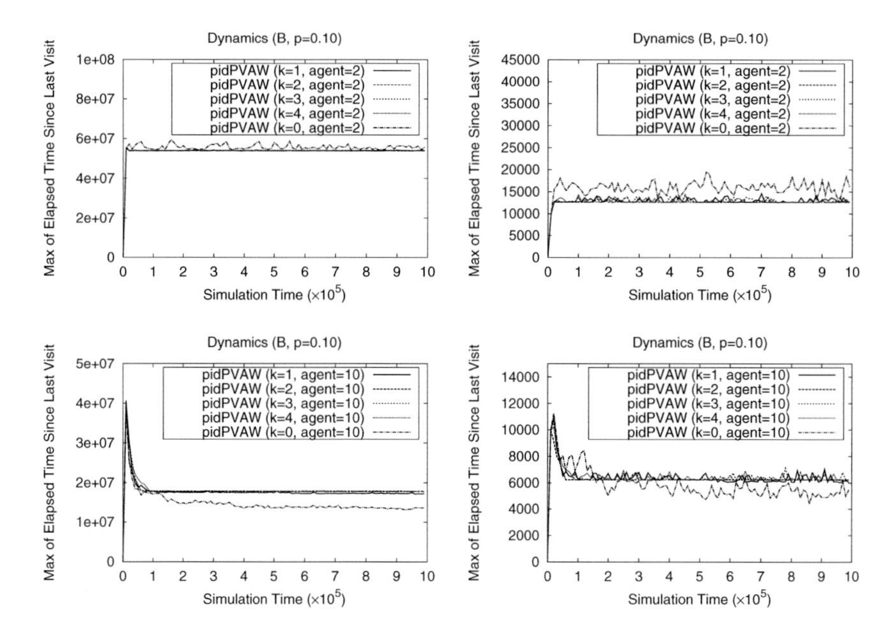

Fig. 6 Simulation result of graph B for Scenario 2. In the two graphs in the top row, the setting $agent = 2$ was used, whereas, in the two graphs in the bottom row, the setting $agent = 10$ was used. For each row, the graph on the left shows s, the sum of the visiting intervals for all nodes, and the graph on the right shows m, the maximum visiting interval for all nodes.

Furthermore, if the number of agents is insufficient, the algorithm can be adapted to scale-free networks. When pidPVAW with $k > 0$ is used an agent is capable to visit a lower degree node that has not been visited by any other agent for a long time. These results show that pidPVAW can be considered more robust to graph topology than pPVAW.

5 Conclusions and Future Work

In this paper, we introduced the parameter k to pidPVAW, which includes the degree of the neighbor nodes. We confirmed that a higher k can lead the agents to the four corner nodes in a square grid; besides, higher k showed better dynamics, when the number of agents is insufficient, in scale-free networks. Thus, the degree of the neighbor nodes is an important parameter, which, if perceived by an agent, should exploited for effective patrolling. Consequently, pidPVAW, which includes the degree of the neighbor nodes, is a highly stable algorithm for the solution of patrolling problems. In the future, we shall investigate further parameter settings and algorithm enhancements to improve pidPVAW performance.

Acknowledgments This work was supported by JSPS KAKENHI Grant Number 26870806.

References

1. Doi, S.: Proposal and evaluation of a robust pheromone-based algorithm for patrolling problem on various network structure. In: Proceedings of 18th Asia Pacific Symposium on Intelligent and Evolutionary Systems, IES 2014, vol. 1, pp. 397–408 (2014)
2. Barabási, A.L., Albert, R., Jeong, H.: Scale-free characteristics of random networks: the topology of the world-wide web. Physica A: Statistical Mechanics and its Applications **281**(1), 69–77 (2000)
3. Machado, A., Ramalho, G., Zucker, J.D., Drogoul, A.: Multi-agent patrolling: an empirical analysis of alternative architectures. In: Proceedings of Third International Workshop MABS 2002 (LNCS 2581), pp. 155–170 (2002)
4. Gabriely, Y., Rimon, E.: Spanning-tree based coverage of continuous areas by a mobile robot. Annals of Mathematics and Artificial Intelligence **31**(1–4), 77–98 (2001)
5. Elmaliach, Y., Agmon, N., Kaminka, G.A.: Multi-robot area patrol under frequency constraints. Annals of Mathematics and Artificial Intelligence **57**(3), 293–320 (2009)
6. Chevaleyre, Y.: Theoretical analysis of the multi-agent patrolling problem. In: Proceedings of IEEE International Conference on Intelligent Agent Technology, pp. 302–308 (2004)
7. Glad, A., Simonin, O., Buffet, O., Charpillet, F., et al.: Theoretical study of ant-based algorithms for multi-agent patrolling. In: 18th European Conference on Artificial Intelligence Including Prestigious Applications of Intelligent Systems (PAIS 2008), ECAI 2008, pp. 626–630 (2008)
8. Portugal, D., Rocha, R.: MSP algorithm: multi-robot patrolling based on territory allocation using balanced graph partitioning. In: Proceedings of ACM Symposium on Applied Computing, pp. 1271–1276 (2010)
9. Santana, H., Ramalho, G., Corruble, V., Corruble, V.: Multi-agent patrolling with reinforcement learning. In: Proceedings of the Third International Joint Conference on Autonomous Agents and Multiagent Systems, AAMAS 2004, vol. 3, pp. 1122–1129 (2004)
10. Lauri, F., Charpillet, F., et al.: Ant colony optimization applied to the multi-agent patrolling problem. In: IEEE Swarm Intelligence Symposium (2006)
11. Almeida, A., Ramalho, G., Santana, H., Tedesco, P., Menezes, T., Corruble, V., Chevaleyre, Y.: Multi-agent patrolling with reinforcement learning. In: Proceedings of 17th Brazilian Symposium on Artificial Intelligence, pp. 526–535 (2004)
12. Elor, Y., Bruckstein, A.M.: Autonomous multi-agent cycle based patrolling. Technion CIS Technical Reports CIS-2009-15 (2009)
13. Elor, Y., Bruckstein, A.M.: Autonomous multi-agent cycle based patrolling. In: Proceedings of 7th International Conference on Swarm Intelligence, LNCS 6234, pp. 119–130 (2010)
14. Doi, S.: Proposal and evaluation of a pheromone-based algorithm for the patrolling problem in dynamic environments. In: Proceedings of the 2013 IEEE Symposium on Swarm Intelligence, pp. 48–55. IEEE SIS 2013 (2013)

An Improved Evacuation Guidance System Based on Ant Colony Optimization

Asuka Ohta, Hirotaka Goto, Tomofumi Matsuzawa, Munehiro Takimoto, Yasushi Kambayashi and Masayuki Takeda

Abstract This paper proposes an evacuation guidance method for use in disaster situations. The method is based on ant colony optimization (ACO). We have implemented the method as ACO-based evacuation system in a simulator and examined the feasibility of the system. Since we cannot depend on the communication infrastructures with a disaster occurs, we make the system utilize mobile ad hoc network (MANET). We expect the ACO-based evacuation system produces quasi-optimized evacuation paths by the cooperation of multiple agents, while MANET provides communication between agents in the environment lacking of network infrastructure. Even though a number of ACO-based guidance systems have been developed, there are still some questions whether evacuees who follow the evacuation paths given by ACO are really safe. We examined how safe following these paths is by simulations, and found that they were not safe in some cases. As a result, in this paper, we propose an improved ACO-based evacuation system that equips deodorant pheromone to actively erase ACO pheromone traces when dangerous locations are found. Our simulation results show the use of deodorant pheromone can improve the safety level of the evacuation guidance system without degrading evacuation efficiency.

Keywords Ant Colony Optimization · Route guidance system · Swarm intelligence · Disaster simulation

A. Ohta(✉) · H. Goto · T. Matsuzawa · M. Takimoto · M. Takeda
Department of Information Sciences, Tokyo University of Science, Tokyo, Japan
e-mail: ask.n.use@gmail.com, t-matsu@is.noda.tus.ac.jp, mune@cs.is.noda.tus.ac.jp

Y. Kambayashi
Department of Computer and Information Engineering, Nippon Institute of Technology, Saitama, Japan
e-mail: yasushi@nit.ac.jp

K. Lavangnananda et al. (eds.), *Intelligent and Evolutionary Systems*,
Proceedings in Adaptation, Learning and Optimization 5,
DOI: 10.1007/978-3-319-27000-5_2

1 Introduction

Because disasters often cause enormous material damage and can result in numerous casualties, studies aimed at minimizing their effects have received significant attention. There are two types of casualties in disaster situations such as earthquakes. The first type casualties are incurred at the moment of the disaster, such as caused by buildings collapse. The second type casualties are caused by the fires that are often up in the aftermath of building breakdowns and people are trapped in hazardous locations. In order to avoid the second type of casualties, post disaster evacuations have to be conducted safely and quickly.

In this study, we consider a system that assists people evacuating from disaster situations, safely and quickly. The system receives information from evacuees and induces them to relocate to safe areas by using ant colony optimization (ACO) [1]. In the ACO method, numerous evacuation trails eventually converge into optimum paths that are used by evacuating people. The problem is that once a path is considered as the optimum path, numerous people will use it, but if the situation changes, the found safe path may become no longer safe. Therefore, in this paper, we conducted numerical experiments on the simulator to prove that the paths given by our approach took evacuees to safe areas without touching dangerous areas. Upon observing the result of the experiments, we propose an improved ACO-based evacuation system using deodorant pheromone to erase once-safe-but-currently-dangerous paths.

The structure of the balance of this paper is as follows. In Section 2, we provide an overview of the ACO method and discuss issues that need to be considered during an evacuation. In Section 3, we describe related works. Section 4 describes the system used for our simulation, while in Section 5 we provide details of experiments on the simulator and discuss the results. Since the results of the simulations show that ACO paths may be efficient but may be unsafe in some cases, we discuss how to extend our system to pursue safety. Section 6 describes this extension and the results of numerical experiments on the extended simulator. The results show that the extension makes the evacuation method safer with little overhead. Finally, Section 7 concludes our discussion.

2 Ant Colony Optimization (ACO)

ACO is an algorithm based on the behavior of ants in the natural world. In the beginning, ants move randomly in order to locate food. When food is found, they gather what they can and return to their nest while depositing volatile pheromone traces on the ground. These pheromone traces serve to guide other ants to the food source found by the first ants that succeeded to find food. Thus, any ants that encounter such a pheromone trace will follow it to reach the food and will also deposit pheromone traces on the path as they return to the nest.

While a number of paths to smaller or greater food sources may exist at first, the pheromone concentration on those paths will change as time passes. When many ants select and use a certain path, its pheromone concentration becomes deepen and the path starts to attract more ants. When fewer ants select a path, its pheromone attraction weakens. Furthermore, pheromone paths tend to converge into one or two because ants will naturally follow the strongest pheromone trail. In this manner, ACO will eventually derive a path that provides an optimum solution between the nest and the food source.

2.1 Base Algorithms

ACO derives a path between the start to end points as follows:

1. Ants begin to explore the path from start point to end point.
2. Ants move only a fixed distance each time (step).
3. When ants encounter selectable multiple solutions like branches, they conduct a local search based on pheromone traces and other information.
4. Ants add pheromone to each selected path, when they reach the end point. Ants then return to the start point to explore again.
5. Pheromone evaporates in constant rate as time passes (steps).
6. Continue to do this until sufficient time has elapsed.

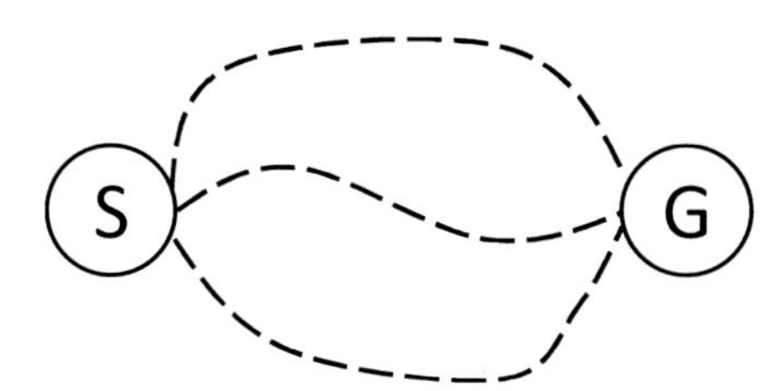

Fig. 1 Possible paths from S to G.

Fig. 1 shows a simple example. Ants begin to explore the path from the start point `S` to the end point `G`. Ants at `S` conduct a local search before selecting the top, middle, or bottom route. On the first trip, the ants select paths in a random manner because no information exists. However, since the middle route is the shortest, ants on the middle reach `G` earliest and add pheromone traces to that route. At this time, other ants at `S` start exploring and conducting local searches. Based on the newly available pheromone information, they are highly probable to select the middle route. Even though, ants select other routes too and other routes receive pheromone traces when ants reach `G`, they tend to receive less pheromone than the middle route.

Thus, when two or more possible routes exist, even though the top, middle, and bottom paths have equal probability of being selected by the first ants, ACO paths tend to converge into the shortest path. Due to the difference in path length, the pheromone traces on the middle route become stronger, while those of the top and bottom routes become weaker. After a sufficient amount of time has elapsed, the ants concentrate on the middle route, which is the optimum solution.

ACO can also dynamically update the optimum path if situations change. For example, should the pheromone traces on one of the optimum routes stop increasing because a part of path has become impassable, the route pheromone traces evaporates, and the ants begin looking for other routes in order to find a path to the goal. After a sufficient amount of time has elapsed, the ants will concentrate on the newly found optimum path.

2.2 Special Note Related to Evacuations

The purpose of our system is to guide evacuees to safe areas along the algorithm of ACO, but not to give the semi-shortest paths using ACO. In other words, we aim to control the evacuees as ants in ACO, and therefore, our research includes showing that this approach works well.

There are differences between the assumptions in evacuation conditions and those in ACO. In the example of Fig. 1, a single start point S and a single end point G exist. In contrast, evacuations often have numerous start points and a number of end points. Furthermore, evacuees those reach safe areas (end points) do not return to their start point, while ants in ACO shuttle repeatedly between start and end points.

For example, one of the related works [2] refers to the ant colony system [3], which is designed to solve the traveling salesman problem (TSP) [4]. In the TSP version, ants can begin exploring from any point because any solution has to eventually include all the points. However, unlike salesmen in the TSP, the evacuees are not required to pass certain points when they go to the safe area.

Since, as we have discussed above, there are significant differences between the assumptions of the evacuation and those of the traditional ACO, it is unclear whether the general optimization technique used with ACO is appropriate for evacuations.

3 Related Works

In the situations of disaster, evacuees must quickly move to safe areas, but oftentimes do not possess sufficient information to evacuate safely. At the minimum, they need information on the location of safe areas, available paths to those locations, and the safety level of those paths. However, safety level of a path can change at any moment during a disaster situation. If a path is unsafe, evacuees may be forced to use an alternative path or switch to another safe area. A number of studies have been conducted to determine how information is exchanged between evacuees or between evacuees and anti-disaster headquarters [5], [6], [7]. We do not consider such headquarters.

A map information sharing system [8] is one of such studies. In this system, some evacuees record road information while moving to shelters, and then exchange the information they collect with each other using a mobile ad hoc network (MANET). Since the situation can change from moment to moment in a disaster area, an ACO-based extension has been proposed to allow evacuees to respond to

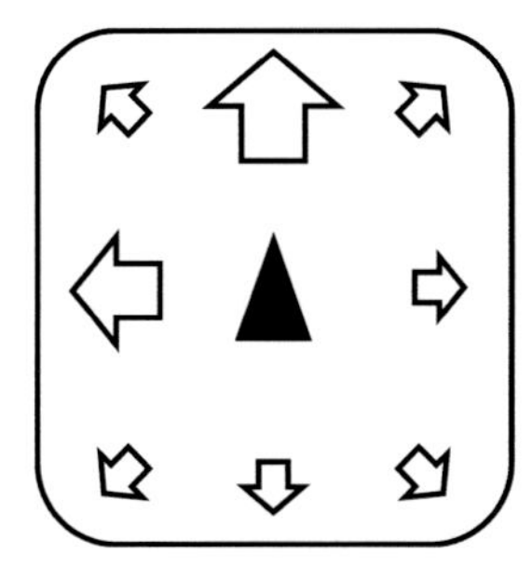

Fig. 2 Device display example. Recommendation degree is indicated by the size of the arrow. Forward and left directions are strongly recommended in this example, which indicate a pheromone path.

such changes [2], [9]. According to the results of agent-based simulations [10], it is possible for evacuees to quickly reach safe areas using such a system.

Another study about evacuation frameworks [11] also collects information from the evacuees, constructs evacuation route using ACO, and anticipates that evacuees will communicate with each other on MANET. Thus, we can conclude that disseminating evacuation guidance via ACO is a good idea, and both of the above papers have shown that it can be done efficiently. More specifically, quasi-optimal evacuation routes can be presented, and routes can be updated dynamically in response to the changes in the situation. Because dangerous routes are deemed suboptimal, the optimum route is assumed to be safe.

In order to propose improvements on ACO-based evacuation system, we examine these ACO features and investigate how effective they are in a general evacuation situation. The next section describes the simulation system we use to investigate a simple evacuation guidance scheme, and identify its shortcomings.

4 Simple Evacuation System Simulation

We have conducted the experiments on the simulated evacuation system as follows. First, we assume that all evacuees have mobile communication devices such as smartphones, and that they would exchange information with each other using those devices. Evacuees use such devices to send information to the system. For example, people can send information such as which paths they have taken, the coordinates of dangerous areas, and the location of safe areas. Each device receives information from the system and displays a recommendation degree for each direction. Fig. 2 shows an example. Evacuees then decide which direction to move based on the provided recommendation.

ACO-based systems are modeled on the behavior of natural ants, which deposit pheromone traces on the ground in order to exchange information. Other ants can perceive the strength of the pheromone traces on the ground. Recommendation degrees correspond to the strength of the pheromone. Thus, the pheromone levels of all coordinates are calculated by the system based on information received from the devices. The system is also equipped with an information storage. In this study, we use a simple configuration that the system and all the connected devices share a single map. The system server receives information from all the connected devices, performs pheromone level calculations, and then propagates responses to all the connected devices.

5 Evacuation Guidance Simulation

In this section, we report the results of simulated evacuations in order to determine the effectiveness of ACO as a method of providing evacuation guidance.

Fig. 3 Simulation start.

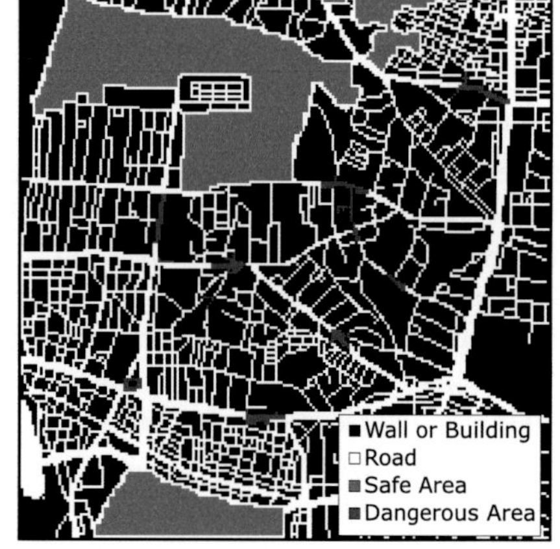

Fig. 4 Simulation end.

The map shown in Fig. 3 and Fig. 4, which shows the Shinjuku District of Tokyo, was used for this simulation. As can be seen in these figures, there are two safe areas on the map. These correspond to actual shelters. The Map uses a 200 $\times$ 200 grid. Each cell is marked with one of the two following attributes:

- Passable Area
  - Road
  - Safe Area
- Impassable Area
  - Wall or Building
  - Dangerous Area

In our simulation, evacuees are placed at random locations in passable areas on the map. They are then directed to move to one of the safe areas, only through passable areas, while using the ACO-based guidance system.

It should be noted here that the evacuees cannot see the map and must make decisions based on the attributes of their current coordinates. They can also learn pheromone trace information regarding the eight cells surrounding their location from the system, but have no information on the other attributes of those cells. Thus, it is inevitable that some will attempt to move into impassable areas.

In this simulation, each step proceeds as follows. Evacuees first obtain pheromone information on the eight cells surrounding them, and then move to one of those cells based on that pheromone information. When evacuees attempt to move to an impassable area, they are informed of the attributes of the destination cell, and must stay in their present location. That means they have failed to move during this step. Once evacuees are in a safe area, they have completed the evacuation process and no longer move. The system then updates pheromone information for all cells. More specifically, it adds pheromone traces to the cell in the evacuee's path if it includes information on an evacuee that has completed the evacuation, and reduces pheromone information for all other cells.

In this simulation, the dangerous area expands as time passes. The simulation begins in the map shown in Fig. 3. The dangerous area then expands when 50%, 65%, and 80% of all evacuees complete their evacuation. When 80% of the evacuees have successfully evacuated, the dangerous area are expanded as shown in the map of Fig. 4. Only roads can be added to the dangerous areas, and no other changes are permitted.

5.1 Pheromone Information Updates

The pheromone value update process proceeds as follows. The pheromone value $\tau_{ij}(t)$ denoted by time t and coordinates (i, j).

$$\tau_{ij}(t+1) = (1-\rho)\tau_{ij}(t) + \sum_{k \in G_t} \Delta\tau_{ij}^k \tag{1}$$

G_t is a set of evacuee that completed the evacuation at time t. $\tau_{ij}(t)$ is decreased by evaporation rate ρ in every step. When evacuee k has completed evacuation, $\tau_{ij}(t)$ is increased by $\Delta\tau_{ij}^k$.

$\Delta\tau_{ij}^k$ is denoted by α and T_k. α is the amount of added pheromone. T_k is a set of coordinates that evacuee k has passed. It includes the shortest paths for evacuation as well as all other available routes, because pheromone traces are used as a means of sharing paths that evacuees have used successfully. It is possible for evacuees to change directions to a shorter shared path, but that might increase the number of evacuees who cannot discover any pheromone traces because the total number of shared paths decrease.

$$\Delta\tau_{ij}^k = \begin{cases} \alpha & (i, j) \in T_k \\ 0 & otherwise \end{cases} \tag{2}$$

Pheromone trace values are limited by upper bound value τ_{max} and lower bound value τ_{min}.

$$0 < \tau_{min} \leq \tau_{ij}(t) \leq \tau_{max} \tag{3}$$

This simulation is based on the idea proposed by MAX-MIN Ant System (MMAS) [12]. The lower bound value τ_{min} leaves the open possibility that the evacuee will move to any coordinates that have minimum pheromone value. The upper bound value τ_{max} prevents to fall into a local optimum generated from the difference of an extreme pheromone value.

5.2 Moving Direction Decision

In each step, evacuees move to one of the eight surrounding cells from the current cell. In this system, evacuees check pheromone information and then use that information to determine which direction to move.

Hence, in this simulation, the pheromone value is treated as a movement probability. The probability $p_{xy}(t)$ that an evacuee k moves to coordinates $(x, y) \in X^k(t)$ at time t is as shown in equation (4). $X^k(t)$ is a coordinates set into which evacuee k can move.

$$p_{xy}(t) = \frac{\tau_{xy}(t)}{\sum_{(i,j)\in X^k(t)} \tau_{ij}(t)} \tag{4}$$

5.3 Results and Discussion

We have conducted numerical experiments on the evacuation simulation system using the parameters shown in Table 1. We have executed the other simulation in which the evacuees walked randomly (did not use pheromone information) for comparison purposes. Both simulations were repeated 500 times. We have recorded the average value for each observation in the simulations. The table in Fig. 5 shows the average results.

Table 1 Parameters

Parameter	Value
Number of evacuees	1000
Volatile rate of pheromone ρ	0.0005
Amount of pheromone adding α	1.0
Lower bound of pheromone τ_{min}	1.0
Upper bound of pheromone τ_{max}	30.0

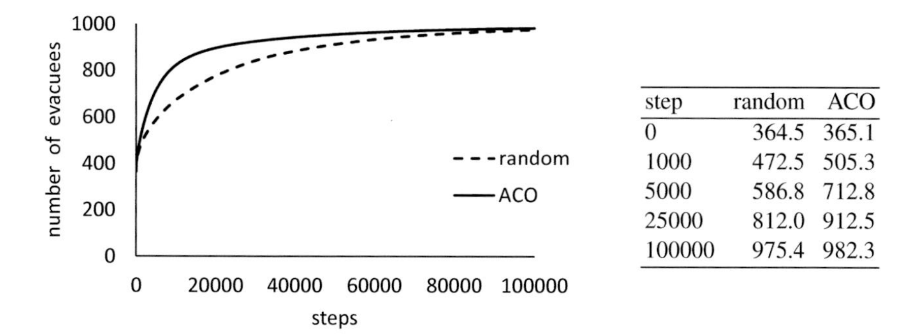

step	random	ACO
0	364.5	365.1
1000	472.5	505.3
5000	586.8	712.8
25000	812.0	912.5
100000	975.4	982.3

Fig. 5 Number of evacuees who have completed evacuation.

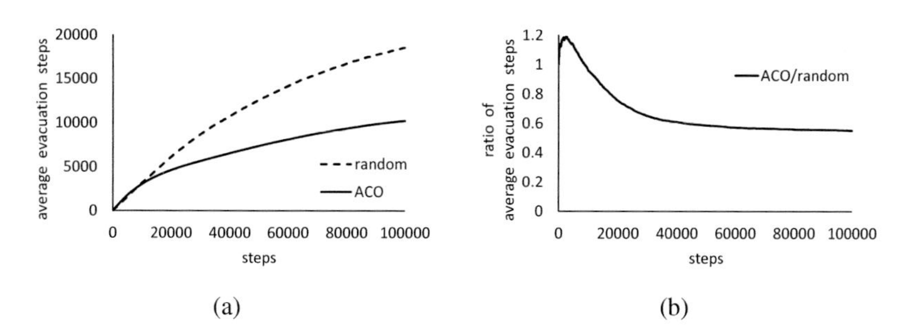

Fig. 6 (a): The horizontal axis shows the time transition, while the vertical axis shows the number of average steps needed for successful evacuation at that time. (b): Ratio of ACO and random walking.

First, we analyze the evacuation speed. Fig. 5 shows the relationship between the number of evacuees who have completed evacuation and the elapsed time. About 360 evacuees have completed evacuation at step 0, because they are placed at safe area from the beginning by the initial random placement.

In this simulation, the evacuees move until they reach safe areas. Therefore all of them eventually complete their evacuation even though they do not use the system. In the ACO, on the other hand, more evacuees complete their evacuation in the early stages. Fig. 6 shows that the number of steps necessary for evacuation has decreased by 40% in the ACO compared with that in the case of the random walk. Thus, we can observe that it is possible to evacuate more efficiently by using the system. Next, we will analyze the safety level of the evacuation.

We use the number of times that evacuees touched the dangerous area as the safety level index. Fig. 7 shows the results of the experiments. The vertical axis shows how many times all the evacuees touch the dangerous area. As we can see in the figure, more evacuees touch the dangerous area in the early stage when the ACO was used. This means that the use of ACO alone does not significantly improve evacuation safeness.

Even though we expect the ACO-based simulation is flexible enough to adapt the changes of situation and provide high safe level, based on the simulations discussed

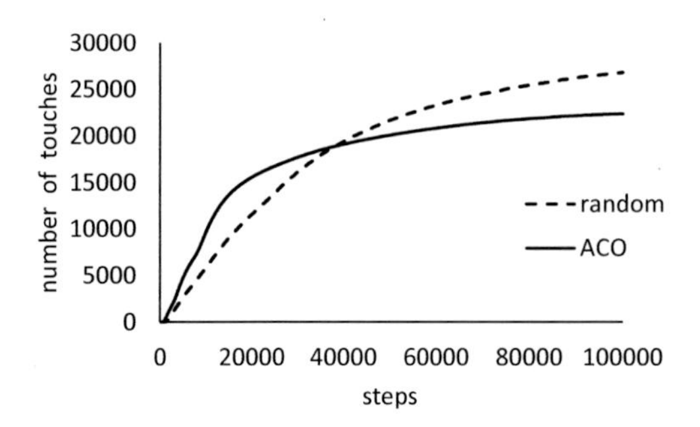

step	random	ACO
0	0.0	0.0
1000	7.2	131.6
5000	2555.4	4629.1
25000	13902.0	16764.1
100000	26814.2	22378.2

Fig. 7 Number of times that evacuees touched the dangerous area.

above, however, ACO does not provide sufficient levels of safety. Below, we will discuss this problem.

First, pheromone traces around dangerous area do not always decrease. People tend to use major roads even though it goes into a few parts of the dangerous area from time to time. In that case, no pheromone decrease would occur because sufficient numbers of evacuees pass over the road and deposit pheromone traces as usual. Since evacuees are moving adjacent to the dangerous area, they can easily touch it.

Next, the ACO changes reflecting to the situation changes occurs somewhat slowly because the evaporation of the pheromone of the old route is slow while they have to discover a new route. In such situations, evacuees would continue being attracted to the vicinity of the dangerous area until the pheromone traces evaporate completely.

One problem that is particularly important in evacuation guidance concerns the route switching speed. In ACO, ants shuttle back and forth between start and end points, while in an evacuation, evacuees stop when they reach the safe area. Therefore, the number of evacuees in an evacuation steadily decreases over time. In order to switch routes, it is necessary for the amount of pheromone on the new route to exceed that of the old route. However, because there are fewer evacuees to add pheromone traces onto a new route as the evacuation progresses, the process takes more time.

6 Extension for Avoiding Dangerous Area

At this point, it is clear that a more proactive mechanism for decreasing pheromone traces around dangerous areas is necessary. To accomplish this, we add new information that can erase pheromone traces into the system. Thus, when an evacuee reaches a dangerous area, new information is produced. We call this a deodorant pheromone.

6.1 Deodorant Pheromone

When evacuees attempt to move to an impassable area, they are checked and return to their previous location. At this time, the evacuees deposit deodorant pheromone

Table 2 Parameters

Parameter	Value
Number of evacuees	1000
Volatile rate of pheromone ρ	0.0005
Amount of pheromone adding α	1.0
Lower bound of pheromone τ_{min}	1.0
Upper bound of pheromone τ_{max}	30.0
Constant of deodorant pheromone τ'	-100
Deodorant rate σ	0.5
Influence range of deodorant pheromone N	2

to the dangerous area coordinates. When evacuees arrive at the safe area, normal pheromone information is sent to the system. The same process is used to send deodorant pheromone information to the system if evacuees have encountered a dangerous area.

Deodorant pheromone $\tau_{ij}(t)' < 0$ works by subtracting normal pheromone traces $\tau_{ij}(t)$. If a deodorant pheromone and an ACO pheromone are present at the same coordinates (i, j), the system treats the pheromone level of those coordinates as $\tau_{ij}(t)'' = \tau_{ij}(t) + \tau_{ij}(t)'$. Thus, when an evacuee finds deodorant pheromone, the normal pheromone $\tau_{ij}(t)$ of surrounding coordinates is decreased as follows:

$$\tau_{ij}(t+1) = (1 - \sigma^{n_{ij}^{k}(t)+1})\tau_{ij}(t) \tag{5}$$

σ is the deodorant rate, and $n_{ij}^{k}(t)$ is the distance at time t between the evacuee k and the coordinates (i, j). The influence of deodorant pheromone declines with distance. In this simulation, the distance $n_{ij}^{k}(t)$ is the number of steps required for k to move to (i, j), the influence range parameter N is used, and the simulation only updates the coordinates satisfying $n_{ij}^{k}(t) \leq N$.

6.2 Results and Discussion

We have conducted two different sets of evacuation simulation using our proposed ACO equipped with a deodorant pheromone and using a normal ACO that does not use deodorant pheromone, and compared the results. We use the parameters shown in Table 2. We have repeated these simulations for 500 times each, and then averaged the results.

In this simulation, deodorant pheromone is set at a constant τ' that does not evaporate. Fig. 8 shows the enhanced safety effect of our proposed ACO. We can observe, in the figure, that the number of times that evacuees touched the dangerous area is suppressed to a very small value by introducing the deodorant pheromone.

Next, we examined the overhead imposed by adding the deodorant pheromone. We found little differences between a normal ACO and our proposed deodorant-added

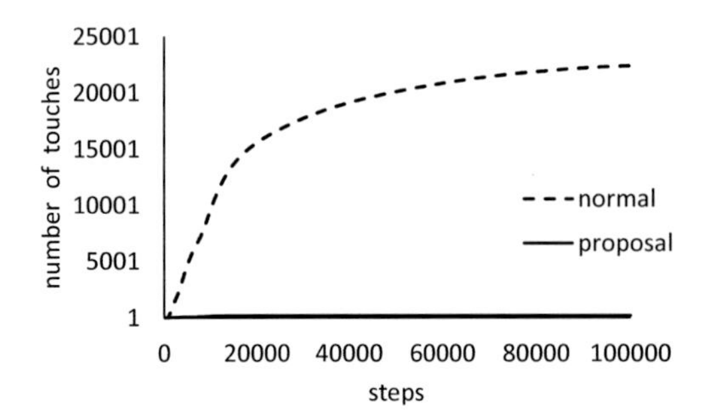

step	normal	proposal
0	0.0	0.0
1000	131.6	15.9
5000	4629.1	84.2
25000	16764.1	183.7
100000	22378.2	187.4

Fig. 8 Number of times evacuees touched the dangerous area.

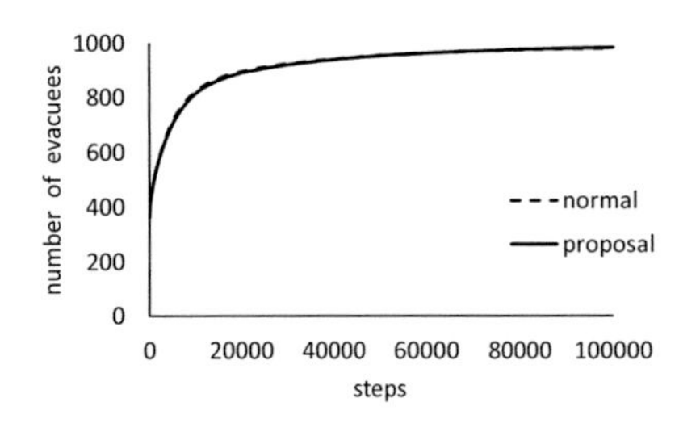

Fig. 9 Number of evacuees who completed evacuation.

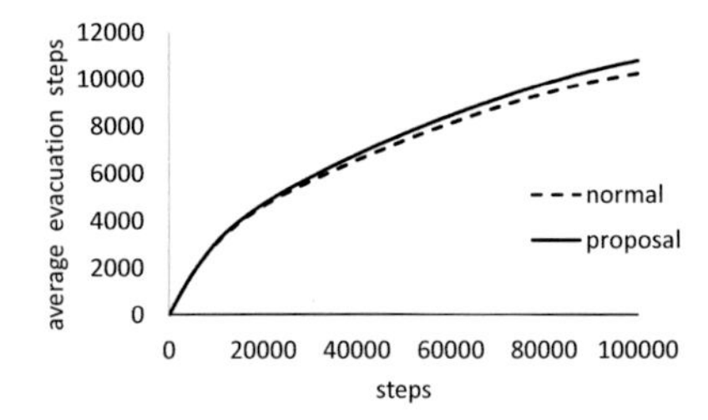

Fig. 10 Number of average steps required for a successful evacuation.

ACO in terms of efficiency, as we observe in Fig. 9 and Fig. 10. Thus, we can conclude that our proposed ACO with deodorant pheromone results in almost no increase in the number of steps required for a successful evacuation.

7 Conclusion

In this paper, we propose an evacuation guidance system based on ant colony optimization (ACO). Evacuees record the routes that they took and share the route information as pheromone deposits with other evacuees once they arrive in the safe area. Because ACO pheromone traces have inducing effects, this system can also induce evacuees to move to safe areas. Our initial simulations, however, show that while ACO guidance is efficient, it is not always safe. In order to increase its safety level, we propose an extension that uses deodorant pheromone as a new guidance mechanism. The results of the second set of simulations show that the number of evacuee induced into safe routes increases without degrading overall ACO efficiency. The reason should be the system that employs deodorant pheromone effectively reduces inducing pheromone traces around dangerous areas.

Acknowledgments This work is supported in part by Japan Society for Promotion of Science (JSPS), with the basic research program (C) (No. 25350482), Grant-in-Aid for Scientific Research.

References

1. Dorigo, M., Maniezzo, V., Colorni, A.: Ant system: Optimization by a colony of cooperating agents. IEEE Transaction on System **26**(1), 29–41 (1996)
2. Asakura, K., Fukaya, K., Watanabe, T.: A map construction system for disaster areas based on ant colony systems. In: 17th International Conference in Knowledge Based and Intelligent Information and Engineering Systems, pp. 494–501 (2013)
3. Dorigo, M., Gambardella, L.M.: Ant colony system: A cooperative learning approach for the traveling salesman problem. IEEE Transactions on Evolutionary Computation **1**, 53–66 (1997)
4. Dorigo, M., Gambardella, L.M.: Ant colonies for the traveling salesman problem. BioSystems **43**, 73–81 (1997)
5. Durresi, A., Paruchuri, V., Barolli, L.: Ad hoc communications for emergency conditions. In: 2011 IEEE International Conference on Advanced Information Networking and Applications (AINA), pp. 787–794 (2011)
6. Iizuka, Y., Iizuka, K.: Disaster evacuation assistance system based on multi-agent cooperation. In: 2015 48th Hawaii International Conference on System Sciences (HICSS), pp. 173–181 (2015)
7. Iizuka, Y., Kinoshita, K., Iizuka, K.: Multiagent approach for effective disaster evacuation. In: Proceedings of the 6th International Conference on Agents and Artificial Intelligence, pp. 223–228 (2014)
8. Asakura, K., Chiba, T., Watanabe, T.: A map information sharing system among refugees in disaster areas, on the basis of ad-hoc networks. In: The 3rd International Conference on Intelligent Decision Technologies, pp. 367–376 (2011)
9. Asakura, K., Watanabe, T.: Construction of navigational maps for evacuees in disaster areas based on ant colony systems. International Journal of Knowledge and Web Intelligence **4**(4), 300–313 (2013)
10. Koichi, A., Watanabe, T.: A movement algorithm for evacuee agents in disaster simulators: towards the development of evacuation guidance systems based on ant colony systems using MANET. In: Intelligent Interactive Multimedia Systems and Services, Springer, Smart Innovation, Systems and Technologies, vol. 40, pp. 269–378 (2013)
11. Avilés, A., Takimoto, M., Kambayashi, Y.: Distributed evacuation route planning using mobile agents. In: Transaction on Computational Collective Intelligence XVII. LNCS, vol. 8790, pp. 128–144 (2014)
12. Stützle, T., Hoos, H.H.: MAX-MIN ant system. Future Generation Computer System **168**, 889–914 (2000). Elsevier

Computational Models Based on Forgiveness Mechanism for Untrustworthy Agents

Ruchdee Binmad and Mingchu Li

Abstract In online communities like e-marketplaces, the success of business transactions only refers to the cooperation between reputable business agents. While untrustworthy agents will never have an opportunity and are enforced to leave the systems even they are potential to cooperate. In this study, we propose computational models of an exploration strategy based on forgiveness mechanism for potential untrustworthy agents to recover their reputation. The implementation of this mechanism is centralized in nature. Therefore, it can incorporate with existing reputation systems to improve the efficiency of online trading.

Keywords Reputation · Forgiveness mechanism · Untrustworthy agents

1 Introduction

In human societies, there are three ways to learn about other people i.e., direct experience, observation, and reputational information [1]. When learning through the direct experience and observation is not available, the last modality becomes essential. In dynamic open systems like online communities, a number of reputation systems and strategies have been proposed in order to deal with distinct aspects of the interactions between software agents. Reputation is one of the important concepts in making more intelligent decision for predicting potential partner's behaviour at least partially or approximately [2]. Reputation systems help selecting trustworthy partners with the ultimate aim is to achieve their goals or maximize their payoffs. Reputation is a gradually learning process based on the outcome of a long-term relationships. Reputation of an agent increases as a result of cooperative interactions.

R. Binmad(✉) · M. Li
School of Software Technology, Dalian University of Technology,
Dalian City 116621, Liaoning Province, People's Republic of China
e-mail: ruchy.tts@gmail.com, mingchul@dlut.edu.cn

K. Lavangnananda et al. (eds.), *Intelligent and Evolutionary Systems*,
Proceedings in Adaptation, Learning and Optimization 5,
DOI: 10.1007/978-3-319-27000-5_3

However, it can also be destroyed in an instant by either intentional dishonest or mistaken behaviours. Once reputation is lost, it may be costly or it may take a long time to rebuild it. In fact, the success of online business transactions only refer to the cooperation between reputable business partners. While untrustworthy agents will never have an opportunity and are enforced to leave the systems even they are potentially capable of cooperating.

Like in business environments, agents always encounter the dilemma of whether to keep interacting with the same trustworthy agents or to keep experimenting by trying other agents with whom they haven't had much experience so far (i.e., explore in order to discover better providers) [3]. In other words, the impact of future welfare is a key factor for choosing which trading partners to interact with no matter how trustworthy they are. According to Braynov et al. [4], an efficient market is unnecessary to require the actual level of trustworthiness of interacting agents but rather the accuracy of each individual agent estimates. Untrustworthy agents can possibly transact as efficiently as trustworthy agents supposing that they hold accurate estimates of one another. Therefore, a market in which agents are trusted to the degree they deserve to be trusted is as efficient as a market with complete trustworthiness.

Consequently, this gives rise to a question about any possibility that untrustworthy agents can build up their reputation [5] as reputation systems incorporating mechanism that allow untrustworthy agents to correct their intentional or unintentional dishonest behaviours have received far less attention. In this study, we propose forgiveness mechanism, an exploration approach based on the evaluation of five forgiveness motivations i.e., intent, history, apology, severity, and importance, from the viewpoints of agents and communities concerned. The result of the mechanism can be utilized in decision making for choosing and recovering potential untrustworthy agents.

The rest of the paper is organised as follows. In section 2, we present related work on enhanced reputation systems with regard to the addition of forgiveness component. This is followed by a background on forgiveness in Section 3. Section 4 is dedicated to our forgiveness mechanism. Last section, we conclude our study along with future work.

2 Related Work

Allowing untrustworthy agents to build up their reputation, two main prosocial motivations are required: forgiveness and regret. A combinatorial framework of trust, reputation, and forgiveness has been proposed by the study of Vasalou et at. [6] called DigitalBlush System. The system which inspired by human forgiveness, uses expressions of shame and embarrassment to elicit potential forgiveness by others in the society. In more detail, offender's natural reactions after shame and embarrassment (i.e. the blush) can prompt sympathy or forgiveness from the victim. However, misinterpretation of emotional signals can be more problematic than they are not applied. In subsequent works of Vasalou et al. [10][13], when trust breaks down, the trustworthiness of the offender will be detected by identifying a number of motivation constituents [10]. If the result is positive, the victim will be presented with those motivation constituents to consider before reassigning a reputation score to the offender. Specifically, this intervention mechanism intends to alleviate the victim's possibly negative attributions, while at the same time it aims to prevent the unintentional/infrequent offender from receiving an unfair judgement. In [13], they investigate trust repairing in one-off online interactions by conducting an experiment that hypothesizes and shows that systems designed to stimulate forgiveness can restore a victim's trust in the offender.

The concept of regret has been proposed by Marsh et al. [15] as a cognitive inconsistency. Regret can occur from truster, trustee, or both counterparts. A truster feels regret because a positive trust decision is betrayed by a trustee. In other words, a truster's regret occurs when their expectations of the interaction toward a trustee were violated and the correspondent betrayal produced severe damage on trust. A trustee feels regret because a negative trust decision is erroneous. It means a trustee expresses regret for what they have done whether it was wrongdoings or not. Both truster and trustee can feel regret as a missed opportunity for what they did not do. Forgiveness and regret are considered as implementable properties to formalize the incorporation of trust defining a computational model.

Forgiveness factor has been used in [16][17] as an extended component of classical reputation model. It is an optimistic view of reconciliation based on the fact that individuals are more likely to forgive someone who committed an offence that seems distant, rather than close, in time. In other words, an agent should always forgive after a sufficiently large time passed without any interaction. Moreover, an agent should assign a reputation value to its partner initially or increasing to the highest possible value.

3 Forgiveness

Undoubtedly, the violations of norms and regulations are inevitable and unavoidable in human societies. As a result, punishments are implemented either emotionally (e.g., experiencing embarrassment) or practically (e.g., prosecution) as a social protective mechanism to maintain a sense of standard order within the community. However, some transgressions can also be forgivable [6]. From the perspective of religious beliefs, the key word in learning to forgive is the willingness to forgive [7]. Forgiveness is the way out of darkness and into the light. The very important issue when considering forgiveness is it somehow can abandon all sense of security. Even so, allowing forgiveness does not conclude other parties agreed with what transgressor has done. The value of forgiveness can be learned by first eliminating the unwillingness to change our belief systems about human nature. This is also supported by the study of Haselhuhn et al. [8] in which individuals who believe that moral character can change over time are more likely to trust their counterpart following an apology and trustworthy behaviour than are individuals who believe that moral character is fixed.

Forgiveness is a consequence of prosocial motivational changes which heals the one's initial negative motivations towards the transgressor (i.e. revenge, avoidance) with positive motivations [9]. Additionally, issuing forgiveness can encourage the transgressor's voluntary reparative actions. More importantly, applying punishment to the transgressor for an unintentional action will result in displeasure and low-compliancy behaviours making forgiveness more difficult [6][10]. The change of motivation is depended on a number of factors, which help to alleviate the soreness of the victim, for example, the severity of the fault occurrence, transgressor's intent, expression of apology and regret, reparative actions, transgressor's prior interactions, and the importance of the transgressor in community.

In more detail, the victim first assesses the severity of current transgression before considering forgiveness. That means a more serious damage has less possibility to forgive than minor damage [11]. Moreover, even the transgression does not cost much damage, but if it occurs more frequently, forgiveness would be assuredly impossible [12]. A transgressor may violate the victim's trust unintentionally. In such case, unintentional act can lead to more positive attitude than intentional one [11][13]. Furthermore, sincere apology and regret can restore a more favourable impression and a perception of trust towards the transgressor. Besides, the transgressor's expression of apology and regret can evoke more empathy which is in turn more likely to grant forgiveness [9]. In business environment, reparative responses (e.g., a discount for the next purchase or a free stuff as compensation) are the most effect means of retaining partner's reputation in order to show that the transgressor takes responsibility of the mistake [13][14]. Additionally, the outcome of past behaviours or previous interactions can help decide whether or not to forgive the transgressor [9]. In other words, poor historical experiences decrease the likelihood of forgiving the current transgression. Lastly, if the only transgressor has a prominent service or product which is necessarily required by other parties, forgiveness tends to be granted to fulfil the requirement of transaction even knowing that the outcome of future transactions might not be maximized. Similarly, the importance of relationship or situation can also override the negative disposition leading to forgiveness which is expected to help restore cooperation between partners after a transgression [15][18][19].

It is worth noting that forgiving a single transgression cannot override someone's attitude as a whole [20]. Specifically, while a current violation may be forgiven, the transgressor's trust towards other the past violations may still impede. Therefore, it is not necessary to consider forgetting or condoning as a part of forgiveness. There are many benefits fostered by forgiveness. Relationship, for example, after individuals grant forgiveness to someone who committed an offence, their willingness to sacrifice negative motivation can improve and maintain relationship satisfaction [21]. Furthermore, social interactions with other parties also improves as they become more supportive and altruistic in general [22]. Also, individuals' status and power can be compromised, when someone has transgressed towards them [23]. The act of forgiveness in this sense can be considered as a sense of justice providing individuals with an opportunity to reassert their status and position of power.

However, apart from highlighting the benefits of forgiveness, Luchies et al. [24] argue that forgiveness can be disadvantageous in some circumstances especially when it applies to an untrustworthy and disagreeable person. Specifically, a reasonable and agreeable person who acted offensively always apologizes sincerely, takes responsibility, and even compensates in some sense. As a result, the relationship between the victim and the transgressor can be maintained. Moreover, forgiveness tends to be appropriate and foster self respect. On the other hand, if the person who acted offensively continues to be untrustworthy and disagreeable, the victim is unlikely to benefit from allowing forgiveness to this person leading to the reduction of self respect. Therefore, forgiveness in this context is rather indignity and humiliation.

4 Forgiveness Mechanism

In this section, we present a novel forgiveness mechanism which is used to explore potential untrustworthy agents who have the required capabilities to fulfil future transactions. The implementation of the mechanism is centralized in nature. It is activated by a specialized agent called the "Forgiveness Facilitator" in a certain time intervals.

4.1 Forgiveness Factors

We start by analysing factors that motivate forgiveness by alleviating the victim's negative responses. We modify a motivation-driven conceptualization of forgiveness which is identified by Vasalou et al. [10] where positive motivations are collectively evaluated to formulate forgiveness. These following factors are used to find the prospective untrustworthy agents who are potential to reestablish their reputation.

- **Intent (PM_{in}).** An offence is more or less forgivable depending on a victim's attribution of a transgressor's intention. In case of intentional act, a transgressor has committed his/herself to deliberately harm a victim resulting in the formation of a victim's harsh dispositional judgments. As a consequence of such intentional offence, forgiveness is unlikely for a victim to be conducted. However, an action of unintentional or infrequent offence can lead to more positive judgements and more likely to foster forgiveness [11]. Also noting that lack of information about the transgressor's intention can presumably make the victim's perception of the transgressor's actions as being intentional even they are unintentional.
- **History (PM_{hi}).** The productive past interactions of the transgressor at both dyadic and corporate level can foster benevolence, which are a key component for trust building. Such benevolence can also increase the tendency towards forgiveness. Conversely, the offensive historical experiences of the transgressor decrease the likelihood of positive motivations which result in the negative inclination to forgive [13].
- **Apology (PM_{ap}).** The transgressor's expression of a truthful apology as a form of affective recovery effort can enhance the victim's perceptions of interactional justice and improve post-recovery satisfaction [25]. Apart from an interpersonal apology, a corporate apology for negative consequences and its willingness to

assume relevant responsibility can lead to a favorable impression that the transgressor and his/her organization are problem solving-oriented rather than inclined to conceal reality. Additionally, forgiveness is related to the time of expressing apology as it shows a sense of taking responsibility. In other words, the transgressor who apologizes immediately after a transgression takes place is more likely to be forgiven than who apologizes later.

- **Severity (PM_{se}).** Forgiveness is more easily granted when the transgression is perceived as less severe [11]. While more severity of the transgression lead to less positive judgments. However, the severity assessment requires some serious consideration as the transgression might not impact only the present, but also the past and the inevitable future. If the consequences of the current transgression are forgiven, then its future consequences are also continually granted forgiveness. Moreover, the transgressor's past offences are compared to the current event. Frequency and severity of past offences can also impact one's inclination to forgive [12].
- **Importance (PM_{im}).** If the only transgressor can provide a crucial product or service which is necessarily most required by the victim or other consumers, forgiveness tends to be granted in order to fulfil the transaction's requirement and avoid deadlock even knowing that the outcome of future transactions might not be maximized. Also the importance of relationship and its quality between the victim and the transgressor can override the victim's negative disposition towards an offence resulting in positive judgements which is expected to help restore cooperation between them [15][18][19]. Furthermore, the vital role played by the transgressor in the community that directly or indirectly influents the victim and other members in some other way can lead to more positive motivations.

4.2 Sources of Forgiveness

We applies the concept of community [26] to categorize agents related to a specific domain e.g., car dealer, book dealer, real estate agency, travel agency, and so on. However, each community member can possibly have different expectation of attributes (e.g., price, quality, and delivery time) of the same product or service. For purposes of simplification, we assume there are two communities: consumer community and service provider community. The consumer community consists of consumer agents which buy the product or use the service provided by service provider agents. While the service provider community consists of service provider agents whose the functionalities are to serve, maintain, and deliver the good quality of product or service to consumer agents.

In this study, we assume a transgression occurs when the service provider breaks transactional agreement by not delivering the product or delivering product of low quality. The consumer as a victim suffers not only from not receiving the product as promised but also wasting of time for making a complaint and spending unnecessary cost to find a new service provider. In many cases, it is not only the consumer who directly suffers the transgression but also consumer and service provider community members who would be indirectly affected by the transgression.

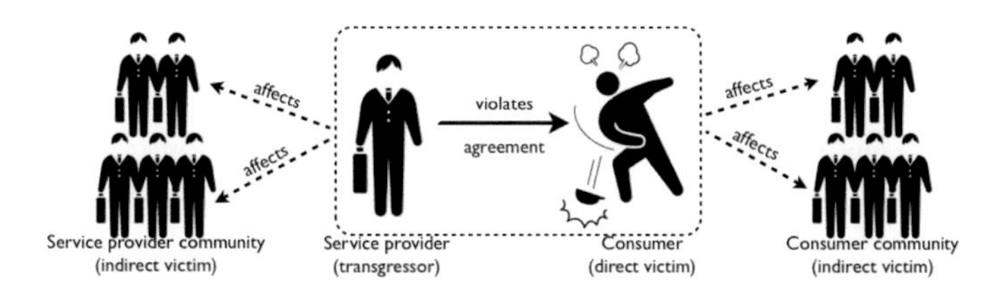

Fig. 1 Direct and indirect sources of forgiveness

As shown in Figure 1, violating contractual terms or agreements by a transgressor can possibly affect both direct and indirect victims. Specifically, a direct victim is the one who commits to do a transaction with a transgressor and the first one who directly affected by a transgression. While an indirect victim is the one or group of members in community who does not transact with a transgressor directly or is not the direct target but still suffers because of that transgression. Therefore, forgiveness is issued from both direct and indirect sources: (i) the victim of a transgression, (ii) the victim(consumer) community, and (iii) the transgressor(service provider) community. The victim community forgives a transgressor based on the fact that the future transactions between a transgressor and other members in the victim community are possible. Similar to human societies, forgiveness from the victim community can be considered as a declaration of amnesty. Furthermore, a transgressor asks for forgiveness from his/her own community for a transgression that negatively affects the whole community trust. The forgiveness from the transgressor community can be motivated to overcome distrust in or restore harmony to the relationship among community members. Hence, the overall forgiveness value is the aggregation of forgiveness evaluation from both direct and indirect victims.

4.3 Computational Model of Forgiveness

Table 1 presents a mapping between positive motivations described above and the sources of forgiveness. Each source of forgiveness has different point of view to evaluate and then make decision whether or not to forgive the transgressor. We suppose agent y violates trust of agent x which also has an affect on trust of agent x's (denote as X) and agent y's (denote as Y) community members. The following subsections detail how each forgiveness factor can be calculated and how each forgiveness source provides their forgiveness value.

Computation of Positive Motivations

In this subsection, we provide computational models for evaluating positive motivations by taking into account the information provided by agents (i.e., agent x and agent y) and communities (i.e., community X and community Y).

Table 1 Positive motivations for evaluating forgiveness from the point of view of the victim, the victim community, and the transgressor community

Positive motivations (PM)	**Victim (F_v)**	**Victim community (F_{vc})**	**Transgressor community (F_{tc})**
Intent (PM_{in})	●		
History (PM_{hi})	●	●	
Apology (PM_{ap})	●	●	●
Severity (PM_{se})	●	●	●
Importance (PM_{im})	●	●	●

- **Intent.** The numbers of transactions between interacting agents are used to evaluate an agent's intent. More specifically, if the number of transactions processed by agent y is high, it means that agent y has high experience. This can lead to the fact that high experience agent y would transact with high intention. Therefore, the agent y's intent of violating agent x's trust can be computed by:

$$PM_{in}(x, y) = 1 - \frac{(N_y^{all+} + N_y^{all-}) - (N_x^{y+} + N_x^{y-})}{(N_y^{all+} + N_y^{all-}) + (N_x^{y+} + N_x^{y-})}, \quad (1)$$

where N_y^{all+} and N_y^{all-} are the total number of successful and defective transactions of agent y respectively. N_x^{y+} and N_x^{y-} are the number of successful and defective transactions between agent x and agent y respectively.
- **History.** The result of past interactions between agent x and agent y can be calculated by considering the number of transactions as:

$$PM_{hi}(x, y) = \frac{N_x^{y+} - N_x^{y-}}{N_x^{y+} + N_x^{y-}}, \quad (2)$$

If $PM_{hi}(x, y) > 0$, historical interactions between them are considered to be productive. Otherwise they are offensive ($PM_{hi}(x, y) \leq 0$).
- **Apology.** As an apology related with time, we first define recency factor (RF) [27]:

$$RF(a) = e^{\frac{\Delta t(a)}{\lambda}}, \quad (3)$$

where $RF(a)$ is a recency factor of apology a. $\Delta t(a)$ is the difference between the time that the transgression takes place (t_o) and the time that the offender apologizes (t_a). More different value between these two times is significantly less positive judgements. The parameter $\lambda \in [0, 1]$ is a decay rate of the apology offer. The small λ indicates inclination relying more on the early time of expressing apology. On the other hand, increasing λ indicates more acceptable on the late apology. The overall apology value is the aggregation of an interpersonal and a corporate apology as follows:

$$PM_{ap}(x, y, Y) = \frac{(a_y^x \times RF(a_y^x)) + (a_Y^x \times RF(a_Y^x))}{RF(a_y^x) + RF(a_Y^x)}, \tag{4}$$

where $RF(a_y^x)$ and $RF(a_Y^x)$ are the recency factors of apology offered to agent x by agent y and community Y respectively. a_y^x and a_Y^x are the honesty of apology offered by agent y and community Y respectively and range from 0 to 1.

- **Severity.** The severity of a transgression is evaluated by considering the current utility that agent x has lost from transacting defectively with agent y:

$$PM_{se}(x, y) = (1 - \frac{U_x^{y+} - U_x^{y-}}{U_x^{y+} + U_x^{y-}}) \times PM_{im}(x, y), \tag{5}$$

where U_x^{y+} is the expected utility that agent x could have been gained from transacting with agent y and U_x^{y-} is the utility that agent x has lost from the transgression committed by agent y. $PM_{im}(x, y)$ is the importance of product/service offered by agent y to agent x.

- **Importance.** We evaluate the importance of product/service offered by agent y to agent x by calculating the number of utility gain that agent x obtained from transacting with agent y compared with utility gain that agent x obtained from all transactions:

$$PM_{im}(x, y) = \frac{\sum_{y_i=1}^{N_x^y} U_x^{y_i}}{\sum_{m \in all(x)} \sum_{m_i=1}^{N_x^m} U_x^{m_i}} \tag{6}$$

where U_x^y is the total utility gain obtained from transactions between agent x and agent y and U_x^m is the total utility gain obtained from transactions between agent x and all other agents in the set $all(x)$.

Forgiveness Values from Different Sources

In order to seek the possibility of issuing forgiveness for the transgression committed by agent y, agent x as a direct victim, community X and community Y as indirect victims evaluate all possible factors according to positive motivations as described below.

- **Victim of a Transgression.** All positive motivations are assessed from the point of view of the victim, agent x, as follows:

$$\begin{aligned} F_v(x, y, Y) = &PM_{hi}(x, y) + PM_{ap}(x, y, Y) + PM_{im}(x, y) - \\ &PM_{in}(x, y) - PM_{se}(x, y), \end{aligned} \tag{7}$$

where $F_v(x, y, Y)$ is the forgiveness value evaluated by agent x for the violation made by agent y which is a member of Community Y. $PM_{in}(x, y)$ is the intent assessment of the transgression reported by both agent x and agent y. $PM_{hi}(x, y)$ is the result of past interactions between agent x and agent y. $PM_{ap}(x, y, Y)$ is the apology offered by agent y towards agent x incorporating the apology offered by community Y towards agent x. $PM_{se}(x, y)$ is the severity of the transgression made by agent y and reported by agent x. $PM_{im}(x, y)$ is the importance of agent y to agent x either in the form of product/service offered or relationship.

- **Victim Community.** Positive motivations from the view point of victim community are different from that of the victim. The victim community's forgiveness can be evaluated according to the expression:

$$F_{vc}(x, X, y, Y) = PM_{ap}(X, y, Y) + PM_{im}(X, y) + \sum_{\substack{i \in X \\ i \neq x}} (PM_{hi}(i, y) - PM_{se}(i, y)), \quad (8)$$

where $F_{vc}(X, y, Y)$ is the forgiveness value aggregated from the members of community X for the transgression made by agent y which is a member of Community Y. $PM_{hi}(i, y)$ is the historical interactions between agent i of community X and agent y. $PM_{se}(i, y)$ is the assessment of severity of the transgression made by agent y towards agent i. Both $PM_{hi}(i, y)$ and $PM_{se}(i, y)$ will be different for each member in the community. $PM_{ap}(X, y, Y)$ is the apology expressed by agent y towards community X incorporating the apology expressed by community Y towards community X. $PM_{im}(X, y)$ is evaluated by considering the importance of product/service that agent y offered to community X.

- **Transgressor Community.** Apart from victim community, transgressor community also suffers from the transgression committed by the transgressor. Some positive motivations are collected and aggregated to formulate transgressor community's forgiveness as defined below:

$$F_{tc}(Y, y) = PM_{ap}(Y, y) + PM_{im}(Y, y) - \sum_{\substack{j \in Y \\ j \neq y}} (PM_{se}(j, y)), \quad (9)$$

where $F_{tc}(Y, y)$ is the forgiveness value calculated from the members of community Y for the transgression committed by agent y. $PM_{se}(j, y)$ is the severity assessment of the transgression made by agent y towards agent j of its own community Y. $PM_{ap}(Y, y)$ and $PM_{im}(Y, y)$ are the apology and the importance of product/service that agent y offered to its own community Y respectively.

In case that the community is large consisting of a huge number of members. It is impossible to request forgiveness assessment from all community members as it is costly and time-consuming. The community leader or a set of members with high reputation can be used to represent the evaluation of positive motivations of the entire community. For example, we assume $X^+ = \{x_1^+, x_2^+, ..., x_m^+\}$, where X^+ is a set of high reputation members of community X. The victim community's forgiveness can then be as follows: $F_{vc}(x, X^+, y, Y) = \sum_{\substack{i \in X^+ \\ i \neq x}} (PM_{hi}(i, y) - PM_{se}(i, y)) + PM_{ap}(X, y, Y) + PM_{im}(X, y)$. Similar to the transgressor community, if we assume $Y^+ = \{y_1^+, y_2^+, ..., y_n^+\}$, where Y^+ is a set of high reputation members of community Y. The forgiveness value calculated by the transgressor community can be expressed as follows: $F_{tc}(Y^{+}, y) = PM_{ap}(Y, y) + PM_{im}(Y, y) - \sum_{\substack{j \in Y^{+} \\ j \neq y}}(PM_{se}(j, y))$.

Furthermore, all forgiveness assessment values from agent x as the victim (F_v), the victim community X (F_{vc}), and the transgressor community Y (F_{tc}), will be transformed by applying the normalized inverse tangent function which is monotonically increased in a range between 0 and 1 [28]. As a result, the victim's ($F_v^{'}$), the victim community's ($F_{vc}^{'}$), and the transgressor community's ($F_{tc}^{'}$) functions can be formulated as shown in Equation (10), (11), and (12) respectively:

$$F_v^{'} = \frac{atan(F_v(x, y, Y) - \alpha) + atan(\alpha)}{\pi/2 + atan(\alpha)}, \quad (10)$$

$$F_{vc}^{'} = \frac{atan(F_{vc}(x, X, y, Y) - \alpha) + atan(\alpha)}{\pi/2 + atan(\alpha)}, \quad (11)$$

$$F_{tc}^{'} = \frac{atan(F_{tc}(Y, y) - \alpha) + atan(\alpha)}{\pi/2 + atan(\alpha)}, \quad (12)$$

where $\alpha > 0$ is a specific constant called forgiveness increasing factor. The overall forgiveness value for the transgression committed by agent y is the aggregation of results both subjective view (at the individual level) and objective view (at the community level):

$$F_{total}(y) = \omega_v F_v^{'} + \omega_{vc} F_{vc}^{'} + \omega_{tc} F_{tc}^{'}, \quad (13)$$

where ω_v, ω_{vc}, and ω_{tc} are the weight factors reflecting the major victim of the transgression and the major contributor to the forgiveness assessment. The summation of all weight factors is equal to 1, that is $\omega_v + \omega_{vc} + \omega_{tc} = 1$. In our case, we set $\omega_{tc} < \omega_{vc} < \omega_v$, meaning that even there are collective views on positive motivations from community level, the main decision for granting forgiveness is still made by the victim.

4.4 *Zone of Forgivability*

The result of forgiveness mechanism should also have some limits or boundary values reflecting the fact that the transgression that might be forgiven should not be completely forgotten [20][29]. Marsh et al. [15], introduce the concept of *the Limit of Forgivability* as a minimum baseline of trust value for determining the worth of the transgressor entering into redemption strategies. However, the concept does not state how much the boundary values after granting the transgressor's forgiveness would be, which means it is possible that the transgressor's trust can be fully reinstated.

In this study, our forgiveness boundary values are indicated as *the Zone of Forgivability* shown in Figure 2. More specifically, the minimum boundary value of forgivability can be determined as:

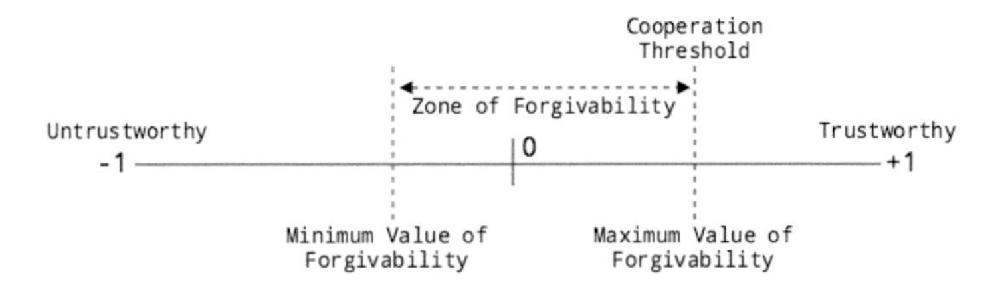

Fig. 2 The zone of forgivability

$$Min_{forgivability} = Max_{untrust} + F_{threshold}, \tag{14}$$

where $Max_{untrust}$ is the possible maximum value of an untrustworthy agent (which is -1) and $F_{threshold}$ is a constant forgiveness threshold. A transgressor is considered to be potential if the aggregation of existing reputation value and forgiveness assessment exceeds the minimum value of forgivability, that is $R^t_{old} + F_{total}(t) \geq Min_{forgivability}$. Whereas the cooperation threshold is used as the maximum value of forgivability which means if the calculated forgiveness value exceeds the maximum boundary value, then the cooperation threshold will be used as a new reputation:

$$R^t_{new} = \begin{cases} R^t_{old} + F_{total}(t) & if \quad R^t_{new} < C_{threshold} \\ C_{threshold} & if \quad R^t_{new} \geq C_{threshold} \end{cases} \tag{15}$$

where $C_{threshold}$ is a constant cooperation threshold. It is worth nothing that the Zone of Forgivability still impede a potential transgressor to be directly considered for selection if the result of forgiveness mechanism less than the cooperation threshold. However, it is not completely rejected but rather a baseline for interacting agents to incorporate with some additional information e.g., cost and quality into its decision making [30].

5 Conclusion and Future Work

In this study, we provide computational models for untrustworthy agents the opportunity to build up their reputation in different types of relationships. We propose forgiveness mechanism which is the evaluation of five positive motivations: intent, history, apology, severity, and importance, of a transgressor based on different viewpoints of the victim, the victim community, and the transgressor community. Moreover, based on the fact that even the violation of norms can be forgiven, it should not be completely forgotten. Therefore, the Zone of Forgivability is presented indicating the minimum and maximum boundary values of forgivability. The outcome of the proposed forgiveness mechanism can be utilized in decision making for improving the efficiency of an online marketplace.

There are some issues worth to be further addressed in future work. For example, incorporating incentive mechanism to encourage the victim and the members of victim community to cooperate with the transgressor in future transactions. Furthermore, risk assessment is considered to be necessary when interacting with untrustworthy agents especially in risky environments. Therefore, incorporating risk assessment with forgiveness mechanism can make the process of finding potential untrustworthy agents more robust. Moreover, a number of experiments are required to evaluate the effectiveness of the proposed forgiveness mechanism through the comparison of different dynamic online marketplace environments.

Acknowledgments This paper is supported by Nature Science Foundation of China (NSFC) under grant number 61272173.

References

1. Frith, C.D., Frith, U.: How We Predict What Other People Are Going To Do. Brain Research **1079**, 36–46 (2006). doi:10.1016/j.brainres.2005.12.126
2. Giardini, F., Conte, R., Paolucci, M.: Reputation. In: Edmonds, B., Meyer, R. (eds.) Simulating Social Complexity: A Handbook (Understanding Complex Systems), pp. 365–399 (2013). doi:10.1007/978-3-540-93813-2
3. Teacy, W.T.L., Chalkiadakis, G., Rogers, A., Jennings, N.R.: Sequential decision making with untrustworthy service providers. In: 7th Int. Conf. on Autonomous Agents and Multiagent Systems (AAMAS 2008), pp. 755–762. International Foundation for Autonomous Agents and Multiagent Systems (2008)
4. Braynov, S., Sandholm, T.: Contracting with Uncertain Level of Trust. Computational Intelligence **18**(4), 501–514 (2002). doi:10.1111/1467-8640.00200
5. Ramchurn, S.D., Huynh, D., Jennings, N.R.: Trust in Multi-Agent Systems. The Knowledge Engineering Review **19**(1), 1–25 (2005)
6. Vasalou, A., Pitt, J.: Reinventing forgiveness: a formal investigation of moral facilitation. In: Hermann, P., Issarny, V., Shiu, S. (eds.) Trust Management. LNCS, pp. 146–160 (2005)
7. Jampolsky, G.G.: Forgiveness: The Greatest Healer of All. Beyond Words Publishing Inc. (1999)
8. Haselhuhn, M.P., Schweitzer, M.E., Wood, A.M.: How Implicit Beliefs Influence Trust Recovery. Psychological Science: A Journal of the American Psychological Society/APS **21**(05), 645–648 (2010). doi:10.1177/0956797610367752
9. McCullough, M.E., Pargament, K.I., Thorensen, C.E.: Forgiveness: Theory, Research and Practice. Guilford Press (2001)
10. Vasalou, A., Pitt, J., Piolle, G.: From theory to practice: forgiveness as a mechanism to repair conflicts in CMC. In: Trust Management. LNCS, vol. 3986, pp. 397–411 (2006)
11. Boon, S., Sulsky, L.: Attributions of Blame and Forgiveness in Romantic Relationships: A Policy Capturing Study. Journal of Social Behaviour & Personality **12**, 19–26 (1997)
12. Buss, A.H.: Self-Consciousness and Social Anxiety. W.H. Freeman, San Francisco (1980)
13. Vasalou, A., Hopfensitz, A., Pitt, J.: In Praise of Forgiveness: Ways for Repairing Trust Breakdowns in One-Off Online Interactions. International Journal of Human-Computer Studies **66**, 466–480 (2008). doi:10.1016/j.ijhcs.2008.02.001
14. Xie, Y., Peng, S.: How to Repair Customer Trust After Negative Publicity: The Roles of Competence, Integrity, Benevolence, and Forgiveness. Psychology and Marketing **26**(07), 572–589 (2009). doi:10.1002/mar

15. Marsh, S., Briggs, P.: Examining trust, forgiveness and regret as computational concepts. In: Golbeck, J. (ed.) Computing with Social Trust. Human-Computer Interaction Series, pp. 9–43 (2009)
16. Burete, R., Bădică, A., Bădică, C.: Reputation model with forgiveness factor for semi-competitive E-business agent societies. In: Communications in Computer and Information Science, vol. 88, pp. 402–416 (2010)
17. Burete, R., Bădică, A., Bădică, C., Moraru, F.: Enhanced Reputation Model with Forgiveness for E-Business Agents. International Journal of Agent Technologies and Systems **3**, 11–26 (2011). doi:10.4018/jats.2011010102
18. McCullough, M.E., Rachal, K.C., Sandage, S.J., Worthington, E.L., Brown, S.W., Hight, T.L.: Interpersonal Forgiving in Close Relationships: II. Theoretical Elaboration and Measurement. Journal of Personality and Social Psychology **75**(06), 1586–1603 (1998). doi:10.1037/0022-3514.75.6.1586
19. Burnette, J.L., McCullough, M.E., Van Tongeren, D.R., Davis, D.E.: Forgiveness Results from Integrating Information about Relationship Value and Exploitation Risk. Personality and Social Psychology Bulletin **38**, 345–356 (2012). doi:10.1177/0146167211424582
20. Exline, J., Worthington Jr., E.L., Hill, P., McCullough, M.E.: Forgiveness and Justice: A Research Agenda for Social and Personality Psychology. Personality and Social Psychology Review **7**(04), 337–348 (2003)
21. Maio, G.R., Thomas, G., Fincham, F.D., Carnelley, K.D.: Unraveling the Role of Forgiveness in Family Relationships. Journal of Personality and Social Psychology **94**(02), 307–319 (2008)
22. Karremans, J.C., Van Lange, P.A.M., Holland, R.W.: Forgiveness and Its Associations with Prosocial Thinking, Feeling, and Doing beyond the Relationship with the Offender. Personality and Social Psychology Bulletin **31**(10), 1315–1326 (2005)
23. Wenzel, M., Okimoto, L.G.: How Acts of Forgiveness Restore a Sense of Justice: Addressing Status/Power and Value Concerns Raised by Transgressions. European Journal of Social Psychology **40**(03), 401–417 (2010)
24. Luchies, L.B., Finkel, E.J., McNulty, J.K., Kumashiro, M.: The Doormat Effect: When Forgiving Erodes Self-Respect and Self-Concept Clarity. Journal of Personality and Social Psychology **98**(05), 734–749 (2010)
25. Smith, A.K., Bolton, R.N., Wagner, J.: A Model of Customer Satisfaction With Service Encounters Involving Failure and Recovery. Journal of Marketing Research **36**, 356–372 (1999)
26. Malik, Z., Bouguettaya, A.: Reputation Bootstrapping for Trust Establishment Among Web Services. IEEE Internet Computing **13**(01), 40–47 (2009)
27. Huynh, T.D., Jennings, N.R., Shadbolt, N.R.: An Integrated Trust and Reputation Model for Open Multi-Agent Systems. Autonomous Agents and Multi-Agent Systems **13**, 119–154 (2006). doi:10.1007/s10458-005-6825-4
28. Zhang, H., Duan, H., Liu, W.: RRM: An Incentive Reputation Model for Promoting Good Behaviors in Distributed Systems. Science in China Series F: Information Sciences **51**(11), 1871–1882 (2008). doi:10.1007/s11432-008-0141-y. SPEC.ISS
29. Vasalou, A., Riegelsberger, J., Joinson, A.: The application of forgiveness in social system design. In: SIGCHI Conference on Human Factors in Computing Systems, pp. 225–228 (2009). doi:10.1145/1518701.1518738
30. Griffiths, N.: A fuzzy approach to reasoning with trust, distrust and insufficient trust. In: Klusch, M., Rovatsos, M., Payne, T.R. (eds.) Cooperative Information Agents X. LNCS, pp. 360–374. Springer (2006)

A New Adaptive Genetic Algorithm for Community Structure Detection

Yilmaz Atay and Halife Kodaz

Abstract Community structures exist in networks which has complex biological, social, technological and so on structures and contain important information. Networks and community structures in computer systems are presented by graphs and subgraphs respectively. Community structure detection problem is *NP-hard* problem and especially final results of the best community structures for large-complex networks are unknown. In this paper, to solve community structure detection problem a genetic algorithm-based algorithm, *AGA-net,* which is one of evolutionary techniques has been proposed. This algorithm which has the property of fast convergence to global best value without being trapped to local optimum has been supported by new parameters. Real-world network which are frequently used in literature has been used as test data and obtained results have been compared with 10 different algorithms. After analyzing the test results it has been observed that the proposed algorithm gives successful results for determination of meaningful communities from complex networks.

Keywords Combinatorial optimization · Community structure detection · Complex networks · Evolutionary computation · Genetic algorithm · Modularity

1 Introduction

Understanding networks provides us very important information about the extraction of meaningful information from complex systems. In eliciting meaningful information from these networks the importance of structures which are named as community structures is huge. The graph structures are used to present the real-world networks. Community structures or clusters can be considered as subgraphs which are partially or completely independent from each other in graph structures.

Y. Atay · H. Kodaz(✉)
Department of Computer Engineering, Selcuk University, Konya, Turkey
e-mail: {yilmazatay,hkodaz}@selcuk.edu.tr

K. Lavangnananda et al. (eds.), *Intelligent and Evolutionary Systems*,
Proceedings in Adaptation, Learning and Optimization 5,
DOI: 10.1007/978-3-319-27000-5_4

As an example, tissues or organs which have the same role in the human body can be considered as clusters [1]. Community structure detection (CSD) is important to an understanding of the biological, economic, social, technological and so on networks. These networks can be synthetic or real-world networks. For real-world networks we can give some example like economic structure networks [2], food networks [3], networks of chemical interaction between proteins and molecules in cells [4-6] and social networks like networks of determination of friendship in groups, relation analysis networks and networks of detection of terrorist attacks [7].

Objects and connections in networks are presented with nodes and edges respectively. Graph structures which are used to represent the above given networks are referred to as simplest form of undirected networks [8].

Community mining problem (CMP) refers to discovery of meaningful subgraph in many complex networks data [9]. In this paper, many real-world data were analyzed by CMP and the obtained results are given in the experimental results section.

Many methods have been developed for detection of community structures in complex networks. These methods give successful results according to many properties yet generalizations about obtaining the best result cannot be made. Performances of the algorithms in literature are very low on large networks. In addition, for detection of community, many algorithms need prior knowledge like community number. Optimal grouping in network is a very difficult problem. Therefore, CSD problem is a *nondeterministic polynomial time - hard* problem [10, 11].

Usually to solve complex problems like CSD two different methods are proposed. These are exact and (meta-)heuristic methods [12]. From these two methods (meta-)heuristic method can offer more convenient solutions for difficult and complex problems than exact methods. Algorithms like memetic and genetic are covered by (meta-)heuristic methods and are also known as bio-inspired algorithms [12]. These algorithms use various community calculating measures according to their own methods in the CSD problems. The most common calculation measure used recently which is recommended by Girman and Newman is *modularity Q* measure [13, 14].

So far the most well-known community detection algorithm is Girman-Newman (*GN*) algorithm [13, 15]. Fast Newman (*FN*) algorithm is an algorithm based on the maximum *modularity Q* [16]. Similarly, another algorithm based on maximum *modularity Q* is called Fast Unfolding algorithm [17]. In addition to these, algorithms like Random Walks [18], Eigenvectors [19], Label Propagation (*LP*) [20] with Spin Glass Type Potts method [21] and *LTE* (Local Tightness Expansion) algorithm [22] are used in the literature. *FN* [16], community detection algorithm for large networks which is proposed by Clauset et al. [23], Extremal Optimization [24] and other algorithms like this have $O(e^3)$ complexity in terms of time complexity. Here the *e* refers to the number of edges [25].

Time complexity increases in a huge amount as the size of the network increases. In small or regular size networks community detection can be done very easily with algorithms given above but as the network size get larger existing algorithms are inadequate in terms of both performance and success. Also when inclusion of prior knowledge to these algorithms become mandatory, discovery of new and efficient algorithms are inevitable. Due to this need, CSD problems are tried to solve with algorithms like genetic algorithms, particle swarm optimization algorithm, ant colony optimization algorithm, memetic algorithms, and differential algorithm. In this paper genetic algorithm which is one of the above algorithms constitutes the basic structure of the proposed algorithm. Genetic algorithm is already very successful in terms of computation, time complexity and solution convergence in *NP-hard* problems and it is almost used in most problems in literature. For the first time Tasgin et al. [26] used genetic algorithms in CSD problem. The method that they developed was named as *GATB* (or *GATHB* [27]) [25]. After that both genetic and other algorithms started to be used widely in CSD problems. In particular, to find the optimal Q value in the most economical way, many methods have been developed by making several changes on methods of genetic algorithm like mutation, crossover, selection and so on.

Shi et al. tried to solve the CSD problem by using genetic algorithm based *GACD* [28] algorithm. They tested their own method with real-world networks which are used quite a lot in the literature and compare the results with *GN* [13], *GN Fast* [23] and *GATB* [25] results. When the obtained results were analyzed, it was stated that genetic algorithm-based algorithms such as *GACD* and *GATB* were quite effective to solve CSD problems. There are many advantages of the developed methods based on genetic algorithm. For example *GATB* [25] has a time complexity of $O(e)$ and does not require any prior knowledge for CSD. In this paper, a new approach based on genetic algorithms has been proposed and has been named as *AGA-net*. *AGA-net* has a time complexity of $O(e)$ and does not require any prior knowledge. The proposed algorithm is based on *adaptive* design of genetic algorithm to reach the most appropriate solution in less time for CSD problem. *AGA-net* was tested in networks given in section 3 and the obtained results were compared with some existing algorithms in literature (see section 3).

1.1 Community Structure Detection

When any given network presented by graph structure, obtained community structures can be considered as subgraphs which have quality or quantity like maximum common feature in itself, number of interactions, positional similarities and so on. Nodes which are the elements of these structures should have maximum interaction and common properties with its own community nodes and less interaction and common properties with other community nodes. Group of people who have strong relationship in social environment, colony of living creatures in environmental networks who feed on each other and cluster of computers having maximum data exchange cooperation can be examples related with CSD.

Let the given $G(V,E)$ graph structure represent undirected and unweighted network. Here the graph G has V set of nodes (vertices) and E set of edges (links).

$$V = \{v_i \mid i = 1, 2, 3, \ldots, n\} \text{ and } E = \{e_j \mid j = 1, 2, 3, \ldots, m\}$$

Here; *i*, *j*, *n* and *m* represent the node index, edge index, number of node and total edge number respectively. Let define adjacency matrix as *Adj* with *nxn* size. And let *Adj* matrix show the relationship of the elements of set *V* by the elements of the set *E*. *Adj* adjacency matrix is generated by Equation (1) [23].

$$Adj = \begin{cases} 1 & \text{if } i.\text{and } j.\text{nodes are connected,} \\ 0 & \text{otherwise.} \end{cases} \quad (1)$$

Modularity Q for graph *G* is given in Equation (2). This fitness function has been proposed by Newman and Girvan in their work by the name of *Finding and Evaluating Community Structure in Networks* [15].

$$Q = \frac{1}{2\times m} \sum_{ij} \left(Adj_{(i,j)} - \frac{k_i \times k_j}{2\times m} \right) \times \delta(C_i, C_j) \quad (2)$$

Where Q is named as *modularity Q* and expresses the objective function to be maximized. $Adj_{(i,j)}$, represents the adjacency matrix of given *G* graph. *m* demonstrates the total number of edge in network and calculated by Equation (3). k_i demonstrates the degree of ith node, k_j demonstrates the degree of jth node and as an example k_i can be calculated by Equation (4). C_i and C_j demonstrate the ith and j^{th} node community respectively. $\delta(C_i, C_j)$ is a function which demonstrates the ith and j^{th} node whether exist in the same community. $\delta(C_i, C_j)$ function is calculated by Equation (5).

$$\mathrm{m} = \frac{1}{2} \sum_{ij} Adj_{(i,j)} \quad (3)$$

$$k_i = \sum_j Adj_{(i,j)} \quad (4)$$

$$\delta = \begin{cases} 1 & \text{if } C_i = C_j \\ 0 & \text{if } C_i \neq C_j \end{cases} \quad (5)$$

Detection of community structure according to fitness value was done by Tasgin et al. in 2007 by the name of *Community Detection in Complex Networks using Genetic Algorithms* [26]. In the specified paper, proposed algorithm was named as *GATHB* [26, 27]. After publication of this paper, many evolutionary algorithms were applied to CSD problems. The *AGA-net* algorithm that we proposed has also used the same objective function as *GATHB* algorithm which is given in Equation (2).

1.2 Genetic Algorithm

GAs were first described by John Holland in the 1960s and further developed by Holland and his students and colleagues at the University of Michigan in the 1960s and 1970s. Holland's goal was to understand the phenomenon of "adaptation" as it occurs in nature and to develop ways in which the mechanisms of natural adaptation might be imported into computer systems. Holland's 1975 book Adaptation in Natural and Artificial Systems (Holland, 1975) presented the GA as an abstraction of biological evolution and gave a theoretical framework for adaptation under the GA [29]. GA is a population based algorithm and can be modal without requiring any prior knowledge or assumptions. Thus this algorithm can be adapted to many problems and has a general-purpose structure property.

2 The Proposed Algorithm

In this paper to solve CSD problem, *AGA-net* algorithm has been proposed which is based on genetic algorithm. As every node of network in CSD problem has a limited number of neighbors therefore the probability of selected neighbor to be selected again is very high. This situation, searching for the best solution, it may cause to be entered to a vicious cycle in the various networks. This problem has been solved by the help of genetic operators given in section 2.4. Thus for rapid convergence to best solution, better solutions has been selected by elitism while entering in to vicious circle is also prevented by crossover and mutation mechanism. The proposed algorithm has the property of convergence of best global *modularity Q* without being trapped in local best solution. *AGA-net* also has a linear time complexity. In addition to basic parameters and operators of standard genetic algorithm, specific changes for CSD problems and new parameters have been included. The proposed algorithm has been named as *Adaptive Genetic Algorithm* (*AGA-net*). The adaptive phrase used herein indicates that every mechanism of algorithm can be adapted to all networks. The proposed algorithm can be operated for all networks on CSD problem without being depended to any internal or external data, with its new specific parameters. Proposed algorithm's steps are given below in detail under separate headings.

2.1 Genetic Representation

The proposed algorithm uses locus-based adjacency representation (*LAR*) structure for graph based representation [30]. Each gene in chromosome holds two different information (*communityID* and *populationID*). Information about these is given in Fig. 1. The first information stores randomly selected neighbor node from i^{th} node neighbors. The second information keeps community knowledge (*communityID*) of ith node for communities generated by the first information. An example of 8-node network has been given in Fig. 1(a), Fig. 1(b) shows an example of chromosomes generated according to the given network and Fig. 1(c) provides community structures generated from given chromosome information. Obtained community structures have been given in different colors.

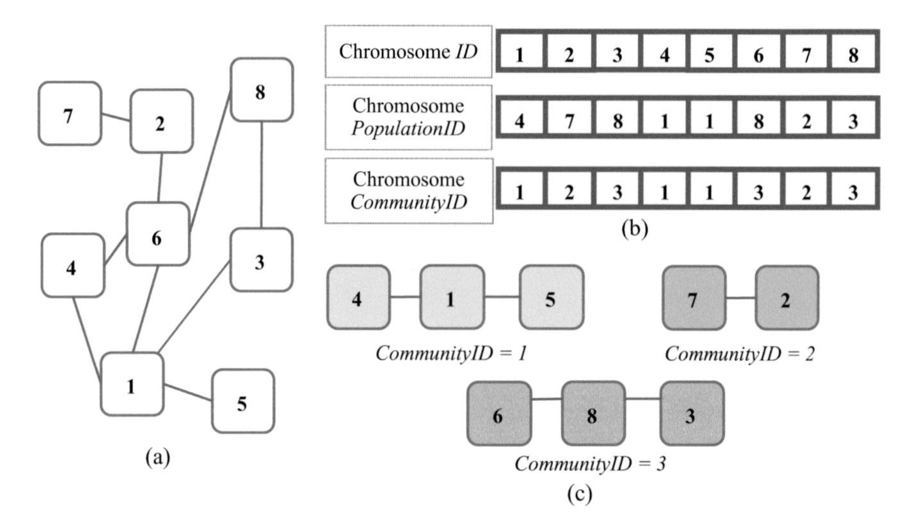

Fig. 1 Examples of a network with 8 node, a chromosome and obtained community structures

The chromosome given in Fig. 1(b) is represented by 3 different array which keeps information about *ID, populationID and communityID*. The first array keeps node sequence number, the second keeps selected neighbor node and the third array keeps information about the node's community.

2.2 Population Initialization

The proposed algorithm produces chromosomes as the size of population in initialization process. Each gene in the chromosome represents a node. The second array from the arrays given in Fig. 1(b) selects randomly neighboring node according to *ID*. After all population formed in this way the 3rd array is formed according to 2nd array which is given in Fig. 1(b). The 2nd array which is given in Fig. 1(b) provides forming of necessary community's list during *modularity Q* calculation.

While determining *CommunityID* of a gene inside chromosomes there should be neighborhood of neighbor gen with existing gen. According to this principle solution space becomes restricted and it saves time.

2.3 Fitness Function

In this paper, *modularity Q* has been used as fitness function. This measure has been first used by Newman and Girvan [15] in 2004. The function has been given in Equation (2). CSD problem can be considered as combinatorial optimization problem according to given objective function. The objective function in the best graph clustering reaches maximum *Q* value. *Q* value varies in the range of -1 to +1.

2.4 Genetic Operators

Elitism, selection, crossover and mutation operations have been used in proposed algorithm. Each operator parameters used in the process is adapted to CSD problem to achieve the most suitable solution. Unlike standard genetic algorithm new parameters have been included in elitism, crossover and mutation operators. Operators and parameters proposed by *AGA-net* algorithm are presented in detail below.

Elitism. This operator is used at two stage of the algorithm. In the first stage, it is selected to transfer chromosomes at the rate of *elitismRate (%)* which has the best Q value in the population to the next generation. At the second stage, new chromosomes with the better Q values change place with bad chromosomes at the same rate. Here *elitismRate* ensure the elimination of the worst chromosome from solution cluster. This parameter has been used in small rates to not reduce the chromosome diversity.

Selection. The process of the production of a new generation individual selection process was carried out with the roulette wheel selection (RWS) [31]. In the proposed algorithm selection process according to RWS method are done as follows.

- The fitness value of each chromosome is calculated and sum of all chromosomes fitness value in the population is calculated by Equation (6).

$$TfitnessQ = \sum_{t=1}^{popSize} fitnessQ_t \tag{6}$$

- The selection probability of each chromosome is calculated by Equation (7).

$$P_t = fitnessQ_t / TfitnessQ \tag{7}$$

- The cumulative total is calculated for each chromosome and cumulative probability is determined by Equation (8).

$$Q_t = \sum_{k=1}^{t} P_k \tag{8}$$

- A random number between 0 and 1 is generated. A chromosome is selected according to the generated number's Q_t range. So, chromosomes to be transferred to the next generation are selected.

Crossover. Two different parameters associated with this operator by the names of *crossover rate* (*CR*) and *crossover choice* (*CC*) has been defined. Of these the *CR* parameters will be subjected to the individuals crossover process in the population and determine the number of subjected process. The *CC* parameter will then provide the production of change control sequence for pairs of chromosome subjected to crossover process. The sequence is generated such that the *CC* value would be 0 if it is smaller than the generated random number and 1 if it is bigger. The occurred crossover process has been given in Fig. 2.

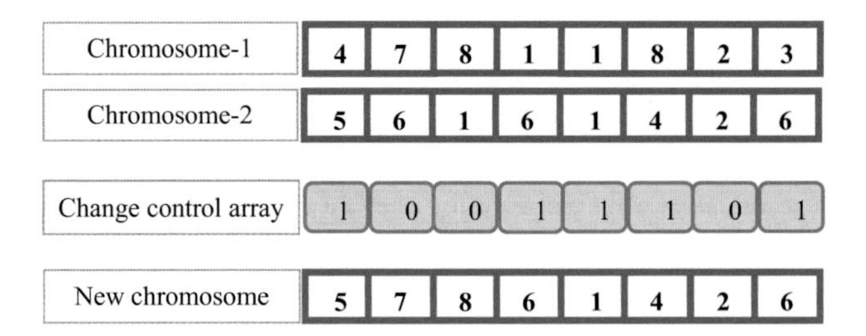

Fig. 2 Crossover operation of *AGA-net* algorithm

Mutation. In the proposed algorithm, mutation process is carried out in two cases, *one-point* and *multi-point*. The first one is single point mutation and the second one is multiple point mutations. Also two parameters have been used in mutation process. The first parameter is *mutation rate* (*MR*) and the second parameter is *multi-point mutation rate* (*multiP*). *MR* parameter is selected in a small ratio and it will determine whether the incoming chromosome mutates or not. And the *multiP* parameter allows the selection of one of the single or multiple mutations option. If the value of this parameter is less than randomly generated number, single point mutation but if it is equal or greater than the randomly generated number multi-point mutation applies. Representative examples showing this process has been given in Fig. 3. Here, each selected genes are mutated by the neighborhood condition (refer to Fig. 1(a)).

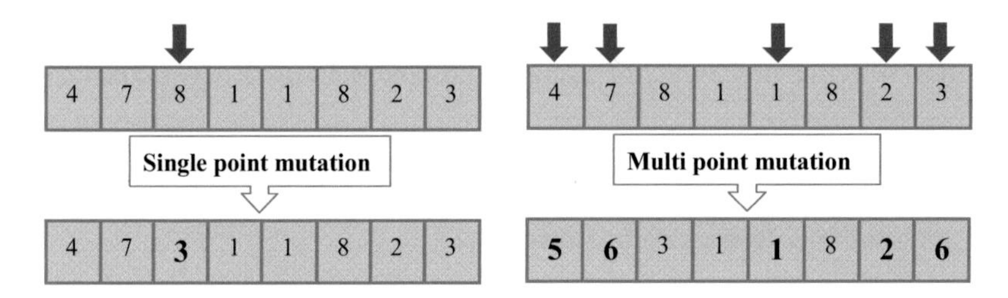

Fig. 3 Single-point and multi-point mutations

3 Experimental Results

In this section the *AGA-net* algorithm has been tested on 5 real networks which are (*Z*) Zachary's Karate Club [32], (*D*) Dolphins Social Network [33], (*A*) American College Football [13], (*B*) Books about US Politics [14] and (*C*) Cattle Protein Interactions (*IntAct*) [34]. These networks are organized as undirected and unweighted. Each node in the network is identified by an *ID*. For example the first node of Cattle Protein Interactions (*IntAct*) network which is *AATM_BOVIN* [35] has been identified by *ID* number 1. Networks and their properties used in the experiments are given in Table 1. All experiments have done on a computer which has the following specifics:

Microsoft Windows 7 (*x64*) OS environment, Intel (R) Core (TM) *i7-3632QM* CPU @ *2.20* GHz processor and *4* GB of RAM.

Table 1 Networks and their properties used in the experiments

Networks[*]	*Number of Nodes (V)*	*Number of Edges (E)*
[*Z*] Zachary's Karate Club	34	78
[*D*] Dolphins Social Network	62	159
[*A*] American College Football	115	613
[*B*] Books about US Politics	105	441
[*C*] Cattle Protein Interactions (IntAct)	268	303

* The given node and edge numbers have been obtained after turning networks to undirected and unweighted by subtracting self-loops featured nodes.

The algorithm was run 50 times for networks in Table 1. The maximum Q, average Q and standard deviation was recorded as test results. The number of population used in the experiments varies according to the size of the network. Population numbers used for networks according to Table 1 order are 20, 30, 100, 80 and 50 respectively. Also for other parameters the following values have been used; for *elitismRate* 0.05, for *crossover rate* 0.8, for *crossover choice* 0.5, for *mutation rate* 0.2 and *multi point mutation* 0.4. These parameters have been tested with experiments inside their borders and the effect of these values to the algorithm has been tested through trial and error. The best results were achieved in this parameter value for all networks.

The proposed algorithm's experimental results were compared with algorithms which are given in Table 2. The results obtained are given in Table 2. In the table the best Q values are indicated in bold and for ease of reading the decimals have been rounded to three digits. For example, the best Q value of Z network was approximately 0.419789612097304 and it was rounded to 0.420. Average *modularity Q* values and standard deviation values for *AGA-net* algorithm are given in Table 3.

Table 2 Comparison of the results according to *modularity Q* values

Networks / *Algorithms*	*Z*	*D*	*A*	*B*
DECD [10]	0.416	-	**0.605**	-
Eigenvector [19]	0.393	0.491	0.488	0.467
GACD [37]	**0.420**	**0.529**	0.604	**0.527**
GATHB [8]	0.402	0.522	0.551	0.518
GN [13]	0.401	0.519	0.599	0.510
FN [16]	0.381	0.510	0.550	0.502
MA-COM [39]	**0.420**	**0.529**	**0.605**	**0.527**
MA-Net [8]	**0.420**	**0.529**	**0.605**	**0.527**
MENSGA [38]	**0.420**	0.527	0.604	0.526
MOGA-Net [36]	0.416	0.505	0.515	0.518
Proposed method (AGA-net)	**0.420**	**0.529**	**0.605**	**0.527**

Apart from the networks given in Table 2, maximum *modularity Q* value obtained for network named as Cattle Protein Interactions (IntAct) is 0.720. After analyzing present studies and as for as it is known this network (*C*) has not been used in CSD problem before. So, this network was not included in the list of comparison given in Table 2.

Table 3 The average *modularity Q* and standard deviation values

Networks	*Z*	*D*	*A*	*B*	*C*
Average Q values	0.420	0.528	0.601	0.527	0.719
Standard deviations	± 0	± 0.000790	± 0.002986	± 0.000137	± 0.000017

Best global and local *modularity Q* values were recorded in each run for all networks. According to obtained results, the average and best *Q* values are almost the same. Also a low level of standard deviation values indicates that the proposed algorithm gave consistent results. For example the standard deviation for *Z* network is zero therefore both the average and the best *Q* values are the same which is 0.420. There are not much different between the obtained results for other networks.

While observing the community numbers generated according to the best *Q* values for *Z*, *D*, *A*, *B* and *C* networks the following community numbers 4, 5, 10, 5 and 40 have been obtained respectively.

When table 2 is examined, till now the *AGA-net* algorithm has reached the best *Q* modularity value for four networks (*Z*, *D*, *A* and *B*) which have been used in literature before. When genetic algorithm-based method proposed by us is compared with other genetic algorithm-based methods given in Table 2 (*GACD* and *GATHB*), it is seen that our proposed method has given better results than other two methods. The most important reason for this is the other two methods mentioned above have limited genotype production mechanism while the crossover and mutation mechanisms used in our proposed method narrows down the solution space to converge to optimal solution.

4 Conclusions

In this paper, CSD problem which is often used in complex networks analysis is discussed and the meaningful information from real-world network has been determined. To test the accuracy of the recommended algorithm the obtained results have been compared with state-of-the-art methods in the literature. In the experiments, four social networks and a biological network have been used. In addition to existing operators and parameters of standard genetic algorithm the *AGA-net* algorithm has been supported by the proposed genetic operators and parameters which are given in section 2.4. These operators and parameters have provided fast convergence of proposed algorithm to global best *Q* value. When analyzing the experimental results it is observed that the *AGA-net* algorithm has obtained the best *modularity Q* values known so far for all networks. These results indicate that *MA-Net* [8] and *MA-COM* [39] algorithms yielded similar results with the proposed algorithm. While the proposed algorithm has obtained the same results with *GACD* [37] algorithm in 3 networks (*Z*, *D* and *B*), it has obtained better result in *A* network. Analogously the proposed algorithm has obtained the same results in *A* network with *DECD* [10] algorithm and better result in *Z* network. And also the proposed algorithm with compare to *MENSGA* algorithm [38] has obtained the same result in *Z* network and better results in other networks.

As a result, the success and consistency of the proposed algorithm can be understood from both comparison and standard deviation tables. In addition, *AGA-net* does not require any prior knowledge and works fast. Thus, with the proposed mechanism almost 20 percent time saving has been provided for each network. This algorithm has been designed in a way which does not consider the size of nodes and edges number therefore it can be applied to all synthetic and real-world networks. After analyzing the test results it has been observed that the proposed algorithm gives successful results for determination of meaningful communities from complex networks.

Acknowledgments This study was supported by Scientific Research Project of Selcuk University and TUBITAK (The Scientific and Technological Research Council of Turkey, 2211-C Domestic Doctoral Scholarship Program Intended for Priority Areas, No. 1649B031402383).

References

1. Fortunato, S.: Community Detection in Graphs. Physics Reports **486**(3), 75–174 (2010)
2. Jackson, M.O.: Social and Economic Networks, vol. 3. Princeton University Press, Princeton (2008)
3. Dunne, J.A., Williams, R.J., Martinez, N.D.: Food-Web Structure and Network Theory: The Role of Connectance and Size. Proceedings of the National Academy of Sciences. **99**(20), 12917–12922 (2002)
4. Gavin, A.-C., et al.: Proteome survey reveals modularity of the yeast cell machinery. Nature **440**, 631–636 (2006)
5. Krogan, N.J., et al.: Global landscape of protein complexes in the yeast saccharomyces cerevisiae. Nature **440**(7084), 637–643 (2006)
6. Lee, J., Hidden, L.J.: information revealed by optimal community structure from a proteincomplex bipartite network improves protein function prediction (2013). http://www.ncbi.nlm.nih.gov/pubmed/23577106
7. Marcus, S., Moy, M., Coffman, T.: Social network analysis. Mining Graph Data, 443–467 (2007)
8. Naeni, L.M., Berretta, R., Moscato, P.: MA-net: a reliable memetic algorithm for community detection by modularity optimization. In: Proceedings of the 18th Asia Pacific Symposium on Intelligent and Evolutionary Systems, vol. 1, pp. 311–323. Springer International Publishing (2015)

9. Liu, D., Jin, D., Baquero, C., He, D., Yang, B., Yu, Q.: Genetic Algorithm with a Local Search Strategy for Discovering Communities in Complex Networks. International Journal of Computational Intelligence Systems **6**(2), 354–369 (2013)
10. Jia, G., et al.: Community detection in social and biological networks using differential evolution. In: Learning and Intelligent Optimization, pp. 71–85. Springer Berlin, Heidelberg (2012)
11. Brandes, U., et al.: On Modularity Clustering. IEEE Transactions on Knowledge and Data Engineering **20**(2), 172–188 (2008)
12. Parpinelli, R.S., Lopes, H.S.: A Computational Ecosystem for Optimization: Review and Perspectives for Future Research. Memetic Computing **7**(1), 29–41 (2015)
13. Girvan, M., Newman, M.E.J.: Community Structure in Social and Biological Networks. Proceedings of the National Academy of Sciences **99**(12), 7821–7826 (2002)
14. Newman, M.E.J.: Modularity and Community Structure in Networks. Proceedings of the National Academy of Sciences **103**(23), 8577–8582 (2006)
15. Newman, M.E.J., Girvan, M.: Finding and Evaluating Community Structure in Networks. Physical Review E **69**(2), 026113 (2004)
16. Newman, M.E.J.: Fast Algorithm for Detecting Community Structure in Networks. Physical Review E **69**(6), 066133 (2004)
17. Blondel, V.D., Guillaume, J.L., Lambiotte, R., Lefebvre, E.: Fast Unfolding of Communities in Large Networks. Journal of Statistical Mechanics: Theory and Experiment **2008**(10), P10008 (2008)
18. Pons, P., Latapy, M.: Computing communities in large networks using random walks. In: Computer and Information Sciences-ISCIS 2005, pp. 284–293. Springer Berlin, Heidelberg (2005)
19. Newman, M.E.J.: Finding Community Structure in Networks using The Eigenvectors of Matrices. Physical Review E **74**(3), 036104 (2006)
20. Raghavan, U.N., Albert, R., Kumara, S.: Near Linear Time Algorithm to Detect Community Structures in Large-Scale Networks. Physical Review E **76**(3), 036106 (2007)
21. Ronhovde, P., Nussinov, Z.: Local Resolution-Limit-Free Potts Model for Community Detection. Physical Review E **81**(4), 046114 (2010)
22. Huang, J., Sun, H., Liu, Y., Song, Q., Weninger, T.: Towards Online Multiresolution Community Detection in Large-Scale Networks. PloS one **6**(8), e23829 (2011)
23. Clauset, A., Newman, M.E.J., Moore, C.: Finding Community Structure in Very Large Networks. Physical Review E **70**(6), 066111 (2004)
24. Duch, J., Arenas, A.: Community Detection in Complex Networks using Extremal Optimization. Physical Review E **72**(2), 027104 (2005)
25. Tasgin, M., Bingol, A.: Communities detection in complex networks using genetic algorithm. In: Proc. of the European Conference on Complex Systems (ECSS 2006) (2006)
26. Tasgin, M., Herdagdelen, A., Bingol, H.: Community Detection in Complex Networks using Genetic Algorithms. arXiv preprint arXiv:0711.0491 [physics.soc-ph] (2007)
27. Li, J., Song, Y.: Community Detection in Complex Networks using Extended Compact Genetic Algorithm. Soft Computing **17**(6), 925–937 (2013)
28. Shi, C., Wang, Y., Wu, B., Zhong, C.: A New Genetic Algorithm for Community Detection, pp. 1298–1309. Springer, Heidelberg (2009)
29. Mitchell, M.: Genetic algorithms: An overview. Complexity **1**(1), 31–39 (1995)
30. Park, Y., Song, M.: A genetic algorithm for clustering problems. In: Proceedings of the Third Annual Conference on Genetic Programming, pp. 568–575 (1998)

31. Sastry, K., Goldberg, D., Kendall, G.: Genetic Algorithms: In: Search methodologies, pp. 97–125. Springer US (2005)
32. Zachary, W.W.: An Information Flow Model for Conflict and Fission in Small Groups. Journal of Anthropological Research, 452–473 (1977)
33. Lusseau, D.: The Emergent Properties of a Dolphin Social Network. Proceedings of the Royal Society of London B: Biological Sciences **270**(Suppl 2), S186–S188 (2003)
34. IntAct Protein-Protein Interaction Network of Cattle. http://biit.cs.ut.ee/graphweb/exampleInput/Cattle_protein_interactions_(IntAct).txt (last accessed date June 1, 2015)
35. UniProtKB, Bos taurus (Bovine) - P12344 (AATM_BOVIN). http://www.uniprot.org/uniprot/P12344 (last accessed date May 15, 2015)
36. Pizzuti, C.: A Multiobjective Genetic Algorithm to Find Communities in Complex Networks. IEEE Transactions on Evolutionary Computation **16**(3), 418–430 (2012)
37. Shi, C., Yan, Z., Wang, Y., Cai, Y., Wu, B.: A Genetic Algorithm for Detecting Communities in Large-Scale Complex Networks. Advances in Complex Systems **13**(01), 3–17 (2010)
38. Li, Y., Liu, G., Lao, S.Y.: A Genetic Algorithm for Community Detection in Complex Networks. Journal of Central South University **20**, 1269–1276 (2013)
39. Gach, O., Hao, J.-K.: A memetic algorithm for community detection in complex networks. In: Coello, C.A.C., Cutello, V., Deb, K., Forrest, S., Nicosia, G., Pavone, M. (eds.) PPSN 2012, Part II. LNCS, vol. 7492, pp. 327–336. Springer, Heidelberg (2012)

A Logical Model of Communication Channels

Matteo Cristani, Francesco Olivieri and Katia Santacà

Abstract A channel is a logical spcae where agents make announcements publicly. Examples of such objects are forums, wikis and social networks. Several questions arise about the nature of such a statement as well as about the attitude of the agent herself in doing these announcements. Does the agent know whether the statement is true? Is this agent announcing that statement or its opposite in any other channel? Extensions to Dynamic Epistemic Logics have been proposed in the recent past that give account to public announcements. One major limit of these logics is that announcements are always considered truthful. It is however clear that, in real life, incompetent agents may announce false things, while deceitful agents may even announce things they do not believe in. In this paper, we shall provide a logical framework, called Multiple Channel Logic, able to relate true statements, agent beliefs, and announcements on communication channels. We discuss syntax and semantics of this logic and show the behaviour of the proposed deduction system. Lastly, we shall present a classification of agents based on the above introduced behaviour analysis.

Keywords Rules · Agents · Multiple channel logic

1 Introduction

In several recent approaches to reasoning about web processes, two different matters tend to intersect: the attitudes of the agents using the web for communication activities, and the beliefs of those agents. Consider, in particular, a situation in which a set of agents use many different communication channels, such as blogs, social networks, forums. It may happen that these agents use those channels in an incoherent

M. Cristani(✉) · F. Olivieri · K. Santacà
Department of Computer Science, University of Verona, Verona, Italy
e-mail: {matteo.cristani,francesco.olivieri,katia.santaca}@univr.it

K. Lavangnananda et al. (eds.), *Intelligent and Evolutionary Systems*,
Proceedings in Adaptation, Learning and Optimization 5,
DOI: 10.1007/978-3-319-27000-5_5

way: for instance one agent may announce one statement in a channel while omitting it in another one, or she can announce a statement in a channel and the opposite statement in another one.

In multiple agent settings, there are two different forms of incoherence. We can look at different agents announcing opposite statements, or to one agent announcing opposite things. Both types of incoherence lie on the same ground: contradictory statements are made. The former type involves different agents making such statements (possibly on the same channel), while the latter type sees one single agent making contradictory announcements (possibly on distinct channels, for certain cases).

In this paper we only look at incoherences generated byn announcements made by one single agent, within a multiple agent logical framework.

We then investigate the set of possible communication attitudes of the agents. A communication attitude is the relation between reality facts and agent beliefs, or the relation between beliefs and announcements the agent makes. It is rather common that agents communicating on channels result not always competent on the matter they are talking about. Some agents can also be insincere, or they can hide some (possibly private) information. The attitudes agents assume when communicating, or the ability to know true facts about the reality are important aspects of the communication processes. When we observe agents communicating, we typically have *prejudices* about their attitudes, where prejudices value an agent behaviour before observing what she announces.

In this paper we investigate how to combine public announcements with beliefs that are also not necessarily aligned with the reality. Somebody can have a false belief or she can believe something that is not known as true or false. Moreover agents can announce things they do not believe, or avoid to announce things they actually believe.

To clarify what we mean with this research boundaries, we provide a general example of announcements and their relations with truth and beliefs.

Example 1. Alice, Bob and Charlie travel quite often for work. They are also passionate about good food and love visiting nice restaurants. To choose the best hotels and restaurants in town, they use as channels the social networks C_1 and C_2, where they are also active users by posting reviews and feedbacks. During their last business trip, they all stayed at the hotel H and they ate at the restaurants R_1, R_2 and R_3. Once back home, Bob posts a review on channel C_1 *announcing* that hotel H was dirty (saying $dirty(H)$), whilst on channel C_2 he announces that H was clean (saying $\neg dirty(H)$). Alice agrees with his announcement on C_1 but does not post any comments on C_2, while Charlie announces $\neg dirty(H)$ both on C_1 and C_2.

We assume that hotel H being clean or dirty is not a matter of opinions but a *provable* fact. It follows that we may draw some conclusions on the statements announced by Alice, Bob and Charlie on C_1 and C_2. First, Bob is not truthful since he announces $dirty(H)$ on C_1 and $\neg dirty(H)$ on C_2. It may be the case that he believes in only one of the two announces (and possibly in none of them) and, consequently, he is lying in one of the two channels.

In this paper, we assume atemporal channels, namely announcements are made in a channel and hold forever and eversince, with respect to the moment in which the announcement is made. When an agent observes a particular channel and another agent contradicts herself in that channel the observer finds it out. Consequently we assume that agents make coherent announcements in every single channel, though it is possible that they make opposite announcements in distinct channels. In temporal channels, an agent might announce a statement, and, later, announce the opposite statement. Provided that she is not lying, this implies a belief revision process she passed through in order to change her point of view.

In general, we include agents that can do more real-life things than just truthful and sincere announcements. They can lie, but it is also possible that they just announce things they simply are not informed about. An agent can make an announcement that corresponds to her belief, or she can claim the opposite of her belief. Moreover she can behave in a combination of the three above basic attitudes on different channels. On the same way, an agent can believe things that are true, that are false, and that are neither true or false.

We assume *consistent* agents, that is agents that either believe in the truthfulness of a given statement, or believe in the truthfulness of the opposite statement; naturally, we allow that an agent may be not competent on a certain topic and, as such, believe in neither of them, but never to believe in both at the same.

Back to the example, while Bob and Charlie announce on every channel, Alice decides to express her opinions just on channel C_1. Finally, Alice and Charlie are consistent with themselves even if not with one another. Therefore, they can be both sincere but one of them is not competent. We shall assume that *competent* agents do *generally* know any topic discussed in every channel, while – admittedly a strong assumption – if the agent is ignorant in at least one topic, then she is considered incompetent.

The structure of the paper is as follows. In Section 2 we define the logical language. In Section 4 we introduce the semantics of the logic and in Section 3 we provide a inference rules of the framework. Section 5 provides the prove of soundness for the introduced logic. Section 6 provides a concise analysis of the related works in literature and we conclude the paper in Section 7 with a summary of the results obtained in the investigation and a proposal of a few further investigations.

2 Multiple Channel Logic

In this section, we present our logical formalism, the *Multiple Channel Logic* (hereafter *MCL*); *MCL* is specified in terms of language, semantics and inference rules.

MCL is a three-layered labelled, modal logical framework. The first layer of *MCL* is a propositional calculus. The second layer is a multi-modal calculus, where we can use three distinct modalities: one modality of *belief*, permitting to assert that an agent believes in a proposition, one modality for stating that a given proposition is asserted by an agent in every channel, and one modality to state that an agent asserts a proposition in at least one channel. The last two are henceforth named *communication modalities*.

The second layer does not allow skolemization, permitting, in particular, only to assert that when an agent believes in a proposition, she cannot believe in the opposite, and that she cannot simultaneously believe in something and not believe in it. The same holds for the communication modalities, combined in a dual fashion. A proposition can not be asserted by an agent in every channel when its opposite is asserted in one channel, and, on the other hand, we cannot assert a proposition in every channel and not asserting it in one channel (and vice versa). Not asserting anywhere a proposition does not imply that the opposite of this proposition is asserted somewhere (and vice versa). When we deal with the deduction rules, in section 3, we shall mark those rules that guarantee this forms of duality.

Within the third layer, we habilitate *agent tagging* with the explicit purpose of allowing assertion of *prejudices* about agent communicative attitudes. In particular, we tag an agent as sincere, collaborative, and other positive or negative tags. The statement of an agent tag associated to a given agent is named a *prejudice*. Predjudices are employed as means to make certain deduction rules apply.

An *MCL* theory $\mathcal{T}$ is a triple $\mathcal{T} = \langle \mathcal{W}, \mathcal{A}, \mathcal{R} \rangle$, where $\mathcal{W}$ is the logical language, $\mathcal{A}$ is the set of axioms, and $\mathcal{R}$ is the set of inference rules, in our specific case, rewriting rules. When the set $\mathcal{A}$ is empty, then we call $\mathcal{T}$ a *calculus*. For a given *MCL* theory M we denote by $\mathcal{R}_M$ the set of axiomn of M. To represent the set of rules $\mathcal{R}$, that is common to any *MCL* theory, we also use, for symmetry with $\mathcal{A}_M$ the notation $\mathcal{R}_M$.

We employ the alphabet $\sum = \mathcal{L} \cup \mathcal{C} \cup \mathcal{M} \cup \mathcal{S} \cup \Lambda \cup \mathcal{T}$ where:

$\mathcal{L}$ is a finite non-empty set of propositional letters $\mathcal{L} = \{A_1, A_2, \ldots, A_n\}$,
$\mathcal{C}$ is the set of connectives $\mathcal{C} = \{\neg, \wedge, \vee \sim, -, \}$,
$\mathcal{M}$ is the set of modalities $\mathcal{M} = \{\mathrm{B}, \mathrm{T}_\Box, \mathrm{T}_\Diamond\}$,
$\mathcal{S}$ is the set of logical signs $\mathcal{S} = \{(,), [,], \bot\}$,
Λ is a finite non-empty set of agents labels $\Lambda = \{\lambda_1, \ldots, \lambda_m\}$,
$\mathcal{T}$ is the set of agent tags $\mathcal{T} = \{Co, S, SCl, WCl, O\}$.

A propositional formula φ is defined by $\varphi := A \mid \neg\varphi \mid \varphi \wedge \varphi \mid \varphi \vee \varphi$ where A denotes a letter. The second layer of *MCL* is a modal logic where the modal operators are B for beliefs, while $\mathrm{T}_\Box$ and $\mathrm{T}_\Diamond$ are the operators for the communication channels. We use $\mathrm{T}_\Box$ ($\mathrm{T}_\Diamond$) to denote that an agent *tells* the embedded formula in every (at least one) communication channel. A modal formula is $\mu ::= \mathrm{B}[\lambda : \varphi] \mid \mathrm{T}_\Box[\lambda : \varphi] \mid \mathrm{T}_\Diamond[\lambda : \varphi] \mid \sim\mu$ where φ denotes a propositional formula. The modal formula for belief $\mathrm{B}[\lambda : \varphi]$ is intended to denote that the agent λ believes φ. For the purpose of this paper, the intended notion of belief embeds the notion of knowledge as meant in Epistemic Logic.

The modal formula $\mathrm{T}_\Box[\lambda : \varphi]$ denotes that the agent λ announces φ everywhere, namely, when λ announces φ in a channel C, then she announces φ in any channel C' that is accessible from C. When we provide the semantics of *MCL* we shall relate the notion of accessibility to the notion of *observation*. A channel C' is accessible from a channel C when the observer of C, also observes C'.

On the third layer of the logical framework we make use of the agent tags. An agent tag formula has one of the two formats $\alpha ::= +(X)\lambda \mid -(X)\lambda$ where $X \in \{Co, S, SCl, WCl, O\}$ is an agent tag, and λ is the label representing the agent. The intended meaning of the tags is *competent* for Co, *sincere* for S, *strongly (weakly) collaborative* for SCl and WCl, respectively and *omniscient* for O.

3 A Deduction System for MCL

In this Section we provide a set of inference rules for *MCL*. These rules are redundant. In the stream of rules below, we prove that some of the rules can be reduced.

The rules can be of three distinct types:

- **Introduction rules** use the elements appearing in the antecendent for building those elements that appear in the subsequent;
- **Elimination rules** de-construct elements appearing in the antecedent into elements in the subsequent;
- $\perp$ **rules** introduce a contradiction, deriving $\perp$ in the subsequent.

The first group of rules, defined below, manage inference on the propositional layer of *MCL*. We adapted the classical presentation of Prawitz [1] for propositional calculus to our needs. We shall then introduce a few other specific $\perp$ rules while providing the single contexts for the inference rules of the propositional layer, for the belief layer and finally for the announcement layer.

Without loss of generality we assume that axioms in a *MCL* theory *M* are all written in *Conjunctive Normal Form*, namely as conjunctions of disjuctions of positive and negative literals.

Below, Introduction is shortened to In., Elimination to El., D.N. for Double Negation, Left is shortened to L and Right to R, and finally Non Contradiction to N.C. The "Ex falso sequitur quodlibet in beliefs" rule is shortened to EFSQ, and Modus Ponens is shortened to MP

R.1 $\dfrac{\varphi \quad \psi}{\varphi \wedge \psi}$ [In. $\wedge$] R.2 $\dfrac{\varphi \wedge \psi}{\varphi}$ [R el. $\wedge$]

R.3 $\dfrac{\varphi \wedge \psi}{\psi}$ [L el. $\wedge$] R.4 $\dfrac{\varphi}{\varphi \vee \psi}$ [In. $\vee$]

R.5 $\dfrac{\varphi \vee \psi}{\psi}$ [R el. of $\vee$] R.6 $\dfrac{\varphi \vee \psi}{\varphi}$ [L el. of $\vee$]

R.7 $\dfrac{\varphi}{\neg\neg\varphi}$ [In. of D.N.] R.8 $\dfrac{\neg\neg\varphi}{\varphi}$ [El. of D.N.]

RC.1 $\dfrac{\varphi \quad \neg\varphi}{\perp}$ [N.C. principle] RC.2 $\dfrac{\perp}{\varphi}$ [EFSQ]

The last operation we need in this group to provide for inferential mechanism is the *Modus Ponens* rule, that we introduce as follows:

$$\text{MP.1}\ \frac{\varphi \quad (\neg\varphi \vee \psi)}{\psi}\ \text{[Propositional modus ponens]}$$

The second group of rules manage inference on the layer of beliefs. The last rule manages the behaviour of combined negations $\sim$ and $\neg$. This is a *quasi*-skolemization, in the sense that it introduces the concept that someone cannot believe that a fact is true and simultaneously not believe that the negation of that fact is false.

$$\text{R.9}\ \frac{\mathrm{B}[\lambda : \varphi] \quad \mathrm{B}[\lambda : \psi]}{\mathrm{B}[\lambda : \varphi \wedge \psi]}\ \text{[In. } \wedge \text{ in B]} \qquad \text{R.10}\ \frac{\mathrm{B}[\lambda : \varphi \wedge \psi]}{\mathrm{B}[\lambda : \varphi]}\ \text{[R el. } \wedge \text{ in B]}$$

$$\text{R.11}\ \frac{\mathrm{B}[\lambda : \varphi \wedge \psi]}{\mathrm{B}[\lambda : \psi]}\ \text{[L el. } \wedge \text{ in B]} \qquad \text{R.12}\ \frac{\mathrm{B}[\lambda : \varphi]}{\mathrm{B}[\lambda : \varphi \vee \psi]}\ \text{[In. } \vee \text{ in B]}$$

$$\text{R.13}\ \frac{\mathrm{B}[\lambda : \varphi]}{\mathrm{B}[\lambda : \neg\neg\varphi]}\ \text{[In. of D.N. in B]} \qquad \text{R.14}\ \frac{\mathrm{B}[\lambda : \neg\neg\varphi]}{\mathrm{B}[\lambda : \varphi]}\ \text{[El. of D.N. in B]}$$

$$\text{R.15}\ \frac{\mathrm{B}[\lambda : \varphi]}{\sim \mathrm{B}[\lambda : \neg\varphi]}\ \text{[Coherence of disbeliefs]}$$

We provide the $\perp$ rules for the belief layer. The first rule is used for belief of contradiction, while the second rule is used for belief contradiction. A belief contradiction occurs when the claim of belief is contradicted by the claim of corresponding disbelief. We also have a belief contradiction rule corresponding to "Ex falso sequitur quodlibet in beliefs". We then introduce the Modus Ponens rule for beliefs.

$$\text{RC.3}\ \frac{\mathrm{B}[\lambda : \varphi] \quad \mathrm{B}[\lambda : (\neg\varphi)]}{\perp}\ \text{[B of contr.]} \qquad \text{RC.4}\ \frac{\mathrm{B}[\lambda : \perp]}{\mathrm{B}[\lambda : \varphi]}\ \text{[B EFSQ]}$$

$$\text{RC.5}\ \frac{\mathrm{B}[\lambda : \varphi] \quad \sim \mathrm{B}[\lambda : \varphi]}{\perp}\ \text{[B contr.]} \qquad \text{MP.2}\ \frac{\mathrm{B}[\lambda : \varphi] \quad \mathrm{B}[\lambda : \neg\varphi \vee \psi]}{\mathrm{B}[\lambda : \psi]}\ \text{[B MP]}$$

We obtain the $\perp$ when we believe in φ and either we believe in the opposite, or we do not believe in it.

The third group of rules is introduced to manage inference on $\mathrm{T}_{\Box}$ and $\mathrm{T}_{\Diamond}$ modalities. The first subgroup, formed by the rules R.16–R.20, supplies the five classical rules for introduction, left and right elimination for $\wedge$, introduction and elimination for double negation for $\mathrm{T}_{\Box}$. The rules R.28 and R.29 represent, respectively, Coherence of missing announcements (COM) and Coherence of provided announcements (COP), and correspond to the quasi-skolemization of the modalities $\mathrm{T}_{\Box}$ and $\mathrm{T}_{\Diamond}$. Seriality is shortened below to Ser.

$$\text{R.16}\ \frac{\mathrm{T}_{\Box}[\lambda:\varphi]\quad \mathrm{T}_{\Box}[\lambda:\psi]}{\mathrm{T}_{\Box}[\lambda:\varphi\wedge\psi]}\ [\text{In.}\wedge\text{ on }\mathrm{T}_{\Box}] \qquad \text{R.17}\ \frac{\mathrm{T}_{\Box}[\lambda:\varphi\wedge\psi]}{\mathrm{T}_{\Box}[\lambda:\varphi]}\ [\text{R el.}\wedge\text{ on }\mathrm{T}_{\Box}]$$

$$\text{R.18}\ \frac{\mathrm{T}_{\Box}[\lambda:\varphi\wedge\psi]}{\mathrm{T}_{\Box}[\lambda:\psi]}\ [\text{L el.}\wedge\text{ on }\mathrm{T}_{\Box}] \qquad \text{R.19}\ \frac{\mathrm{T}_{\Box}[\lambda:\varphi]}{\mathrm{T}_{\Box}[\lambda:\varphi\vee\psi]}\ [\text{In.}\vee\text{ in }\mathrm{T}_{\Box}]$$

$$\text{R.20}\ \frac{\mathrm{T}_{\Box}\varphi}{\mathrm{T}_{\Box}[\lambda:\neg\neg\varphi]}\ [\text{In. of D.N. on }\mathrm{T}_{\Box}] \qquad \text{R.21}\ \frac{\mathrm{T}_{\Box}[\lambda:\neg\neg\varphi]}{\mathrm{T}_{\Box}[\lambda:\varphi]}\ [\text{El. of D.N. on }\mathrm{T}_{\Box}]$$

$$\text{R.22}\ \frac{\mathrm{T}_{\Diamond}[\lambda:\varphi\wedge\psi]}{\mathrm{T}_{\Diamond}[\lambda:\varphi]}\ [\text{R el.}\wedge\text{ on }\mathrm{T}_{\Diamond}] \qquad \text{R.23}\ \frac{\mathrm{T}_{\Diamond}[\lambda:\varphi\wedge\psi]}{\mathrm{T}_{\Diamond}[\lambda:\psi]}\ [\text{L el.}\wedge\text{ on }\mathrm{T}_{\Diamond}]$$

$$\text{R.24}\ \frac{\mathrm{T}_{\Diamond}[\lambda:\varphi]}{\mathrm{T}_{\Diamond}[\lambda:\varphi\vee\psi]}\ [\text{In.}\vee\text{ in }\mathrm{T}_{\Diamond}] \qquad \text{R.25}\ \frac{\mathrm{T}_{\Diamond}\varphi}{\mathrm{T}_{\Diamond}[\lambda:\neg\neg\varphi]}\ [\text{In. of D.N. on }\mathrm{T}_{\Diamond}]$$

$$\text{R.26}\ \frac{\mathrm{T}_{\Diamond}[\lambda:\neg\neg\varphi]}{\mathrm{T}_{\Diamond}[\lambda:\varphi]}\ [\text{El. of D.N. on }\mathrm{T}_{\Diamond}] \qquad \text{R.27}\ \frac{\mathrm{T}_{\Box}[\lambda:\varphi]}{\mathrm{T}_{\Diamond}[\lambda:\varphi]}\ [\text{Ser. }\mathrm{T}_{\Box}\text{ on }\mathrm{T}_{\Diamond}]$$

$$\text{R.28}\ \frac{\mathrm{T}_{\Box}[\lambda:\varphi]}{\sim\mathrm{T}_{\Diamond}[\lambda:\neg\varphi]}\ [\text{COM}] \qquad \text{R.29}\ \frac{\mathrm{T}_{\Diamond}[\lambda:\varphi]}{\sim\mathrm{T}_{\Box}[\lambda:\neg\varphi]}\ [\text{COP}]$$

If agent λ announces φ on every channel, it is straightforward that she announces it on at least one channel (R.27). If λ announces φ on every channel, then in no channel she may announces the opposite (R.28). Lastly, if λ announces φ on at least one channel, then she cannot announce the opposite on every channel (R.29).

We now present the $\perp$ rules for announcements. Announcement contradictions are shortened below to AC.

$$\text{RC.6}\ \frac{\mathrm{T}_{\Diamond}[\lambda:(\varphi\wedge\neg\varphi)]}{\perp}\ [\text{AC on }\mathrm{T}_{\Box}] \qquad \text{RC.7}\ \frac{\mathrm{T}_{\Box}[\lambda:(\varphi\wedge\neg\varphi)]}{\perp}\ [\text{AC}]$$

$$\text{RC.8}\ \frac{\mathrm{T}_{\Box}[\lambda:\varphi]\quad \mathrm{T}_{\Diamond}[\lambda:\neg\varphi]}{\perp}\ [\text{AC on }\mathrm{T}_{\Diamond}]$$

RC.8 is derived from R.28–R.29: announcing φ on every channel contradicts with announcing $\neg\varphi$ somewhere. Again we provide a modus ponens rule for $\mathrm{T}_{\Box}$ and $\mathrm{T}_{\Diamond}$. The first is Channel existential Modus Ponens (CEMP), the second is Channel Universal Modus Ponens (CUMP).

$$\text{MP.3}\ \frac{\mathrm{T}_{\Box}[\lambda:\varphi]\quad \mathrm{T}_{\Diamond}[\neg\varphi\vee\psi]}{\mathrm{T}_{\Diamond}:[\lambda:\psi]}\ [\text{CEMP}] \qquad \text{MP.4}\ \frac{\mathrm{T}_{\Diamond}[\lambda:\varphi]\quad \mathrm{T}_{\Box}[\neg\varphi\vee\psi]}{\mathrm{T}_{\Diamond}:[\lambda:\psi]}\ [\text{CUMP}]$$

We need to manage introduction and elimination of missing beliefs and missing announces. We summarise the above by using the expression μ to denote a modal formula of *MCL*.

$$\text{R.30}\ \frac{\mu[\lambda:\varphi]}{\sim\sim\mu[\lambda:\varphi]}\ [\text{In. of double}\sim] \qquad \text{R.31}\ \frac{\sim\sim\mu[\lambda:\varphi]}{\mu[\lambda:\varphi]}\ [\text{El. of double}\sim]$$

We can introduce the rules for the agent tags. We have two rules that relate reality and beliefs, based on the expressed prejudice of omniscience and competence. Moreover, we have two pairs of rules for weak and strong collaboration tags, and one rule for sincereness which relates beliefs and communication channels.

All the above mentioned rules are employed for positive tags. For every positive tag, we have the dual rule for the negative counterpart.

$$\text{R.32}\ \frac{\varphi \quad +(O)\lambda}{\mathrm{B}[\lambda : \varphi]}\ [\text{Ext.}\ +(O)] \qquad \text{R.33}\ \frac{\mathrm{B}[\lambda : \varphi] \quad +(Co)\lambda}{\varphi}\ [\text{Ext.}\ +(Co)]$$

$$\text{R.34}\ \frac{\mathrm{B}[\lambda : \varphi] \quad +(Wcl)\lambda}{\mathrm{T}_{\diamond}[\lambda : \varphi]}\ [\text{Ext.}\ +(WCl)] \qquad \text{R.35}\ \frac{\mathrm{B}[\lambda : \varphi] \quad +(SCl)\lambda}{\mathrm{T}_{\Box}[\lambda : \varphi]}\ [\text{Ext.}\ +(SCl)]$$

$$\text{R.36}\ \frac{\mathrm{T}_{\diamond}[\lambda : \varphi] \quad +(S)\lambda}{\mathrm{B}[\lambda : \varphi]}\ [\text{Ext.}\ +(S)] \qquad \text{R.37}\ \frac{\sim \mathrm{B}[\lambda : \varphi] \quad \varphi}{-(O)\lambda}\ [\text{In.}\ -(O)]$$

$$\text{R.38}\ \frac{\mathrm{B}[\lambda : \varphi] \quad \neg\varphi}{-(Co)\lambda}\ [\text{In.}\ -(Co)] \qquad \text{R.39}\ \frac{\mathrm{B}[\lambda : \varphi] \quad \sim \mathrm{T}_{\diamond}[\lambda : \varphi]}{-(Wcl)\lambda}\ [\text{In.}\ -(WCl)]$$

$$\text{R.40}\ \frac{\mathrm{B}[\lambda : \varphi] \quad \sim \mathrm{T}_{\Box}[\lambda : \varphi]}{-(Scl)\lambda}\ [\text{In.}\ -(SCl)] \qquad \text{R.41}\ \frac{\mathrm{T}_{\diamond}[\lambda : \varphi] \quad \sim \mathrm{B}[\lambda : \varphi]}{-(S)\lambda}\ [\text{In.}\ -(S)]$$

An agent is omniscient if she knows every true formula. An agent is competent is every formula she knows is true. Notice that in the case of omniscience, the set of formulae believed true by the agent is a superset of the (actually) true formulae: the agent knows all what is true but she may also believe in some formulae proven neither true, nor false. On the contrary, in case of competence, the agent's beliefs are a subset of the true formulae; as such, all the agent's beliefs are proven to be true but there may be true formulae "out" of her knowledge base.

If an agent believes that φ is true and she is weakly collaborative, then she will announces φ in at least one channel. Sincerity relates the communication of an agent with her beliefs. As such, a sincere agent that tells φ on a channel, then she believes φ to be true. We might be tempted to formulate sincerity from "the other agents' perspective" and state that occurs whenever an agent announces φ somewhere while $\neg\varphi$ anywhere. The proposed formulation of R.36 has the advantage that a sincere agent cannot contradict herself. This would lead to a contradiction due to λ being sincere and the application of R.36 to both $\mathrm{T}_{\diamond}[\lambda : \varphi]$ and $\mathrm{T}_{\diamond}[\lambda : \overline{\varphi}]$, and the subsequent application of RC.3.

Sincerity and collaboration do not derive one another. Assume three channels and that agent λ believes both formulae φ and ψ to be true. Suppose that λ announces φ on the first channel, $\neg\varphi$ on the second one, and ψ on the third one. In this setting, λ is collaborative but not sincere. Suppose now that λ solely announces ψ on the first channel. In this case, λ is sincere but not collaborative. It is straightforward to notice that R.35 subsumes R.34 through seriality. In addition, it implies a subtle form of sincerity. A strongly collaborative agent cannot announce anything she does not believe to be true, even if she might announce something she believes to be neither true nor false. (Indeed, whenever the strongly collaborative λ has $\sim\mathrm{B}\varphi$ as well as $\sim\mathrm{B}\neg\varphi$, nothing prevent her to announce either φ or $\neg\varphi$ somewhere.)

Trivially, an agent: (R.37) is non-omniscient whenever she does not believe true an actually true formula, (R.38) is incompetent whenever she believe true a false formula. An agent is not weakly (strongly) collaborative if she knows something which she does not announce in at least one channel. Finally, an insincere agent announces, in at least one channel, a formula she does not believe to be true. In our formulation, it is not necessary that the agent believes in the truthfulness of a formula φ while announcing $\neg\varphi$ to be considered insincere.

R.38 depends on R.37. In fact, given the premises of R.37, by applying R.15 to $\mathrm{B}[\lambda : \varphi]$ we obtain $\sim\mathrm{B}[\lambda : \neg\varphi]$.

Finally we introduce the rules that provide contradictions between prejudices (positive and negative). The set of rules can be summarised by the expressions $+(P)\lambda$ as the assertion of the prejudice P on λ, and $-(P)\lambda$ as the negation of the prejudice P on λ.

RC.9 $\dfrac{+(P)\lambda \quad -(P)\lambda}{\bot}$ [Prejudice contradiction]

4 The Semantics of *MCL*

In *MCL* the announcement of a formula by one agent cannot be bound to appear on a single, specific channel. We can only bind an announcement to appear in either all channels, or at least one. We therefore employ a semantics for the modalities that follows Kripke's modelling guidelines.

In order to build the semantics of a *MCL* theory we provide the interpretation of the signature, of first-layer formulae, of second-layer formulae and of agent tags. Accordingly, the semantics of a *MCL* theory is a tuple $\mathcal{M} = \langle \mathcal{W_C}, \mathcal{W_A}, \mathcal{W_R}, \mathcal{W_L}, \mathcal{R_C}, \mathcal{I} \rangle$. where $\mathcal{W_C}$ is the domain of *Channels*, $\mathcal{W_A}$ is the domain of *Agents*, $\mathcal{W_R}$ is the Boolean twofold interpretation domain $\{true, false\}$, $\mathcal{W_L}$ is the the domain of the *Letters*, a finite non-empty set, as numerous as the letters in the signature of *MCL*, $\mathcal{R_C}$ is the accessibility relation between elements of $\mathcal{W_C}$, $\mathcal{I}$ is the *Interpretation* function.

Given a *MCL* theory L, the interpretation function maps every propositional letter of the signature of L in one element of $\mathcal{W_R}$, every agent letter in one element of $\mathcal{W_A}$, every channel letter in one element of $\mathcal{W_C}$, and finally every agent tag in a subset of $\mathcal{W_A}$. The accessibility relation is defined in $\mathcal{W_C}$. The accessibility relation captures the idea that a channel C_1 is related to a channel C_2 *iff* the external observer representing the theory can observe C_2 whenever she observes C_1. The accessibility relation is assumed therefore to be reflexive and transitive, while we do not make any assumption about symmetry.

The interpretation of $\bot$ is $\mathcal{I}(\bot) = false$. The truth of a propositional formula follows the classic interpretation of $\wedge$, $\vee$ and $\neg$ operators.

Interpretations are models when they provide consistent evaluations for the propositional layer, the beliefs, the announcements and the relationships between the above determined by the expression of predjudices.

In particular, we assume that for a model the following hold:

1. The interpretation of the set of propositional axioms is consistent with the interpretation of letters;
2. The interpretation of each agent belief set is a consistent propositional theory with respect to the interpretation of pairs formed by agent and propositional letters;
3. The set of announcements for each agent in every single channel is a consistent propositional theory with respect to the interpretation of triples formed by agent, channel and propositional letters;
4. For every modal formula $T_{\Box}[\lambda : \varphi]$ asserted axiomatically in L and for every channel C in which φ is announced by λ, φ is announced by λ in every channel related from C by the accessibility relation;
5. For every modal formula $T_{\Diamond}[\lambda : \varphi]$ asserted axiomatically in L and for every channel C in which φ is announced by λ, φ is announced by λ in at least one channel related from C by the accessibility relation;
6. For every *omniscient* agent λ and every formula φ that is interpreted true, then φ is also a belief of λ;
7. For every *competent* agent λ and every formula φ believed by λ, then φ is interpreted true;
8. For every *sincere* agent λ, if a formula φ is announced in one channel by λ, then φ is a belief of λ;
9. For every *strongly collaborative* agent λ, and every belief φ of λ, λ announces φ in every channel;
10. For every *weakly collaborative* agent λ, and every belief φ of λ, λ announces φ in at least one channel.

As usual, when a *MCL* theory L has a model, then L is called *satisfiable*. Conversely, when it has no model is called *unsatisfiable*. A set of axioms containing the symbol $\perp$ is unsatisfiable.

When we value semantics as defined above, we name such a model a *MCL*-model, and we say that this holds for a *MCL*-semantics.

To explain how the interpetation of a theory works we introduce here an example.

Example 2. Consider a set of at least two agents. The construction of the model is performed as follows: starting from a set of axioms $A_1, A_2, \ldots A_n$, corresponding to the interpetation of the letters $A_1^{\mathcal{I}} \ldots A_n^{\mathcal{I}}$.

An interpretation is, in fact, a set of literals assumed true: $A_1, A_3, \overline{A_5}, A_6$.

An assignments for the beliefs of the agents is something like:

$$\lambda_1 \; A_1, \overline{A_3}, A_6$$

$$\lambda_2 \quad \overline{A_1}, A_6$$

$$\lambda_3 \qquad A_2$$

and we can define for each C, where C is a channel, an assignement as in the scheme below.

$$\begin{array}{cccc} \lambda_{1,C1}\ A_1,\ A_2\ \lambda_{2,C1} & \dots\ \lambda_{3,C1} & \dots\ \lambda_{N,C1} & \dots \\ \lambda_{1,C2}\ A_1,\ A_2\ \lambda_{2,C2} & \dots\ \lambda_{3,C2} & \dots\ \lambda_{N,C1} & \dots \\ \vdots\ \vdots & \vdots\ \vdots & \vdots\ \vdots & \vdots\ \vdots \\ \lambda_{1,Cm}\ A_1,\ A_2\ \lambda_{2,Cm} & \dots\ \lambda_{3,Cm} & \dots\ \lambda_{N,CM} & \vdots \end{array}$$

On the other hand, a channel is a space for interpreting announcements of the agents. The underlying idea, again, is that a single agent in a single channel announces things in a coherent way, being irrational that an agent contradicts herself in a completely observable channel. Thus, the set of axioms corresponding to agent assertion quantifications are mapped to channels by choosing a set of literals for each agent label that, assumed true, makes coherent the statements of the agent mapped in that particular channel.

5 Formal Properties of *MCL*

In this section we prove that *MCL* is sound and complete with respect to the introduced semantics. For the sake of space, the proofs of simplest results are omitted, and we only deal with the main ones.

The task we look at is *consistency checking*. Given a *MCL* theory L, we aim at establishing whether the set of axioms in L are consistent with each other, or, in other terms, whether they can or cannot derive a contradiction.

The first property we prove is that the set of deduction rules introduced for *MCL* theories preserve satisfiability. When employing the deduction rules of *MCL* we transform a theory L into other theories, called *derived from* L. We specifically say that a theory L is deductively closed when applying the deduction rules to L we obtain L. Conversely, when applying rules to a theory L leads to a theory L', and further on, to a theory L^*, that is deductively closed, we name L^* the *deductive closure* of L.

Lemma 1. *Given a satisfiable* MCL *theory L, if L' is derived from L, then L' is satisfiable.*

Proof. The proof is a direct consequence of the rules R.1-R.40 and of the rules MP.1-MP.4, along with the definition of *MCL*-models.

Lemma 2. *Given an unsatisfiable* MCL *theory L, if L' is derived from L, then L' is unsatisfiable.*

Proof. The proof is a direct consequence of Lemma 1 and of rules RC.1-RC.9, along with the definition of *MCL*-models..

An immediate, straightforward consequence of the above mentioned lemmas is claimed in Corollary 1, whose proof is omitted.

Corollary 1. *The deductive closure L^* of a* MCL *theory L is consistent* iff *L is consistent.*

Based on Lemma 1, 2 and Corollary 1, we can conclude Theorem 1.

Theorem 1. *The rules of* MCL *preserve satisfiability.*

Once we have proved that the rules preserve satisfiability, we are now able to claim a soundness result. The soundness result is obtained by means of the rules, the standard interpretation of $\bot$ and the notion of *MCL*-model.

First of all, we claim that when a theory is contradictory, its deductive closure contains $\bot$. Secondly, we prove the inverse: if a *MCL* theory L' contains $\bot$, when another theory L can be transformed by the rules onto L', then L contains either $\bot$ or a contradiction. When, given a theory L', there is not a theory L such that L' is derived from L, we say that L' is *primary*.

Lemma 3. *The deductive closure of a contradictoy theory contains $\bot$.*

Lemma 4. *Every theory L' containing $\bot$ that is not primary, can be derived from a different theory L that either contains $\bot$ or is contradictory.*

As a consequence of the above lemmas we conclude the following theorem.

Theorem 2. *The deduction system of* MCL *is sound.*

In order to prove the completeness of the *MCL* deduction rules with respect to the semantics introduced in Section 4 we firstly show the following lemma.

Lemma 5. *If a* MCL *theory L is satisfiable, then it is consistent.*

Proof. Lemmas 3 and 4 prove that deductively closed satisfiable theories do not contain $\bot$. As a consequence, if a theory had a model, then it would be consistent.

Based on Lemma 5 we can finally prove the following theorem.

Theorem 3. *The deduction system of* MCL *is complete.*

6 Related Work

There is a rather long research stream on the problem of aligning beliefs and announcements, that starts from Dynamic Epistemic Logic. See [2], for a general framework analysis in the early stage of Public Announcement Logics (PAL), and recent development in [3–5]. Although these scholars have deeply dealt with the problem of announcements, they have made a very strong, and clearly oversimplified assumption: agents are always truthful and sincere. As a consequence, an observer on a communication channel always trusts the agents making announcements on that channel. In recent developments on PAL, the possibility that an announcement is made producing changes in beliefs of agents is provided in the form of belief revision operators:

whenever an agent belief contrasts with what announced, she revises her knowledge base [5].

On the other hand, there have been meany scholars who concentrated their attention upon the ways in which agents communicate false announcements, as recently analysed in [6].

Back to early stage investigations on Public Announcements, the idea that someone could forecast others' lies is incorporated within the logic itself [7]. A more recent study focused on communication, and introduces, in a different way with respect to the approach we adopted here, a notion of channel [8]. The framework is Dynamic Epistemic Logic. Authors consider only truthful communications. A message can be sent only if the sender knows that the message is true. Agents cannot lie. Agents communicate by sending messages to a group of other agents. This is distinct from passing on channels where: (i.) The agent can choose on which channels say what, (ii.) the agent has not control on who is looking at those channels, he thus does not have control on who can access her announcement. The authors deal with updating the knowledge of each agent after a message has been sent.

In [9], authors dealt with the problem of how to express a semantics for Agent Communication Logic in order to make non-monotonic inferences on the ground of speech acts. The above mentioned researches have exploited the flaws we referred to in this paper. Some attempts to solve these flaws have been proposed in [10–16]. Although interesting perspectives, the focus of those papers and their aims share little with the purpose of this work.

The most comprehensive investigation about lying agents, from the viewpoint of agent communication logical framework is [6].

First things first, what author considers a lie? You lie to me that p, if you believe that p is false while you say that p, and with the intention that I believe p.

This is the first strong difference between the two approaches. An agent's beliefs are typically private, as such another agent can only guess if an agent believes something or not. Therefore, lying in our framework is based on just what an agent announces; an agent thus lies if she announces p as well as $\neg p$ on (possibly) different channels.

The author claims that a lying agent considers that p is false when she announces it. We value this viewpoint not exhaustive. If an agent has not any knowledge regarding p nor $\neg p$ but she announces p on a channel while $\neg p$ on (possibly) another channel, we say that even in this situation that agent is lying. (In [6], the author calls it *bluffing*, whereas we have named it incompetent.)

Another strong difference is that a lie is successful if the recipient of the lie ends up believing in its truthfulness (provided the type of this recipient agent, that is credulous, sceptical, or revising). Even if we can understand the author aim, we do not agree on it. A lie is a lie, independently on whether, after it has been told, some agents end up in believing in it. If you announce that the authors of this work are females, we shall not believe it, but we know that you are a liar (or a joker in a more sympathetic scenario).

An agent observing the communication channels is not really interested in what the beliefs of another agent are, but rather on *what is said* on the channels under her observation. The information she uses by her deduction process can therefore rely exclusively on those announcements, her knowledge and her own beliefs, and she can combine all such information to make her own prejudices about the other agents.

To complete this short analysis of the reference literature, an important investigation regards complexity of reasoning in PAL. In [17], the author proves two interesting results: Satisfiability in single-agent PAL is NP-complete and in multi-agent PAL is PSpace-complete.

7 Conclusions

In this paper we dealt with the problem of combining beliefs and announcements in a framework that also allows to provide prejudices about agent communication attitudes. The basic results we obtained are: (i.) a formalisation of the modal logic *MCL* which allows to express facts, beliefs and announcements, (ii.) the analysis of a semantics for this logic, and (iii.) the proof of soundness for the logic itself.

As stated in Section 1, we have based our work on the stream of extensions to Dynamic Epistemic Logic, in particular referring to PAL, which was originally proposed by Plaza in [18]. The basis of our approach has been to quit the oversimplified assumption of truthfulness of agents. Issues about truthfulness of agents have often been dealt with in PAL and other agent-based logic approaches, as in [19].

There are several ways in which this research can be taken further. First, we shall investigate completeness of *MCL*. Moreover, we are developing a tableaux approach to the consistency checking, essentially by extending the method used for proving soundness in this paper. We are finally looking at extensions to the logical framework to cover *partially observable channels* that include temporal aspects and access permissions. Agent tags presented in Section 3 might be refined. For instance, an agent can be considered insincere only when announcing the opposite of a belief of hers, while an agent making a statement on which she has no knowledge of truthfulness might be classified as braggart.

Acknowledgements The authors thank Luca Viganò and Margherita Zorzi for hints and discussion. Matteo Cristani and Francesco Olivieri are supported by project MYENERGY by Sardegna Ricerche.

References

1. Prawitz, D.: On the idea of a general proof theory. Synthese **27**(1–2), 63–77 (1974)
2. van Benthem, J., van Eijck, J., Kooi, B.P.: Logics of communication and change. Inf. Comput. **204**(11), 1620–1662 (2006)
3. Balbiani, P., Guiraud, N., Herzig, A., Lorini, E.: Agents that speak: modelling communicative plans and information sources in a logic of announcements. In: 10th International Conference on Autonomous Agents and Multiagent Systems (AAMAS 2011), Taipei, Taiwan, May 2-6, 2011, vol. 1-3, pp. 1207–1208, (2011)

4. Sonenberg, L., Stone, P., Tumer, K., Yolum, P. (eds.): 10th International Conference on Autonomous Agents and Multiagent Systems (AAMAS 2011), Taipei, Taiwan, May 2-6, 2011, Vol. 1-3. IFAAMAS (2011)
5. Balbiani, P., Seban, P.: Reasoning about permitted announcements. J. Philosophical Logic **40**(4), 445–472 (2011)
6. van Ditmarsch, H.: Dynamics of lying. Synthese **191**(5), 745–777 (2014)
7. Batlag, A., Moss, L.S., Solecki, S.: The logic of public announcements and common knowledge and private suspicions. In: Gilboa, I., (ed.) Proceedings of the 7th Conference on Theoretical Aspects of Rationality and Knowledge (TARK-98), Evanston, IL, USA, July 22-24, 1998, pp. 43–56. Morgan Kaufmann (1998)
8. Sietsma, F., van Eijck, J.: Message passing in a dynamic epistemic logic setting. In: Apt, K.R. (ed.): Proceedings of the 13th Conference on Theoretical Aspects of Rationality and Knowledge (TARK-2011), Groningen, The Netherlands, July 12-14, 2011, pp. 212–220. ACM (2011)
9. Boella, G., Governatori, G., Hulstijn, J., Riveret, R., Rotolo, A., van der Torre, L.: Time and defeasibility in FIPA ACL semantics. J. Applied Logic **9**(4), 274–288 (2011)
10. Singh, M.P.: A social semantics for agent communication languages. LNCS, vol. 1916, pp. 31–45. Springer (2000)
11. Verdicchio, M., Colombetti, M.: From message exchanges to communicative acts to commitments. Electr. Notes Theor. Comput. Sci. **157**(4), 75–94 (2006)
12. Gaudou, B., Herzig, A., Longin, D.: A logical framework for grounding-based dialogue analysis. Electr. Notes Theor. Comput. Sci. **157**(4), 117–137 (2006)
13. Gaudou, B., Herzig, A., Longin, D., Nickles, M.: A new semantics for the FIPA agent communication language based on social attitudes. In: ECAI 2006, 17th European Conference on Artificial Intelligence, August 29 - September 1, 2006, Riva del Garda, Italy, pp. 245–249 (2006)
14. Boella, G., Damiano, R., Hulstijn, J., van der Torre, L.W.N.: Role-based semantics for agent communication: embedding of the 'mental attitudes' and 'social commitments' semantics, pp. 688–690. ACM (2006)
15. Boella, G., Damiano, R., Hulstijn, J., van der Torre, L.: A common ontology of agent communication languages: Modeling mental attitudes and social commitments using roles. Applied Ontology **2**(3–4), 217–265 (2007)
16. Nickles, M., Fischer, F.A., Weiß, G.: Communication attitudes: A formal approach to ostensible intentions, and individual and group opinions. Electr. Notes Theor. Comput. Sci. **157**(4), 95–115 (2006)
17. Lutz, C.: Complexity and succinctness of public announcement logic. In: 5th International Joint Conference on Autonomous Agents and Multiagent Systems (AAMAS 2006), Hakodate, Japan, May 8-12, 2006, pp. 137–143 (2006)
18. Plaza, J.: Logics of public communications. Synthese **158**(2), 165–179 (2007)
19. Wooldridge, M.: Semantic issues in the verification of agent communication languages. Autonomous Agents and Multi-Agent Systems **3**(1), 9–31 (2000)

Part II
Data Mining and Its Applications

A New Approach for Wrapper Feature Selection Using Genetic Algorithm for Big Data

Waad Bouaguel

Abstract The increased dimensionality of genomic and proteomic data produced by microarray and mass spectrometry technology makes testing and training of general classification method difficult. Special data analysis is demanded in this case and one of the common ways to handle high dimensionality is identification of the most relevant features in the data. Wrapper feature selection is one of the most common and effective techniques for feature selection. Although efficient, wrapper methods have some limitations due to the fact that their result depends on the search strategy. In theory when a complex search is used, it may take much longer to choose the best subset of features and may be impractical in some cases. Hence we propose a new wrapper feature selection for big data based on a random search using genetic algorithm and prior information. The new approach was tested on 2 biological dataset and compared to two well known wrapper feature selection approaches and results illustrate that our approach gives the best performances.

Keywords Wrapper · Feature selection · Big data

1 Introduction

Over the past few years, the problem of understanding cancer treatment went from basic to one of the most important task in data mining, thanks to expanding knowledge of cancer genomics and the technologies that make such understanding possible [1].

Genomic sequencing is continuously changing the way we understand cancer. Over the time, we have come to a point where the challenge is not so much how to generate large amounts of data, but how to connect the enormous amounts of genomic data churned out by ever-advancing technologies so that they translate into

W. Bouaguel (✉)
LARODEC, ISG, University of Tunis, Tunis, Tunisia
e-mail: bouaguelwaad@mailpost.tn

K. Lavangnananda et al. (eds.), *Intelligent and Evolutionary Systems*,
Proceedings in Adaptation, Learning and Optimization 5,
DOI: 10.1007/978-3-319-27000-5_6

meaningful cancer prevention and treatment strategies. As there are thousands of gene expressions and only a few dozens of observations in a typical gene expression data set, the number of genes d is usually of order 1000 to 10000 while n the number of biological observations is somewhere between 10 and 100 [1]. Such a condition makes the application of many classification methods a hard task.

Feature selection aims at identifying a subset of features for building robust learning models. Since only a small number of genes among tens of thousands show strong correlation with the targeted disease, some works address the problem of defining which is the appropriate number of genes to select [2, 3]. The choice of the best set of pertinent features to retain is a key factor for a successful and effective classification [1]. In general, redundant and irrelevant features can never help to improve the performance of a classifier or a model. However, they are usually added by mistake to the learning process. Let's take the case of cancer diagnosis where we aim to study the link between the symptoms and their class of diseases. For example, If the patient identification (ID) is considered as one of the input features, the classifier my conclude that the class of disease is influenced by the patient ID, which will influence badly the final result. Thus, these kind of features should be removed in order to increase the learning performance.

Usually, a feature selection method try to find a representative subset of features from the original features space. This selected subset should bring the same information of the original feature space and improve the accuracy of a particular application. According to [1] feature selection process may reduce the time complexity of an algorithm and usually facilitate the data understanding.

Feature selection methods can be grouped into two groups: filter and wrapper methods [4]. On one hand, filter methods evaluate features, individually before the learning process and eliminate some. Wrapper methods on the other hand, are an other category of feature selection methods, in which the prediction accuracy of a classifier is used as a threshold to separate the best features from the others. According to [5] wrapper methods generally result in better performance than filter methods because the feature selection process is optimized for the classification algorithm to be used. Typically wrapper approach use some sort of search strategy to generate the candidate subsets. The search strategy is broadly classified as exhaustive (eg. branch & bound), heuristic (eg. forward selection, backward selection), and random search (eg. genetic algorithm (GA)). The search complexity depends of the data dimensionality, it is usually exponential for an exhaustive search and quadratic for a heuristic search and may be linear to the number of iterations for a random search [4]. Hence using random search seems to be to most appropriate choice but the feature space have to be first reduced using some prior information in order to have a linear complexity.

The presence of prior information and additional information about how the features will interact in the classification model have always a great impact on feature selection and on its subsequent application. So whenever possible try to use this information. For example, when the biological relevance of feature can be ascertained, the potentially irrelevant or obvious features can also be eliminated.

Further to enhance the classification accuracy and learning runtime in big data as biological ones, we propose a new wrapper feature selection method that use in a first step prior information to find a minimum set of features in order to reduce the search space then use a random search using genetic algorithm leading to a new set of features such that the resulting probability distribution of the data classes is as close as possible to the original distribution obtained using all features.

This paper is organized as follows. "Wrapper framework" describes the wrapper feature selection approach. "New approach for wrapper feature selection" proposes a two-stage feature selection approach combining prior knowledge and GA. "Experimental investigations" describes the used datasets and the performance metrics. Then, our results are summarized in "Results Analysis" and conclusions are drawn in "Conclusion".

2 Wrapper Framework

Typically a wrapper approach use a generation mechanism to generate candidate subset: The original feature set contains d features, the total number of competing candidate subsets to be generated is 2^d, which is a huge number even for medium-sized d. The ideal feature selection approach is the exhaustive search of the full set of features to find the optimal subset. However, as the number of features increases the exhaustive search becomes rapidly impractical even for a moderate number of features [6]. If we look at different ways in which features subsets are generated among many variations, three basic schemes are available in the literature namely forward selection, backward elimination and random scheme [4].

Forward selection and backward elimination are considered as heuristics. Generally, sequential generation can help in getting a valid subset within a reasonable time but still it cannot find an optimal subset. This is due to the fact that the generation scheme uses a heuristic to obtain an optimal subset by selecting sequentially the best, as in the forward case, or removing the worst as in the backward case. Using such kind of generator will without doubt speed up the selection process. However, if the search falls in a local optima it cannot turn back. In fact the generator has no way to get out of the local optima because what has been removed cannot be added and what has been added cannot be removed. This is a big shortcoming of sequential schemes. To overcome this problem we may use the random generation scheme, to add randomness to the fixed rule of sequential generation and avoid getting stuck at some local optima [7].

Random search works well for search spaces with a high density of good solutions. GA can be considered as a random search algorithm, since randomness is embedded in GA at almost every level [8]. The idea of applying genetic algorithms with wrapper feature selection is not novel. Of these, Yang and Hanovar used genetic algorithm and neural network to investigate feature subset selection [9].

3 New Approach for Wrapper Feature Selection

In this section we propose a novel approach for wrapper feature selection:

- At first, a based on similarity study with the prior knowledge primary dimensionality reduction step is conducted on the original feature space. This step is used to reduce the search space.
- Second, the subset generation step is performed using genetic algorithm.

3.1 Primary Dimensionality Reduction Step: Similarity Study

The first step of our proposed approach is designed specifically to select less redundant features without sacrificing quality. Redundancy is measured by a similarity measure between a preselected set of features and the remaining features in the dataset. In this step we enhance an existing set of preselected features by adding additional features as a complement. In any data mining application we may already have a set of features preselected with prior information. In fact, experts have years of experience on some particular knowledge about which features are more important. This knowledge is generally obtained by years of use of classical feature selection methods. Thus, a possible improvement of any search strategy is to use the prior knowledge and to eliminate redundant features before generating the candidate subsets. Since our goal is to take advantage of any additional information about the feature, we may want to select a set of features complementary to those preselected by experts. Hence, we need to study the effect of using prior information on relevant feature complexity.

First, we split the features set in two sets. The first one regroups a set of features that were assumed to be more relevant according to some prior knowledge. The second set contains the remaining ones. Once the two sets are obtained we conduct a similarity study and a similarity matrix is constructed. In this step the mutual information (MI) is chosen as a similarity measure given its efficiency in providing a solid theoretical framework for measuring the relation between the classes and a feature or more than one feature [10]. Formally, the MI of two continuous random variables X^j and $X^{j'}$ is defined as follows:

$$MI(X^j, X^{j'}) = \int\int p(x^j, x^{j'}) log \frac{p(x^j, x^{j'})}{p(x^j)p(x^{j'})} dx^j dx^{j'}, \quad (1)$$

where $p(x^j, x^{j'})$ is the joint probability density function and $p(x^j)$ and $p(x^{j'})$ are the marginal probability density functions. In the case of discrete random variables, the double integral becomes a summation, where $p(x^j, x^{j'})$ is the joint probability mass function, and $p(x^j)$ and $p(x^{j'})$ are the marginal probability mass functions. MI is an information metric used to measure the relevance of features taking into account the amount of information shared by two features [11]. Large values of MI indicate high correlation between the two features and zero indicates that two features are uncorrelated. Many authors proposed feature selection methods based on MI in different evaluation functions [11, 12].

Finally, we investigate level of similarity of each feature from the remaining set with the features of the first set. If the similarity is over 80%, the evaluated feature is eliminated else it is retained for further examination. More details are given in Fig. 1.

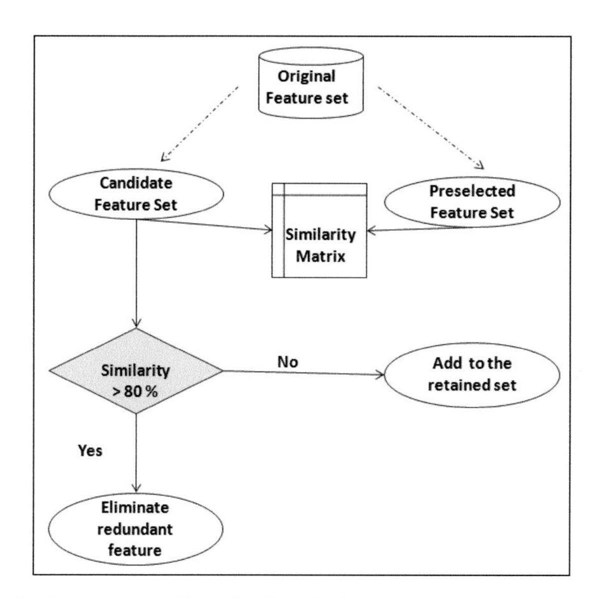

Fig. 1 First reduction process using prior knowledge.

3.2 Random Search Using Genetic Algorithms

Finding the best feature candidates from the reduced set can be seen as an enumeration problem. A random search algorithm refers to an algorithm that uses some kind of randomness or probability in the definition of the method. The term metaheuristic is also commonly associated with random search algorithms. Tabu search, evolutionary programming, ant colony optimization, GA [13, 14] and other random search methods are being widely applied to feature generation problems.

A GA use prior information to guide the search into the best region in the search space.

GA are better than conventional artificial intelligence (AI) algorithm in that it is more robust. GA is one of the artificial intelligence (AI) algorithms. However, unlike these older AI algorithms, a GA perform well even with noisy data or when the inputs changed slightly. Also, a GA may offer significant advantages over more usual search of optimization techniques especially in presence of large feature space.

GAs, are general adaptive optimization search methodologies that were developed by [15] to imitate the mechanism of genetic models of natural evolution and selection. They are a promising alternative to conventional random search methods. They work on the basis of a set of candidate solutions. Each candidate solution is called a "chromosome", and the whole set of solutions is called a "population". The algorithm allows movement from one population of chromosomes to a new population in an iterative way, until acceptable results are obtained. Each iteration is called a "generation". A fitness function assesses the quality of a solution in the evaluation step. The crossover and mutation functions are the main operators that randomly impact the fitness value. Chromosomes are selected for reproduction by evaluating the fitness value. The fitter chromosomes have higher probability to be selected into the recombination pool using the roulette wheel or the tournament selection methods. Fig. 2 depicts the GA evolutionary process mentioned above.

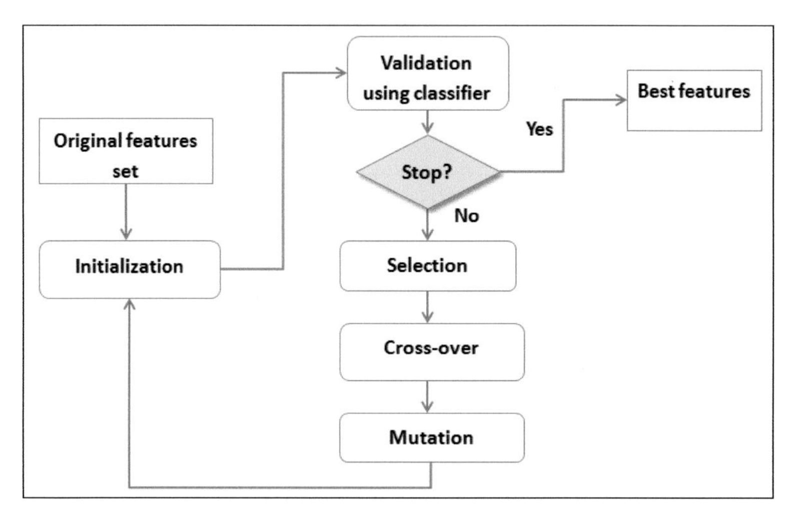

Fig. 2 General scheme for genetic algorithms.

4 Experimental Investigations

The experiments were conducted on Central Nervous System (CNS), a large data set concerned with the prediction of central nervous system embryonal tumor outcome based on gene expression. This data set includes 60 samples containing 39 medulloblastoma survivors and 21 treatment failures. These samples are described by 7129 genes [16]. We consider also the Leukemia microarry gene expression dataset that consists of 72 samples which are all acute leukemia patients, either acute lymphoblastic leukemia (47 ALL) or acute myelogenous leukemia (25 AML). The total number of genes to be tested is 7129 [3]. Table 1 displays the characteristics of the datasets that have been used for evaluation.

Our wrapper approach is used with GA as random search technique wrapped with two different classifiers namely support vector machine (SVM) and decision tree (DT).

For GA, population size is 100, number of generation is 10 as terminating condition, crossover rate is 0.7 and mutation rate is 0.001. In order to study the performance of the proposed approach, several evaluation measure derived from the confusion matrix were used [17]. these evaluation measures are: the percentage of correct positive predictions (Precision), the percentage of positive classified instances that were predicted

Table 1 Datasets summary

Names	CNS	Leukemia
Total instances	60	72
Total features	7129	7129
Number of classes	2	2
Missing Values	No	No

as positive (Recall). Our new approach is compared to forward and backward wrapper feature selection and results are summarized in Tables 2 and 3.

5 Results Analysis

The results in Tables 2 and 3 show that the relevant attributes identified by the various wrappers have indeed improved classification precision and recall of DT and SVM when compared to classification precision and recall with all the inputs. In fact using forward and backward feature selection as a wrapper improved significantly the classification performances compared to using all features, but most of the cases, experimental results show employing wrapper feature selection using GA and prior information enhanced the classification performances.

Table 2 Classification accuracy using wrapper feature selection approach for CNS dataset.

Wrapper approach	Number of Attributes	Precision (%)	Recall (%)
GA+ DT	396	91 (%)	76(%)
GA+ SVM	361	89(%)	82(%)
Forward feature selection	367	69(%)	72(%)
Backward feature selection	370	67(%)	75(%)
With all inputs	7129	49(%)	57(%)

Table 3 Classification accuracy using wrapper feature selection approach for Leukemia dataset.

Wrapper approach	Number of Attributes	Precision (%)	Recall (%)
GA+ DT	392	92 (%)	73(%)
GA+ SVM	373	86(%)	85(%)
Forward feature selection	360	62(%)	68(%)
Backward feature selection	358	63(%)	65(%)
With all inputs	7129	48(%)	59(%)

A closer look at Tables 2 and 3 shows that results are much better within the DT outputs. Actually, DT classifiers are sometimes considered as embedded methods. These kinds of methods essentially perform feature selection within the learning process, which means that they are able to select relevant features on their own: using their own search strategy and splitting mechanism. In other words DT classifiers select relevant features at two different stages. In the first stage features are selected by DT objective function individual and in the second features are selected by wrapper evaluation with GA. In this way, only features that are selected at both stages will form the final feature subset which is very likely to include features of high relevance.

6 Conclusion

In this work we propose a new approach for wrapper feature selection using genetic algorithm for random search and prior information. The motivation is to construct a more robust feature selection model with less complexity than usual search strategies. In a first part we investigated the effect of using prior information on the search space. Then we we conduct a random search on the reduced space of features using genetic algorithm. Results on two biological datasets show the performance of our approach.

References

1. Ben Brahim, A., Bouaguel, W., Limam, M.: 24. In: Combining Feature Selection and Data Classification Using Ensemble Approaches: Application to Cancer Diagnosis and Credit Scoring, pp. 517–532. Taylor & Francis (2014)
2. Schowe, B., Morik, K.: Fast-ensembles of minimum redundancy feature selection. In: Ensembles in Machine Learning Applications: Studies in Computational Intelligence, vol. 373
3. Golub, T.R., Slonim, D.K., Tamayo, P., Huard, C., Gaasenbeek, M., Mesirov, J.P., Coller, H., Loh, M.L., Downing, J.R., Caligiuri, M.A., Bloomfield, C.D.: Molecular classification of cancer: class discovery and class prediction by gene expression monitoring. Science **286**, 531–537 (1999)
4. Liu, H., Yu, L.: Toward integrating feature selection algorithms for classification and clustering. IEEE Transactions on Knowledge and Data Engineering **17**(4), 491–502 (2005)
5. Karegowda, A.G., Jayaram, M.A., Manjunath, A.: Article: Feature subset selection problem using wrapper approach in supervised learning. International Journal of Computer Applications **1**(7), 13–17 (2010). Published By Foundation of Computer Science
6. Chan, Y.H., Wing, W.Y.N., Daniel, S.Y., Chan, P.P.K.: Empirical comparison of forward and backward search strategies in L-GEM based feature selection with RBFNN. In: Proceedings of the International Conference on Machine Learning and Cybernetics (ICMLC), pp. 1524–1527 (2010)
7. Yun, C., Shin, D., Jo, H., Yang, J., Kim, S.: An experimental study on feature subset selection methods. In: Proceedings of the 7th IEEE International Conference on Computer and Information Technology. CIT 2007, Washington, DC, USA, pp. 77–82. IEEE Computer Society (2007)

8. Martínez, H.P., Yannakakis, G.N.: Genetic search feature selection for affective modeling: a case study on reported preferences. In: Proceedings of the 3rd International Workshop on Affective Interaction in Natural Environments. AFFINE 2010, New York, NY, USA, pp. 15–20. ACM (2010)
9. Feature subset selection using a genetic algorithm. In: Liu, H., Motoda, H. (eds.): Feature Extraction, Construction and Selection. The Springer International Series in Engineering and Computer Science, vol. 453
10. Bonev, B.: Feature Selection based on Information Theory. Ph.D. thesis, University of Alicante (2010)
11. Kumar, G., Kumar, K.: A novel evaluation function for feature selection based upon information theory. In: Proceedings of the Canadian Conference on Electrical and Computer Engineering (CCECE), pp. 395–399 (2011)
12. Al-Ani, A., Deriche, M.: An optimal feature selection technique using the concept of mutual information. In: Proceedings of the Sixth International Symposium on Signal Processing and its Applications, pp. 477–480 (2001)
13. Zhang, H., Sun, G.: Feature selection using tabu search method. **35**(3), 701–711 (2002)
14. Ramirez, R., Puiggros, M.: A genetic programming approach to feature selection and classification of instantaneous cognitive states. In: Giacobini, M. (ed.) Applications of Evolutionary Computing. Lecture Notes in Computer Science, vol. 4448, pp. 311–319. Springer, Heidelberg (2007)
15. Holland, J.H.: Adaptation in natural and artificial systems. MIT Press, Cambridge (1992)
16. Pomeroy, S.L., Tamayo, P., Gaasenbeek, M., Sturla, L.M., Angelo, M., McLaughlin, M.E., Kim, J.Y.H., Goumnerova, L.C., Black, P.M., Lau, C., Allen, J.C., Zagzag, D., Olson, J.M., Curran, T., Wetmore, C., Biegel, J.A., Poggio, T., Mukherjee, S., Rifkin, R., Califano, A., Stolovitzky, G., Louis, D.N., Mesirov, J.P., Lander, E.S., Golub, T.R.: Prediction of central nervous system embryonal tumour outcome based on gene expression. Nature **415**(6870), 436–442 (2002)
17. Okun, O.: Feature Selection and Ensemble Methods for Bioinformatics: Algorithmic Classification and Implementations (2011)

An Enhanced Univariate Discretization Based on Cluster Ensembles

Kittakorn Sriwanna, Tossapon Boongoen and Natthakan Iam-On

Abstract Most discretization algorithms focus on the univariate case. In general, they take into account the target class or interval-wise frequency of data. In so doing, useful information regarding natural group, hidden pattern and correlation among the attributes may be inevitably lost. In response, this paper introduces a new pruning method that exploits natural groups or clusters as an explicit constraint to traditional cut-point determination techniques. This unsupervised approach makes use of cluster ensembles to reveal similarities between data belonging to adjacent intervals. To be precise, a cut-point between a pair of highly similar or related intervals will be dropped. This pruning mechanism is coupled with three different univariate discretization algorithms, with the evaluation is conducted on 10 datasets and 3 classifier models. The results suggest that the proposed method usually achieve higher classification accuracy levels, than those of the three baseline counterparts.

Keywords Discretization · Clustering · Cluster ensembles · Data mining

1 Introduction

Discretization is a preprocessing technique that has been used for data reduction in the fields of data mining and machine learning. It transforms a numeric attribute to the discrete or nominal one. This is achieved by replacing the raw values of the

K. Sriwanna(✉) · N. Iam-On
School of Information Technology, Mae Fah Luang University,
Muang, Chiang Rai 57100, Thailand
e-mail: {kittakorn.sri,nt.iamon}@gmail.com

T. Boongoen
Royal Thai Air Force Academy, 171/1 Klongthanhon, Saimai,
Bangkok 10220, Thailand
e-mail: tossapon_b@rtaf.mi.th

K. Lavangnananda et al. (eds.), *Intelligent and Evolutionary Systems*,
Proceedings in Adaptation, Learning and Optimization 5,
DOI: 10.1007/978-3-319-27000-5_7

continuous attribute to non-overlapping interval labels (e.g. , 0-5, 6-10 , etc.). Most of data mining models have been designed to handle different data types. While some are able to cope only with numerical data, others are specifically developed for nominal data. To tackle many real problems with mixed data type, some others attempt to hybridize the two schemes. However, the problem remains with those work only with nominal domains, where discretization is required prior to the learning process. Data patterns are easier to be captured and interpreted once the data were reduced and simplified [1–3]. Also, previous works suggested the improved accuracy obtained by nominal classification methods [4].

Discretization can be classified into various categories; such as supervised vs. unsupervised, univariate vs. multivariate, splitting vs. merging, etc. [5, 6]. Supervised methods consider class information whereas unsupervised ones exploit non-class, natural characteristics embedded in the data. Splitting approach starts from one interval and recursively selects the best cut-point to split the domain into two sub-intervals. On the other hand, merging techniques kick off with the whole set of single-value intervals and iteratively merge adjacent intervals into a larger group. Univariate methods discretize each attribute independently, without considering possible relationships across attributes. However, multivariate methods consider other attributes to determine the best cut-points.

Most of the discretization algorithms are univariate [3]. These include ChiMerge [7] and Chi2 [8], which are also supervised and merging. They use statistical metrics such as χ^2 values to determine the most similar pair adjacent intervals, to be merged. Class-Attribute Interdependence Maximization (CAIM) [9] and Class-Attribute Dependent Discretizer (CADD) [10] are univariate, supervised and splitting methods. A cut-point is acquired through the greedy optimization, with class-attribute interdependence being employed as the discretization criterion. Likewise, Minimum Entropy-Minimum Description Length Principle (Ent-MDLP) [11] and Discretizer 2 (D-2) [12] can also be classified as univariate, supervised and splitting methods. They recursively select the best cut-points to partition the domain under examination, based on class information entropy [13].

Despite previous success, failing to blend interactions among attributes with a univariate discretization may lead to important information loss [3] and inability to reach a global optimal [14]. As a result, multivariate discretization methods such as Independent Component Analysis (ICA) [14], Clustering + Minimum Entropy-Minimum Description Length Principle (CEnt-MDLP) [15] and Hypercube Division Discretizer (HDD) [3] have been proposed to fill this gap. Specific to ICA, it first transforms original attributes to a new attribute space, then discretize the new attribute space using the univariate discretization method, Ent-MDLP. Similarly, CEnt-MDLP initially finds the pseudo-class by clustering the original data. After that the discretization is based on Ent-MDLP with the target class and pseudo-class information. It finds the best cut-point by averaging both the entropies that consider the target class and pseudo-class. In addition, HDD extends CAIM by improving the stopping criterion that encodes associations between attributes.

Given the recent trend of discretization, data clustering [1, 16] has been an attractive alternative to conventional statistics to represent attribute-wise relations and native data properties. Following CEnt-MDLP, Clustering + Rought Sets Discretizer (Cluster-RS-Disc) [17] also seeks optimal set of intervals based on knowledge disclosed by data clusters. The intuition of clustering is to partition a given set of data based on the similarity measured across attributes. As such, the interdependency is prescribed as clusters, where instances within a cluster are more similar to each other than to those in different clusters. However, most clustering algorithms are highly parameterized, and specifying an optimal setting has long been a difficult task. In particular to CEnt-MDLP, many trials of parameter k (i.e., number of clusters) for k-means clustering have been assessed, with the purpose of acquiring the best predictive accuracy. In response, the methodology of cluster ensemble is recently put forward to overcome such barrier. It turns a variety of parameter sets to the competitive edge for the generation of more robust and stable clustering results [18].

Provided this insight, the paper introduces an enhanced univariate discretization based on cluster ensembles, which is the multivariate method that consider the similarity or distance between data points. The proposed algorithm aims to preserve information regarding natural clusters in such a way that it can be coupled with conventional discretization methods. As an active constraint to cut-point determination, it drops any cut-point between intervals whose instance members are of high similarity, or belong to similar clusters. This may help to increase the purity of intervals, hence the quality of nominal domains.

The rest of this paper is organized as follows. Related works with respect to discretization and cluster ensembles are provided in Section 2. Section 3 presents the proposed algorithm, which includes the generation and use of cluster ensembles. Following that, Section 4 contains performance evaluation with a collection of real data sets and different classifiers. Finally, conclusion and future research are shown in Section 5.

2 Related Work

This section provides a brief summary of two topics of: discretization and cluster ensembles. The discretization part introduces three algorithms that are closely related to the proposed method. In the other, basis of cluster ensembles is provided to set the scene for the following sections.

2.1 Discretization

CAIM. The algorithm was proposed by Kurgan and Cios [9]. Its goal is to find the minimum number of discrete intervals, while minimizing the loss of class-attribute interdependency. A greedy approach is exploited for this optimization problem, which is the fundamental of splitting process. It iteratively searches for the best cut-point in order to split the interval into two sub-intervals until the stopping criterion becomes true. The algorithm mostly generates discretization schemes where the number of intervals equal to the number of classes [19, 20].

FUSINTER. This algorithm is a greedy merging method [21]. It employs the strategy similar to that of ChiMerge. It first merges the adjacent intervals that all instances of the intervals are of the same target class. The algorithm keeps on merging adjacent intervals until no improvement can be made or the number of intervals reaches 1. In practice, users must specify two parameters, α and λ, which are significance level and variable tuning, respectively. In this paper, these parameters are used with the default values $\alpha = 0.975$ and $\lambda = 1$. Like any parameterized model, performance is subjected to parameter tuning, usually with the help of prior knowledge or multiple trials.

PKID. The algorithm proposed by Young and Webb [22]. This unsupervised method aims to maintain discretization bias and variance by tuning interval frequency and interval number, especially with Naive-Bayes classifier. The *proportional discretization* (PKID) sets the interval frequency and interval number equally proportional to the amount of training data in order to achieve low variance and low bias.

2.2 Cluster Ensembles

Data clustering is an unsupervised learning, whose goal is to discover structure, natural groupings of data or objects. It categorizes data into groups such that the objects in the same cluster are more similar to each other than to those in different clusters. It has been used for many problem domains, ranging from engineering to health informatics.

Despite the success, there is no single clustering algorithm that performs best for all datasets. This holds true as each technique make specific assumption regarding shape and structure of data clusters [18]. Each clustering algorithm has its own strength and weakness. Moreover, its performance is normally determined by some parameters. Thus, it is extremely difficult for users to decide the optimal parameter setting. Cluster ensembles are able to overcome these limitations. It combines different decisions of various clustering algorithms to achieve superior accuracy [23].

The pairwise similarity [24, 25] is one of the approaches developed in the field of cluster ensembles. It uses co-occurrence relationships between all pairs of objects or

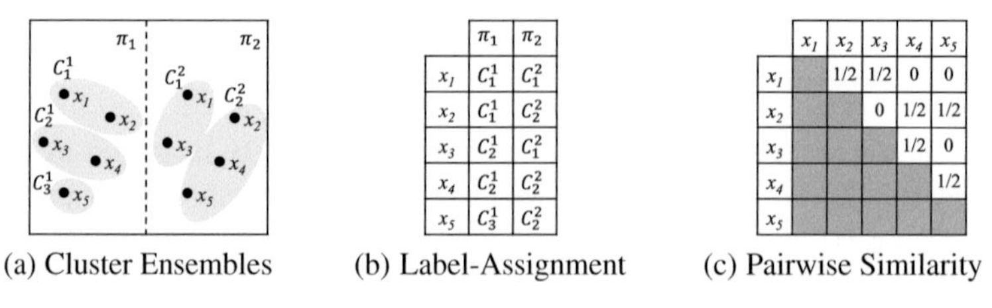

(a) Cluster Ensembles (b) Label-Assignment (c) Pairwise Similarity

Fig. 1 Examples of (a) cluster ensembles and the corresponding ensembles-information matrices: (b) label-assignment, and (c) pairwise similarity. Note that $X = \{x_1, \ldots, x_5\}$, $\Pi = \{\pi_1, \pi_2\}$, $\pi_1 = \{C_1^1, C_2^1, C_3^1\}$, and $\pi_2 = \{C_1^2, C_2^2\}$.

data-points. The example of cluster ensembles and the so-called pairwise similarity is illustrated in Fig. 1. Given the cluster ensemble with two based clusterings that is shown in Fig. 1(a), Fig. 1(b), and Fig. 1(c) depicts two different matrices representing information of the underlying ensemble. The pairwise similarity matrix (see Fig. 1(c) for and example) contains the similarity among all data points. It can be constructed from the label-assignment matrix, given in Fig. 1(b). The similarity value is the probability that the pair of data pints are the same cluster ensembles. Having obtained this similarity matrix, agglomerative or spectral clustering techniques can be applied to create the final clustering result.

3 Proposed Method

Most of the previous discretization algorithms make an extensive use of the criterion exhibiting purity of the target class or equal interval sizes. They simply neglect natural characteristics such as similarity among data points. Therefore, the cut-point selection may overlook a strong inter-interval association, which results in a premature partitioning. Intuitively, encoding data clusters and similarity into the discretization procedure can help to avoid the aforementioned dilemma.

The proposed method is an unsupervised learning that is based on cluster ensembles. For the current research, the homogeneous-ensemble generation of the work [23] is adopted, with k-means being repeatedly applied to the data, each with different value of k. This clustering technique is chosen due to its simplicity and efficiency. Euclidean metric [26] is employed to compute the distance between data-points under examination. Its native instability, with respect to initialization of cluster centers and number of clusters, can be useful to boost diversity, hence the cluster ensembles are quality.

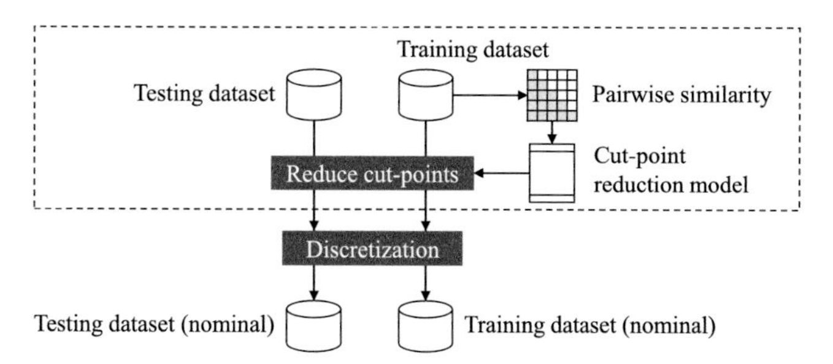

Fig. 2 The framework of the cut-point and data reduction algorithm. It includes three steps of: Create a pairwise similarity matrix, create cutting-point reduction model, and apply models to reduce cut-point and data.

The proposed framework is shown in Fig. 2, as the part encapsulated by the dashed border. It includes three steps of: create a pairwise similarity matrix from a given set of data, create the cut-point reduction model, and reduce potential cut-points and

Input: Training dataset ($train$), testing dataset ($test$).
Output: The reduced data of $train$ and $test$.
1: Create cluster ensembles $\prod = \{\pi_1, \ldots, \pi_m\}$ of $train$.
2: Create pairwise similarity matrix.
3: **for all** numeric attribute A_i **do**
4: Sort A_i in ascending and find all posible cut-points (C).
5: **repeat**
6: **for all** cut-point c_i, where $c_i \in C$ **do**
7: Calculate $IS(c_i)$.
8: **end for**
9: Select the best cut-point c_* that highest $IS(c_*)$.
10: **if** $IS(c_*) \geq \beta$ **then**
11: Remove c_*.
12: **end if**
13: **until** $IS(c_*) < \beta$
14: Store the intervals of the remain cut-points of A_i (IA_i).
15: **end for**
16: Reduce $train$ and $test$ using IA.

Fig. 3 The proposed cut-point reduction for discretization algorithm (CRD).

deliver the resulting data intervals, respectively. The cut-point reduction algorithm is summarized in Fig. 3. Firstly, the pairwise similarity matrix is created from cluster ensemble $\prod = \{\pi_1, \ldots, \pi_m\}$, which consists of m clustering results of k-means, using $k \in \{2, 3, 4\}$. In this study, the maximum k of k-means (K_{max}) was setted to 4, which given high classification accuracy, see Sect 4.2 for the details. Because k-means is nondeterministic algorithm, in this study, each k was run 10 times, $m = 30$. Having accomplished that, the algorithm reduces the cut-point of one attribute at a time by firstly sorting the attribute values in ascending order (see line 4 of Fig. 3). Then, as shown from lines 6 to 8 in Fig. 3, the similarity between intervals $IS(c_i)$ of each cut-pint c_i is calculated by Eq. 1.

$$IS(c) = \frac{2}{|N|(|N|-1)} \sum_{\forall u,v \in N, u \neq v} sim(u, v) \tag{1}$$

where c is a cut-point, $IS(c)$ is an interval similarity at c, $|N|$ denotes the size of set N, which contains data-points in the adjacent intervals of c, and $sim(u, v)$ is a pairwise similarity value of the data-points u and v. For any $\pi_g \in \prod$, the similarity between two data points u and v can be justified by the following:

$$sim_g(u, v) = \begin{cases} 1 & \text{if } u \text{ and } v \text{ are in the same cluster,} \\ 0 & \text{otherwise} \end{cases} \tag{2}$$

Then, the pairwise similarity across all clusterings in $\prod$ can be estimated as:

$$sim(u,v) = \frac{1}{m} \sum_{\forall \pi_g \in \prod} sim_g(u,v) \tag{3}$$

Where $sim(u, v) \in [0, 1]$, with 0 and 1 denoting the least and the most similar cares, respectively.

The interval similarity $IS(c)$ is an average of similarity measures between all data-point pairs. The maximum and minimum values of $IS(s)$ are 1 and 0, respectively. It reflects a probability of two adjacent intervals belonging to the same cluster. From line 9 to line 12 in Fig. 3, the algorithm selects a candidate cut-point c with the highest IS(c), where the intervals ae the most similar and should not be separated. Hence, the selected cut-point c will be dropped, and repeat the aforementioned procedure with the other cut-points. This continues as long as The $IS(c)$ value is equal or greater than a threshold β, otherwise stop the cut-point reduction process. In this study, the β was tested to several values. The value that gives the best result is 0.8, 80% confident of intervals' similarity.

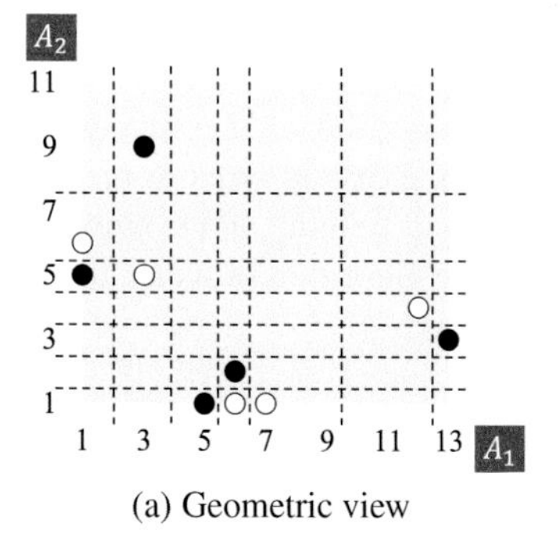

(a) Geometric view

A_1	A_2	Class
1	5	B
1	6	W
3	5	W
3	9	B
5	1	B
6	1	W
6	2	B
7	1	W
12	4	W
13	3	B

(b) Data view

Fig. 4 The toy dataset in the view of: (a) geometric view, and (b) data view. Note that the vertical dashed-line and horizontal dashed-line of sub figure (a) are the cut-point of attribute A_1 and A_2, respectively.

The proposed cut-point reduction for discretization algorithm (CRD) is designed in such a way that it can be coupled with existing discretization algorithms, both supervised and unsupervised methods. For clarity, the example of a toy dataset is given in Fig. 4, which shows the geometry and data view of the dataset. This dataset contains 10 data-points and 2 classes of black (B) and white (W). The vertical and horizontal dashed-lines in Fig. 4(a) represent all possible cut-pints of attributes A_1 and A_2, respectively. Following that, Fig. 5 illustrates the resulting cut-points after applying the CRD method. As shown in Fig. 5(a), CRD algorithm removes the cut-points between data-points that appear to be in the same clusters. The result of data reduction is presented in Fig. 5(b), where each data point was reduced by replacing the

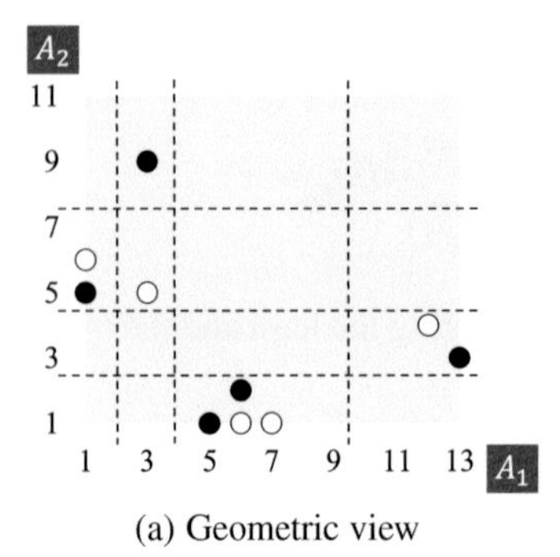

(a) Geometric view

A_1	A_2	Class
1	6	B
1	6	W
3	6	W
3	9	B
6.75	1.75	B
6.75	1.75	W
6.75	1.75	B
6.75	1.75	W
11.25	3.5	W
11.25	3.5	B

(b) Data view

Fig. 5 The result of toy dataset after reducing the cut-point: (a) cut-point reduction view, and (b) the data reduction view

center value of the cut-point as calculated by Eq. 4, where $cut_l(data)$ and $cut_h(data)$ are low and high adjacent cut-point values of the examined data, respectively.

$$data* = \frac{cut_l(data) + cut_h(data)}{2} \quad (4)$$

CRD algorithm is a pre-discretization algorithm. It iteratively reduce the cut-point of all numeric attributes, then reduce the data in order to prevent partitioning the natural group of data. CRD reduces data of training and testing dataset by using the center value of each interval. Finally, the algorithm provides this set of reduced data to a conventional discretization technique.

4 Performance Evaluation

4.1 Experiment Design

The experimental evaluation was conducted on ten datasets taken from the benchmark UCI Machine Learning Database Repository [27]. The main characteristics of these datasets are summarized in Table 1. Three discretization techniques; CAIM, FUSINTER, and PKID are coupled with the proposed CRD algorithm. This results in three new combined method of CAIM+, FUSINTER+, and PKID+, respectively. Note that the aforementioned baseline models are included in the empirical assessment due to their exceptional performance reported in the literature [5].

This investigation evaluates the CRD algorithm with respect to two distinct prospects. On one hand, it is to compare the quality of nominal data generated by CRD, against those derived by baseline techniques. This can be justified by predictive accuracy obtained with C4.5, K-Nearest Neighbors (KNN, the number of neighbours is 3), and Naive Bayes (NB) classifiers. For each of the experimental datasets, the accuracy of any pair of discretization model and classifier is as the average based

Table 1 Description of datasets: number of instances (n), number of attributes (d), number of numeric attributes (d^u), number of nominal attributes (d^o), and number of classes (c).

Dataset	n	d	d^u	d^o	c
Blood	748	4	4	0	2
Bupa	345	6	6	0	2
Column	310	6	6	0	3
Faults	1941	27	25	2	7
Haberman	306	3	3	0	2
Ionosphere	351	34	32	2	2
Iris	150	4	4	0	3
Liver	345	6	6	0	2
Sonar	208	60	60	0	2
Wine	178	13	13	0	3

on 10-fold cross validation. On the other, it is also crucial to obtain a small number of nominal values, such that the learning model can be simple, efficient, easy to understand and analyze. However, too small of a number of nominal intervals may lead to worse classification performance [19]. Specific to this point, the number of intervals achieved by investigated techniques will be compared and discussed later on.

Particular to CRD, the underlying cluster ensembles are generated using the homogeneous generation. As such, k-means with different number of clusters and initializations is exploited to create m = 30 base clusterings, from which the similarities between data points are estimated.

4.2 Experiment Results and Analysis

Based on the classification accuracy, Table 2 compares the accuracy of the original discretizations algorithm and their couplings with CRD algorithm. There are three comparison groups: CAIM vs. CAIM+, FUSINTER vs. FUSINTER+, and PKID vs. PKID+. For each dataset, the highest predictive accuracy in each comparison group is highlighted in **boldface**. According to these measures, CAIM+ usually perform better than its baseline counterpart, i.e., CAIM, with exceptional improvements being observed on Wine, Iris and Bupa datasets. This is similarly seen in the cases of FUSINTER+ and PKID+, which are able to increase the accuracy levels achieved by their original models. For each classifier, the average predictive accuracy across ten datasets is shown in Fig. 6. It reinforces that the use of CRD can boost the accuracy normally achieved by the conventional algorithms of CAIM, FUSINTER and PKID. To this end, the average accuracy of CAIM+ increase by 0.97% from CAIM, FUSINTER+ increase by 2.28% from FUSINTER, and PKID+ increase by 0.63% from PKID, respectively.

Table 2 Average accuracy with standard deviation results on 10 real datasets using C4.5, KNN, and NB classifier. The highest accuracy of each comparing group is highlighted in **boldface**.

Dataset	Classifier	CAIM	CAIM+	FUSINTER	FUSINTER+	PKID	PKID+
Blood	C4.5	76.20±1.2	**78.68±3.0**	76.21±0.4	76.21±0.4	76.21±0.4	76.21±0.4
	KNN	76.21±1.6	**78.68±3.0**	74.19±2.6	**75.62±3.3**	75.27±4.4	**76.11±2.7**
	NB	74.20±4.0	**75.69±4.8**	73.40±3.4	**74.53±4.0**	74.46±5.1	**74.96±5.4**
Bupa	C4.5	65.49±10.0	**66.53±8.2**	57.98±0.9	**58.01±0.9**	57.10±2.6	**57.89±1.7**
	KNN	61.15±11.4	**65.39±9.9**	56.53±8.8	**59.70±9.4**	60.01±5.6	**61.03±6.4**
	NB	65.22±7.0	**66.74±6.1**	60.32±7.0	**63.87±7.7**	**61.45±10.4**	61.24±8.3
Column	C4.5	78.06±8.2	**78.81±10.3**	68.71±5.7	**74.26±6.2**	73.55±6.4	**75.39±7.2**
	KNN	76.45±8.5	**76.65±11.1**	65.81±6.8	**71.84±7.6**	**71.94±5.9**	71.19±7.2
	NB	76.13±7.2	**76.84±11.1**	66.77±8.1	**71.00±6.5**	**75.81±7.3**	73.29±7.9
Faults	C4.5	**71.98±3.2**	71.95±2.2	43.54±11.1	**60.98±6.7**	69.61±2.5	**70.81±3.0**
	KNN	**71.51±2.5**	71.45±2.3	65.07±2.7	**65.48±2.4**	68.01±2.6	**68.44±2.7**
	NB	**67.75±3.4**	67.05±4.3	62.86±3.3	**64.01±3.6**	**67.08±4.2**	66.98±3.9
Haberman	C4.5	75.48±3.9	**76.40±3.8**	73.53±1.0	73.53±0.9	73.53±1.0	73.53±0.9
	KNN	75.81±4.2	**76.23±4.0**	**73.54±2.7**	72.54±2.9	**72.20±5.0**	71.28±3.7
	NB	**75.82±3.8**	74.89±4.0	**73.85±4.9**	73.12±5.9	72.85±5.7	**74.06±5.9**
Ionosphere	C4.5	88.91±5.2	**90.58±3.7**	73.50±7.0	**81.45±5.0**	**88.62±3.7**	86.02±5.8
	KNN	88.61±4.4	**89.21±4.0**	90.04±3.6	**90.55±2.7**	**91.44±4.3**	91.40±2.9
	NB	**90.60±5.7**	89.35±5.1	**89.75±3.6**	88.58±5.6	88.04±5.2	**88.27±4.7**
Iris	C4.5	93.33±5.4	**96.87±4.5**	94.00±5.8	**96.73±4.5**	92.67±9.1	**96.87±4.5**
	KNN	93.33±7.0	**93.40±7.5**	92.00±6.1	**92.53±8.9**	92.67±8.6	**94.87±8.8**
	NB	94.00±6.6	**94.20±8.2**	92.67±5.8	**94.47±7.6**	92.00±6.1	**95.93±6.3**
Liver	C4.5	65.49±10.0	**66.56±7.2**	57.98±0.9	57.98±0.8	**57.10±2.6**	56.37±4.6
	KNN	61.15±11.4	**64.14±9.0**	56.53±8.8	**59.94±5.4**	60.01±5.6	**60.21±9.0**
	NB	65.22±7.0	**65.99±5.2**	60.32±7.0	**63.48±5.3**	61.45±10.4	**61.73±7.9**
Sonar	C4.5	73.52±11.7	**75.88±9.4**	64.86±10.0	**65.21±9.9**	66.86±9.5	**67.28±9.2**
	KNN	**81.69±6.0**	81.61±7.8	**73.60±8.9**	73.50±9.7	**83.17±4.1**	82.65±7.0
	NB	78.88±8.7	**79.62±8.2**	70.26±5.5	**73.31±8.9**	75.57±10.3	**76.46±6.5**
Wine	C4.5	92.71±5.9	**93.16±6.2**	69.05±4.3	**77.40±11.9**	78.14±9.5	**86.88±8.6**
	KNN	94.41±5.2	**95.36±4.3**	**97.19±4.0**	93.39±4.8	**94.35±4.6**	93.21±5.5
	NB	97.78±2.9	**98.33±2.6**	**96.01±4.7**	95.41±5.0	**96.08±3.8**	95.88±4.6
Average		78.24±6.10	**79.21±6.04**	72.34±5.19	**74.62±5.49**	75.58±5.56	**76.21±5.44**

Based on the number of intervals, Table 3 compares the number of intervals created by examined methods. Specific to a dataset, the lowest number of intervals in each comparing group is highlighted in **boldface**. Since CAIM+, FUSINTER+ and PKID+ discretize the reduced data made by the CRD algorithm, the resulting numbers of intervals are not greater than those of CAIM, FUSINTER, and PKID, respectively. Regarding the pair of CAIM and CAIM+, both algorithms often give the same interval sizes. This is due to the fact that CAIM is designed as to create the number of intervals as small as the number of target classes.

The average running time of creating the cut-point reduction model and reducing data shown for each dataset in Table 4. Note that the proposed CRD algorithm is implemented in Java. All experiments are conducted on a workstation with Intel(R) Xeon(R) CPU@2.40 GHz and 4 GB RAM. As shown in table 4, the highest running time is 10.21 seconds for faults dataset. This is due to the size of data points, which

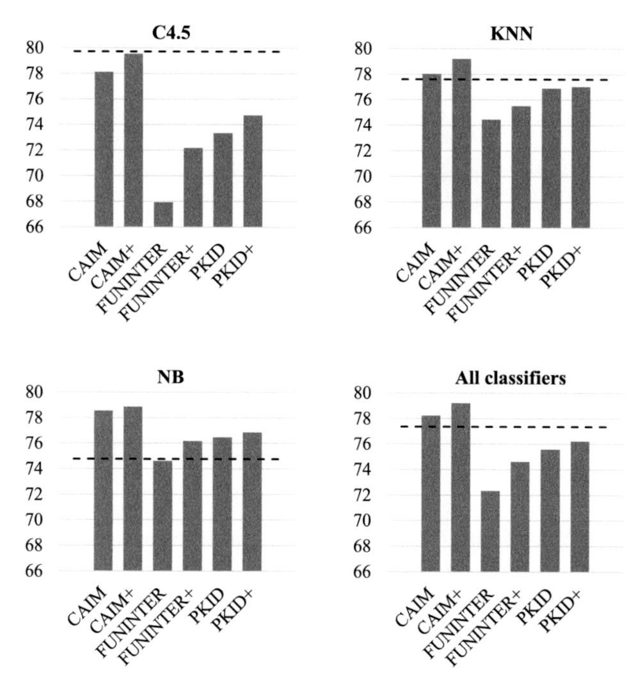

Fig. 6 The average predictive accuracy of C4.5, KNN, NB and all classifiers. Note that, the dashed horizontal lines are the average predictive accuracy without the discretization algorithms.

Table 3 Average number of intervals with standard deviation results on 10 real datasets. The lowest number of intervals of each comparing group is highlighted in **boldface**.

Dataset	CAIM	CAIM+	FUSINTER	FUSINTER+	PKID	PKID+
Blood	2.00 ±0.00	2.00 ±0.00	32.33 ±16.0	**11.74 ±6.72**	16.45 ±5.1	**9.70 ±4.34**
Bupa	2.00 ±0.00	2.00 ±0.00	40.92 ±19.4	**28.82 ±14.26**	16.05 ±2.6	**14.20 ±3.22**
Column	3.00 ±0.00	**2.995 ±0.03**	129.75 ±27.4	**48.70 ±34.76**	17.00 ±0.0	**11.44 ±3.98**
Faults	6.49 ±1.48	**6.48 ±1.48**	471.69 ±412.2	**310.32 ±290.34**	33.69 ±13.0	**30.58 ±12.29**
Haberman	2.00 ±0.00	2.00 ±0.00	25.80 ±11.2	**22.84 ±14.26**	12.70 ±3.3	**10.21 ±5.30**
Ionosphere	2.00 ±0.00	2.00 ±0.00	63.88 ±16.9	**37.80 ±19.73**	16.67 ±2.7	**13.47 ±4.42**
Iris	3.00 ±0.00	3.00 ±0.00	13.73 ±6.4	**7.83 ±5.30**	11.95 ±0.2	**6.39 ±3.58**
Liver	2.00 ±0.00	2.00 ±0.00	40.92 ±19.4	**28.74 ±14.16**	16.05 ±2.6	**14.33 ±3.40**
Sonar	2.00 ±0.00	2.00 ±0.00	85.06 ±9.4	**63.79 ±8.93**	14.00 ±0.0	**13.75 ±0.75**
Wine	3.00 ±0.00	3.00 ±0.00	47.57 ±10.1	**40.34 ±11.97**	13.00 ±0.0	**11.83 ±1.77**
Average	2.749 ±0.15	**2.748 ±0.15**	95.16 ±54.8	**60.09 ±42.04**	16.76 ±3.0	**13.59 ±4.31**

require a large proportion of time to estimate pairwise-similarity values. While CRD algorithm is effective to improve the quality of discretized intervals, hence the resulting classification accuracy, it can be computationally expensive for a very large dataset.

Table 4 The running time results on 10 real datasets of crate data reduction model includes the process of reducing data, repeat 10 runs (10 folds) and average.

Dataset	Run time (in seconds)
Blood	0.673
Bupa	0.510
Column	0.387
Faults	10.210
Haberman	0.221
Ionosphere	1.582
Iris	0.122
Liver	0.498
Sonar	1.640
Wine	0.345

The Maximum *k* of *k*-Means (K_{max}). In order to find the appropriate K_{max} that user can make the best use of the proposed framework, K_{max} was tested to several values. Fig. 7 illustrates a relationship, based on the average of accuracy across all datasets and all classifiers. In the figure, the value of K_{max} is varied from 2 through 15, insteps of 1. An important observation of average accuracy is that the proposed algorithms performed well with K_{max} is 4. Some results of the proposed framework also perform well with K_{max} higher than 4. However, it demand higher time consuming. Therefore, in this study, K_{max} was setted to 4.

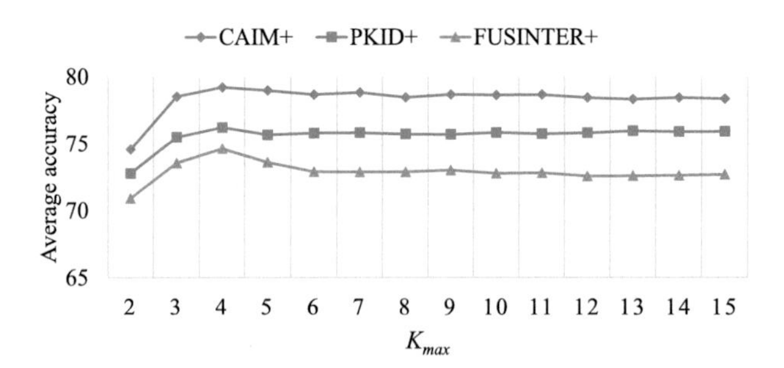

Fig. 7 The relationship of K_{max} and average accuracy across all datasets and all classifiers.

5 Conclusion

This paper presents an enhanced univariate discretization based on cluster ensembles, which proposed the Cut-point Reduction for Discretization (CRD) algorithm. This is a preprocessing of a conventional discretization technique. It aims to remove the cut-point and reduce data before proceeding to the usual discretization procedure. In principle, CRD gets rid of any cut-point that may damage the natural group or cluster of data based on cluster ensembles. The proposed algorithm is an unsupervised method, which can be coupled with both supervised and unsupervised discretization algorithms. The empirical study, with different univariate discretization algorithms, classifiers, and datasets, suggests that CRD can enhanced baseline discretization techniques, in term of predictive accuracy and number of intervals generated. However the algorithm requires a larger proportion of time to estimate pairwise-similarity values. With respect to this limitation, an approximated framework to calculating similarities between data points is an important future work. This widens the application of CRD to a large collection of data that has been frequently encountered in the modern-age society. Another work is to investigate the behaviour of algorithmic parameters, and the application of CRD to various problem domains.

Acknowledgements The authors would like to thank KEEL software [28, 29] for distributing the source code of discretization algorithm.

References

1. Han, J., Kamber, M., Pei, J.: Data Mining: Concepts and Techniques, 3rd edn. Morgan Kaufmann Publishers Inc., San Francisco (2011)
2. Sriwanna, K., Puntumapon, K., Waiyamai, K.: An enhanced class-attribute interdependence maximization discretization algorithm. In: Advanced Data Mining and Applications, pp. 465–476. Springer (2012)
3. Yang, P., Li, J.S., Huang, Y.X.: HDD: a hypercube division-based algorithm for discretisation. International Journal of Systems Science **42**(4), 557–566 (2011)
4. Bay, S.D.: Multivariate discretization for set mining. Knowledge and Information Systems **3**(4), 491–512 (2001)
5. Garcia, S., Luengo, J., Sáez, J.A., López, V., Herrera, F.: A survey of discretization techniques: Taxonomy and empirical analysis in supervised learning. IEEE Transactions on Knowledge and Data Engineering **25**(4), 734–750 (2013)
6. Sang, Y., Li, K.: Combining univariate and multivariate bottom-up discretization. Journal of Multiple-Valued Logic & Soft Computing **20** (2013)
7. Kerber, R.: Chimerge: discretization of numeric attributes. In: Proceedings of the Tenth National Conference on Artificial Intelligence, pp. 123–128. Aaai Press (1992)
8. Liu, H., Setiono, R.: Feature selection via discretization. IEEE Transactions on knowledge and Data Engineering **9**(4), 642–645 (1997)
9. Kurgan, L.A., Cios, K.J.: Caim discretization algorithm. IEEE Transactions on Knowledge and Data Engineering **16**(2), 145–153 (2004)
10. Ching, J.Y., Wong, A.K., Chan, K.C.C.: Class-dependent discretization for inductive learning from continuous and mixed-mode data. IEEE Transactions on Pattern Analysis and Machine Intelligence **17**(7), 641–651 (1995)
11. Fayyad, U., Irani, K.: Multi-interval discretization of continuous-valued attributes for classification learning (1993)
12. Catlett, J.: On changing continuous attributes into ordered discrete attributes. In: Machine learning EWSL 1991, pp. 164–178. Springer (1991)
13. Dougherty, J., Kohavi, R., Sahami, M., et al.: Supervised and unsupervised discretization of continuous features. In: ICML, pp. 194–202 (1995)

14. Kang, Y., Wang, S., Liu, X., Lai, H., Wang, H., Miao, B.: An ica-based multivariate discretization algorithm. In: Knowledge Science, Engineering and Management, pp. 556–562. Springer (2006)
15. Gupta, A., Mehrotra, K.G., Mohan, C.: A clustering-based discretization for supervised learning. Statistics & Probability Letters **80**(9), 816–824 (2010)
16. Parashar, A., Gulati, Y.: Survey of different partition clustering algorithms and their comparative studies. International Journal of Advanced Research in Computer Science **3**(3) (2012)
17. Singh, G.K., Minz, S.: Discretization using clustering and rough set theory. In: International Conference on Computing: Theory and Applications, ICCTA 2007, pp. 330–336. IEEE (2007)
18. Kuncheva, L., Hadjitodorov, S.T., et al.: Using diversity in cluster ensembles. In: 2004 IEEE International Conference on Systems, Man and Cybernetics, vol. 2, pp. 1214–1219. IEEE (2004)
19. Cano, A., Nguyen, D., Ventura, S., Cios, K.: ur-CAIM: improved caim discretization for unbalanced and balanced data. Soft Computing, 1–16 (2014)
20. Tsai, C.J., Lee, C.I., Yang, W.P.: A discretization algorithm based on class-attribute contingency coefficient. Information Sciences **178**(3), 714–731 (2008)
21. Zighed, D.A., Rabaséda, S., Rakotomalala, R.: Fusinter: a method for discretization of continuous attributes. International Journal of Uncertainty, Fuzziness and Knowledge-Based Systems **6**(03), 307–326 (1998)
22. Yang, Y., Webb, G.I.: Discretization for naive-bayes learning: managing discretization bias and variance. Machine Learning **74**(1), 39–74 (2009)
23. Iam-on, N., Boongoen, T., Garrett, S.: LCE: a link-based cluster ensemble method for improved gene expression data analysis. Bioinformatics **26**(12), 1513–1519 (2010)
24. Fred, A.L., Jain, A.K.: Combining multiple clusterings using evidence accumulation. IEEE Transactions on Pattern Analysis and Machine Intelligence **27**(6), 835–850 (2005)
25. Monti, S., Tamayo, P., Mesirov, J., Golub, T.: Consensus clustering: a resampling-based method for class discovery and visualization of gene expression microarray data. Machine Learning **52**(1–2), 91–118 (2003)
26. Xing, E.P., Jordan, M.I., Russell, S., Ng, A.Y.: Distance metric learning with application to clustering with side-information. In: Advances in Neural Information Processing Systems, pp. 505–512 (2002)
27. Bache, K., Lichman, M.: UCI machine learning repository (2013)
28. Alcalá-Fdez, J., Sánchez, L., García, S., del Jesus, M., Ventura, S., Garrell, J., Otero, J., Romero, C., Bacardit, J., Rivas, V., Fernández, J., Herrera, F.: Keel: a software tool to assess evolutionary algorithms for data mining problems. Soft Computing **13**(3), 307–318 (2009)
29. Alcalá, J., Fernández, A., Luengo, J., Derrac, J., García, S., Sánchez, L., Herrera, F.: Keel data-mining software tool: Data set repository, integration of algorithms and experimental analysis framework. Journal of Multiple-Valued Logic and Soft Computing **17**(255–287), 11 (2010)

Recursive Binary Tube Partitioning for Classification

Suebkul Kanchanasuk and Krung Sinapiromsaran

Abstract A classifier aims to categorize instances into well-defined groups based on a model called classifier. One of the most widely used classifiers is a decision tree built using a recursive partitioning algorithm. This paper applies the recursive partitioning technique based on the series of tubes. A tube is identified from three information; 1) a core vector, 2) a tube length and 3) a tube radius. The first component is the core vector generated by the extreme pole and the centroid of the current dataset and the second component is the tube length which is the maximum magnitude of the projections from all instances onto the core vector and the last component is the tube radius which is the maximum distance of the farthest point away from the core vector. Our experiment was performed on synthesized datasets of varying sizes with 2, 4, 6 and 8 attributes. The results showed the improvement over the conditional inference tree and C4.5 tree via the F-measure and G-measure score.

Keywords Classification · Recursive partitioning · Extreme pole · Tube

1 Introduction

The knowledge discovery is a process to uncover interesting knowledge from a large amount of data where the classification is one of the main techniques. It categorizes each instance into a designated class where members share their common characteristics. The classifier built from the classification algorithm is used to categorize a future instance into appropriate class.

Classifier proposed in the literature had been applied to several application domains such as business intelligences, bioinformatics, finances, and telecommunication

S. Kanchanasuk(✉) · K. Sinapiromsaran
Applied Mathematics and Computational Science,
Department of Mathematics and Computer Science, Faculty of Science,
Chulalongkorn University, Bangkok 10330, Thailand
e-mail: suebkul.k@gmail.com, krung.s@chula.ac.th

K. Lavangnananda et al. (eds.), *Intelligent and Evolutionary Systems*,
Proceedings in Adaptation, Learning and Optimization 5,
DOI: 10.1007/978-3-319-27000-5_8

[1–4]. These examples showed the presence of classification algorithm in the real world application. Moreover, there are literatures that showed continuity of the algorithm development. For example, Miller designed a decision tree using kNN and recursive partitioning techniques applying to biological medicine data [5]. Ko et al. proposed a fire detection method using SVM in image and video processing [6]. Farid et.al. modified the decision tree with Naïve Bayes' theorem to classify a biological dataset [7]. Delen applied three classification methods; Artificial neural networks, Decision tree and Logistic regression, in breast cancer survivability domain [8]. Sirisomboonrat and Sinapiromsaran proposed a classification method using multi-attributed lens to detect breast cancer patients [9]. Bunkhumpornpat, et al. proposed an over-sampling technique for the class imbalance problem and used C4.5 as one of the classifiers [10].

C4.5 [11] has been picked as the most used classification algorithm applying to build a decision tree [12]. The method was originally proposed as an improvement of a classification algorithm ID3 [13]. Basically, the C4.5 algorithm builds a binary decision tree based on the maximum gain ratio. The selected attribute always split at a single value for the current dataset. However, the single best attribute for splitting may not be appropriate for all situations.

In addition, the statistical software "R" contains many classifier packages including a party package which contains *ctree* [14, 15]. It uses conditional inference binary trees which embeds tree-structured regression models and applies a statistical hypothesis testing from multivariate linear statistics according to the best attributes ordering position. However, the main drawback is its computation which mostly affected from the complex dataset. Its computation time depends on the number of attributes.

Sirisomboonrat and Sinapiromsaran introduced a recursive partitioning to recognize the pattern of the breast cancer patients [9]. It uses multi-attributed lens, which generates the core vector from the furthest paired instances, to capture the main characteristics of patients. The concept of the core vector can split a region into three partitions, two of which contains only a pure class.

Our approach determines a position and a direction of a core vector which is different from the multi-attribute lens while maintaining the property of a pure class region. A core vector expands to cover a region containing instances of the same class, called a tube. Instances which are not captured by any tube needs to be identified using other models such k-NN. A tube then expands to cover the largest region that guarantee the property of a pure class. For the intersection of two tubes, it needs the recursive partitioning technique to construct tubes to split it into appropriate regions.

The rest of the paper is organized as follows. Section 2 reviews the basic concept of the decision tree induction and mathematical notation. In Section 3, mathematical concepts of recursive binary tube partitioning are presented. An experimental result is discussed in section 4. Finally, Section 5 concludes the paper.

2 Decision Tree Induction

A decision tree consists of a set of nodes and a set of edges. Nodes can be categorized as an internal node or a leaf node. All the internal nodes represent split criteria to partition a dataset into various subsets based on a single characteristic and the root node is just the first internal node to be considered. An edge represents by one of possible outcomes of the test condition from the internal node. A leaf node is labeled by the class.

In the past decade, one of the top ten algorithms in data mining is C4.5 [12] which is commonly regarded for inducing decision. C4.5 attempts to build a decision tree with the gain ratio measure of each feature and branching on the attribute which returns the maximum information gain ratio. At any point during the search, a chosen attribute is considered to have the highest discriminating ability among the concepts whose description are generated.

ID3 and C4.5 use the entropy measurement to find the best split from the attribute that has the highest information gain while C4.5 uses Shannon entropy as an impurity measure. Let *IM(D)* be an overall impurity of the dataset *D* computed by (1) where p_i is the approximated probability of the number of instances in class *i*. The information gain is calculated by (2) where V is a selected attribute and C is a class of instances.

$$IM(D) = \text{Entropy}(D) - \sum_{i}^{|C|}(p_i \log p_i) \tag{1}$$

$$\text{InfoGain}(V) = IM(D) - IM_V(D) \tag{2}$$

The impurity of a collection of subsets based on the V attributes, $IM(D)$, can be computed by (1) where D_v is a partition which contains instances having the value in attribute V as v.

$$IM_v(D) = \sum_{v \in V} \frac{|D_v|}{|D|} \text{Entropy}(D_v) \tag{3}$$

The split information measure is presented in equation (4) where p_v is the probability of an instance with the attribute V having value v. Furthermore, C4.5 deals with this problem using a gain ratio calculated by (5).

$$\text{SplitInfo}_v(D) = -\sum_{v \in V} p_v \log p_v \tag{4}$$

$$\text{GainRatio} = \frac{\text{InfoGain}(V)}{\text{SplitInfo}_v(D)} \tag{5}$$

A limitation of decision tree is the single attribute selection which may not help to split a dataset clustering along the non-orthogonal dimensions.

3 Recursive Tube Partitioning

3.1 *A Classifier: Tube*

The method for generating a tube required four essential components: the extreme pole ($\mathbf{p}$), the centroid of the current partition ($\mathbf{c}$), the tube length (H), and the tube radius (R). The tube length and tube radius are computed from instances to guarantee the tube covering of the same class instances. For a binary classification dataset, two tubes were generated by the instances of each class, called a positive tube and a negative tube. A positive tube covers all positive instances and a negative tube covers all negative instances. In this section, only the process for generating the positive tube will be given while the process for generating the negative tube is performed similarly.

The positional arrangement of a tube concept is explained by a core vector as a direction of a tube. Initially, two extreme poles are identified as the farthest pair of the instances. The vector generated by these two poles is called the core vector. It guarantees to make an acute angle with a vector generated from an extreme pole to each instance [9, 16]. In this context, a new core vector is defined using an extreme pole and a centroid, denoted by $\overrightarrow{\mathbf{pc}}$. As shown in Fig. 1(a), this new core vector has the same property with the core vector generated from two extreme poles.

The tube length is defined as the magnitude of the projection vector onto the core vector. For an instance $\mathbf{x^i}$, the vector ($\overrightarrow{\mathbf{px^i}}$) is generated from the pole $\mathbf{p}$ to $\mathbf{x^i}$. Note that the pole $\mathbf{p}$ is the pole that has the largest distance with respect to the centroid of a dataset. The tube length formula for $\mathbf{x^i}$ is

$$h_{\mathbf{x^i}} = \left\| \mathrm{proj}_{\widehat{\mathbf{pc}}} \overrightarrow{\mathbf{px^i}} \right\| = \widehat{\mathbf{pc}} \cdot \overrightarrow{\mathbf{px^i}} \tag{6}$$

The tube radius is defined as the minimum distance of $\mathbf{x^i}$ to a core vector. An alternative computing of $r_{\mathbf{x^i}}$ is a norm of vector $\overrightarrow{\mathbf{pc}} - \left(\widehat{\mathbf{pc}} \cdot \overrightarrow{\mathbf{px^i}}\right)\widehat{\mathbf{pc}}$, as shown in Fig. 1(b). So the tube radius formula is

$$r_{\mathbf{x^i}} = \left\| \overrightarrow{\mathbf{pc}} - \left(\widehat{\mathbf{pc}} \cdot \overrightarrow{\mathbf{px^i}}\right)\widehat{\mathbf{pc}} \right\| \tag{7}$$

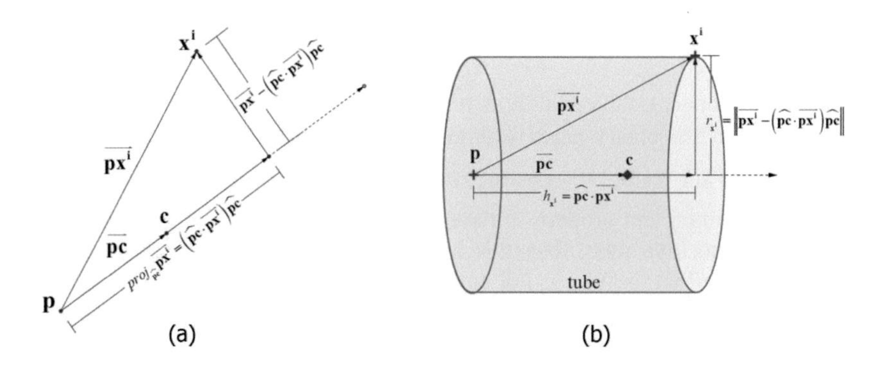

Fig. 1 (a) A length is described by a magnitude of the projection vector $\overrightarrow{\mathbf{px^i}}$ onto $\overrightarrow{\mathbf{pc}}$ and a radius is identified by the length of vector $\overrightarrow{\mathbf{px^i}} - \left(\widehat{\mathbf{pc}} \cdot \overrightarrow{\mathbf{px^i}}\right)\widehat{\mathbf{pc}}$. (b) Illustrated the region of a tube in 3D. The figure shows the region of this tube guaranteed to cover an instance $\mathbf{x^i}$.

As described above, $h_{\mathbf{x^i}}$ and $r_{\mathbf{x^i}}$ are the initial conditions to construct a tube which $\mathbf{x^i}$ was covered by this tube. Our approach determined the length and the radius of this tube which cover all of the instances in Λ_{pos}. The tube length (H) and tube radius (R) were computed by

$$H = \max_{\mathbf{x^i} \text{ in } \Lambda_{pos}} \left(\widehat{\mathbf{pc}} \cdot \overrightarrow{\mathbf{px^i}}\right), \tag{8}$$

and

$$R = \max_{\mathbf{x^i} \text{ in } \Lambda_{pos}} \left(\left\|\overrightarrow{\mathbf{pc}} - \left(\widehat{\mathbf{pc}} \cdot \overrightarrow{\mathbf{px^i}}\right)\widehat{\mathbf{pc}}\right\|\right) \tag{9}$$

```
Algorithm 1: Construct a tube.
Input: dataset
   1. Find the farthest distance pair of instances, ex-
      treme poles.
   2. Find a centroid (c)
```

$$c = \left(\frac{\sum_{i=1}^{N} x_1^i}{N}, \frac{\sum_{i=1}^{N} x_2^i}{N}, \ldots, \frac{\sum_{i=1}^{N} x_{d-1}^i}{N}, \frac{\sum_{i=1}^{N} x_d^i}{N}\right)$$

```
   3. Pick an extreme pole (p) that is the farthest
      distance of c.
   4. Compute length of tube (H) as
```

$$H = \max_{\mathbf{x^i} \text{ in } \Lambda_{pos}} \left(\widehat{\mathbf{pc}} \cdot \overrightarrow{\mathbf{px^i}}\right)$$

```
   5. Compute radius of tube (R)
```

$$R = \max_{\mathbf{x^i} \text{ in } \Lambda_{pos}} \left(\left\|\overrightarrow{\mathbf{pc}} - \left(\widehat{\mathbf{pc}} \cdot \overrightarrow{\mathbf{px^i}}\right)\widehat{\mathbf{pc}}\right\|\right)$$

3.2 Recursive Binary Tube Partitioning (RBTP)

As described above, a tube exhibit as a region which covered all instances of the same class. For the binary classification, the dataset has two disjoint sets; Λ_{pos} and Λ_{neg}. Each set is used to construct two tubes; the positive tube and the negative tube. The important property for these two tubes are that if an instance is not covered by a positive tube, then this instance must be classified as a negative class. Similarly, if an instance is not covered by a negative tube, then this instance must be classified as a positive class. A binary tube consists of a length H and a radius R. As illustrated in Fig. 2(a), the spaces is partition into four possible regions (R0 - R3) where R0 is the region outside both the positive and the negative tubes, R1 is the region inside the positive tube excluding the region of the negative tube, R2 is the region insider the negative tube excluding the region of the positive tube and R3 is the region inside both the positive and the negative tubes. As illustrated in Fig. 2(b), the region R0 in the training phase must be empty however it may not be empty in the testing phase. Note that the region R3 contains the mixed class instances which require additional recursive partitioning split. The recursion is applied until all regions have pure class.

Algorithm 2: Classify class instance.
requirement: $\mathbf{p}_{pos}, \mathbf{c}_{pos}, H_{pos}, R_{pos}$ from Λ_{pos}
$\mathbf{p}_{neg}, \mathbf{c}_{neg}, H_{neg}, R_{neg}$ from Λ_{neg}
input: instance ($\mathbf{x}^{\mathbf{i}}$)

1. Calculate $h_{\mathbf{x}^{\mathbf{i}}}(pos) = \widehat{\mathbf{p}_{pos}\mathbf{c}_{pos}} \cdot \overrightarrow{\mathbf{p}_{pos}\mathbf{x}^{\mathbf{i}}}$
   and $r_{\mathbf{x}^{\mathbf{i}}}(pos) = \left\| \overrightarrow{\mathbf{p}_{pos}\mathbf{x}^{\mathbf{i}}} - \left(\widehat{\mathbf{p}_{pos}\mathbf{c}_{pos}} \cdot \overrightarrow{\mathbf{p}_{pos}\mathbf{x}^{\mathbf{i}}}\right)\widehat{\mathbf{p}_{pos}\mathbf{c}_{pos}} \right\|$
2. Calculate $h_{\mathbf{x}^{\mathbf{i}}}(neg) = \widehat{\mathbf{p}_{neg}\mathbf{c}_{neg}} \cdot \overrightarrow{\mathbf{p}_{neg}\mathbf{x}^{\mathbf{i}}}$
   and $r_{\mathbf{x}^{\mathbf{i}}}(neg) = \left\| \overrightarrow{\mathbf{p}_{neg}\mathbf{x}^{\mathbf{i}}} - \left(\widehat{\mathbf{p}_{neg}\mathbf{c}_{neg}} \cdot \overrightarrow{\mathbf{p}_{neg}\mathbf{x}^{\mathbf{i}}}\right)\widehat{\mathbf{p}_{neg}\mathbf{c}_{neg}} \right\|$
3. Decide region:
   - $\mathbf{x}^{\mathbf{i}} \in R0 \Leftrightarrow \left[\left(\left(h_{\mathbf{x}^{\mathbf{i}}}(pos) > H_{pos}\right) \vee \left(r_{\mathbf{x}^{\mathbf{i}}}(pos) > R_{pos}\right)\right) \wedge \left(\left(h_{\mathbf{x}^{\mathbf{i}}}(neg) > H_{neg}\right) \vee \left(r_{\mathbf{x}^{\mathbf{i}}}(neg) > R_{neg}\right)\right)\right]$
   - $\mathbf{x}^{\mathbf{i}} \in R1 \Leftrightarrow \left[\left(\left(h_{\mathbf{x}^{\mathbf{i}}}(pos) \le H_{pos}\right) \wedge \left(r_{\mathbf{x}^{\mathbf{i}}}(pos) \le R_{pos}\right)\right) \wedge \left(\left(h_{\mathbf{x}^{\mathbf{i}}}(neg) > H_{neg}\right) \vee \left(r_{\mathbf{x}^{\mathbf{i}}}(neg) > R_{neg}\right)\right)\right]$
   - $\mathbf{x}^{\mathbf{i}} \in R2 \Leftrightarrow \left[\left(\left(h_{\mathbf{x}^{\mathbf{i}}}(pos) > H_{pos}\right) \vee \left(r_{\mathbf{x}^{\mathbf{i}}}(pos) > R_{pos}\right)\right) \wedge \left(\left(h_{\mathbf{x}^{\mathbf{i}}}(neg) \le H_{neg}\right) \wedge \left(r_{\mathbf{x}^{\mathbf{i}}}(neg) \le R_{neg}\right)\right)\right]$
   - $\mathbf{x}^{\mathbf{i}} \in R3 \Leftrightarrow \left[\left(\left(h_{\mathbf{x}^{\mathbf{i}}}(pos) \le H_{pos}\right) \wedge \left(r_{\mathbf{x}^{\mathbf{i}}}(pos) \le R_{pos}\right)\right) \wedge \left(\left(h_{\mathbf{x}^{\mathbf{i}}}(neg) \le H_{neg}\right) \wedge \left(r_{\mathbf{x}^{\mathbf{i}}}(neg) \le R_{neg}\right)\right)\right]$
4. Classify class:
   - If $\mathbf{x}^{\mathbf{i}}$ is in R0, $\mathbf{x}^{\mathbf{i}}$ is outlier or no class.
   - If $\mathbf{x}^{\mathbf{i}}$ is in R1, $\mathbf{x}^{\mathbf{i}}$ is a positive class.
   - If $\mathbf{x}^{\mathbf{i}}$ is in R2, $\mathbf{x}^{\mathbf{i}}$ is a negative class.
   - If $\mathbf{x}^{\mathbf{i}}$ is in R3, recursion step with new parameters.

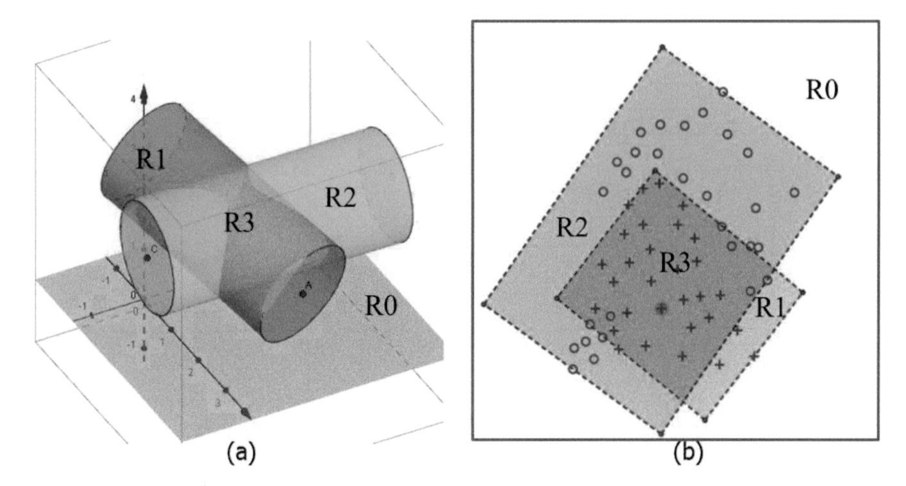

Fig. 2 The four disjoint regions of a binary tube.

4 Experiment Results

4.1 Dataset and Performance Measures

In this paper, *ctree* and *J48* were executed and compared with RBTP. *ctree* is the conditional inference tree of party package [14] and *J48* is C4.5 tree of RWeka package [17, 18]. We implemented RBTP using RStudio with statistical software R. Nine equally assigned class datasets were synthesized with the standard normal distribution, varying sizes from 1000, 2000 and 5000 and varying attributes from 2, 4, and 6 attributes. The performance measures in this study were average of F-measure and G-measure. Additionally, the 10-fold cross validation was applied for these experiments.

4.2 Result

Table 1 highlighted the best performance of all classifiers. RBTP achieves the highest performance of F-measure and G-measure when the dimensions 4. The performance of *J48* and *ctree* are slightly better than RBTP only in the dataset with 5000 instances for 2 and 4 dimensions.

Table 1 F-measure and G-measure of RBTP, *ctree*, and *J48*, respectively.

Number of dimensions	Number of instances	F-measure			G-measure		
		RBTP	ctree	J48	RBTP	ctree	J48
2	1000	**100.00**	99.80	99.80	**100.00**	99.81	99.81
	2000	**99.38**	99.20	98.90	**99.38**	99.20	98.91
	5000	99.18	**99.46**	99.01	99.18	**99.46**	99.02
4	1000	**100.00**	99.90	99.90	**100.00**	99.90	99.90
	2000	**98.09**	96.85	96.55	**98.10**	96.88	96.58
	5000	99.94	**99.98**	**99.98**	99.94	**99.98**	**99.98**
6	1000	**100.00**	99.90	99.80	**100.00**	99.90	99.81
	2000	**99.95**	99.85	99.90	**99.95**	99.85	99.90
	5000	**100.00**	99.98	99.98	**100.00**	99.98	99.98

5 Conclusion

From the experiment results section, RBTP was displayed the better performance than C4.5 tree and conditional inference tree. However, all tests were executed on the synthetic datasets with a small number of attributes. RBTP should be applied to larger datasets and some well-known UCI datasets to see its effectiveness. Moreover, RBTP is not designed to deal with discrete attributes datasets so further improvement is needed.

References

1. Wang, J., Wu, X., Zhang, C.: Support vector machines based on K-means clustering for real-time business intelligence systems. Int. J. Bus. Intell. Data Min. **1**, 54–64 (2005)
2. Che, D., Liu, Q., Rasheed, K., Tao, X.: Decision tree and ensemble learning algorithms with their applications in bioinformatics. In: Arabnia, H.R., Tran, Q.-N. (eds.) Software Tools and Algorithms for Biological Systems SE - 19, pp. 191–199. Springer, New York (2011)
3. Laesanklang, W., Sinapiromsaran, K., Intiyot, B.: Entropy multi-hyperplane credit scoring model (2010)
4. Jiawei, H., Kamber, M.: Data mining: concepts and techniques. Morgan Kaufmann (2001)
5. Miller, D.W.: Results of a New Classification Algorithm Combining K Nearest Neighbors and Recursive Partitioning. J. Chem. Inf. Comput. Sci. **41**, 168–175 (2000)
6. Ko, B.C., Cheong, K.-H., Nam, J.-Y.: Fire detection based on vision sensor and support vector machines. Fire Saf. J. **44**, 322–329 (2009)
7. Farid, D.M., Zhang, L., Rahman, C.M., Hossain, M.A., Strachan, R.: Hybrid decision tree and naïve Bayes classifiers for multi-class classification tasks. Expert Syst. Appl. **41**, 1937–1946 (2014)
8. Delen, D., Walker, G., Kadam, A.: Predicting breast cancer survivability: a comparison of three data mining methods. Artif. Intell. Med. **34**, 113–127 (2005)
9. Sirisomboonrat, C., Sinapiromsaran, K.: Breast cancer diagnosis using multi-attributed lens recursive partitioning algorithm. In: 2012 Tenth International Conference on ICT and Knowledge Engineering, pp. 40–45. IEEE (2012)

10. Bunkhumpornpat, Chumphol, Sinapiromsaran, Krung, Lursinsap, Chidchanok: Safe-level-SMOTE: safe-level-synthetic minority over-sampling TEchnique for handling the class imbalanced problem. In: Theeramunkong, Thanaruk, Kijsirikul, Boonserm, Cercone, Nick, Ho, Tu-Bao (eds.) Advances in Knowledge Discovery and Data Mining SE - 43, pp. 475–482. Springer, Heidelberg (2009)
11. Quinlan, J.R.: Induction of Decision Trees. Mach. Learn. **1**, 81–106 (1986)
12. Wu, X., Kumar, V., Ross Quinlan, J., Ghosh, J., Yang, Q., Motoda, H., McLachlan, G.J., Ng, A., Liu, B., Yu, P.S., Zhou, Z.-H., Steinbach, M., Hand, D.J., Steinberg, D.: Top 10 algorithms in data mining (2007)
13. Quinlan, J.R.: C4.5: Programs for Machine Learning. Morgan Kaufmann Publishers (1993)
14. Hothorn, T., Hornik, K., Zeileis, A.: Party: A Laboratory for Recursive Part (y) itioning (2006). https://cran.r-project.org/web/packages/party/party.pdf
15. Hothorn, T., Hornik, K., van de Wiel, M.A., Zeileis, A.: A Lego System for Conditional Inference (2006)
16. Kaveelerdpotjana, B., Sinapiromsaran, K., Intiyot, B.: Farthest boundary clustering algorithm: half-orbital extreme pole. In: 2013 International Computer Science and Engineering Conference (ICSEC), pp. 168–173 (2013)
17. Hornik, K., Buchta, C., Zeileis, A.: Open-source machine learning: R meets Weka. Comput. Stat. **24**, 225–232 (2009)
18. Kurt, A., Karatzoglou, A., Meyer, D.: Package "RWeka" (2015). https://cran.r-project.org/web/packages/RWeka/RWeka.pdf

Extreme-Centroid Tree for Outlier Detection

Panote Songwattanasiri and Krung Sinapiromsaran

Abstract Outlier detection is one of the knowledge discovery problems that identifies a data point which does not agree with majority data points in a dataset. In the real-world datasets, the majority data points normally line up into patterns that can be captured by some models. In this paper, we propose the new outlier detection algorithm based on the dynamically updated tree model. It composes of two-step processes (1) constructing the extreme-centroid tree from a sampling dataset, and (2) dynamically updated extreme-centroid tree. In the extreme-centroid tree construction step, the root initially identifies two extreme data points from the centroid of a sampling dataset and uses them for splitting data points into groups. It continues splitting until the terminal criterion is met. A leaf node with a single data point is assigned as a suspected outlier in this process. The suspected outliers are trimmed from the tree model and sent back to the rest of a dataset. In the dynamically updated extreme-centroid tree step, a data point from the rest of a dataset will be inserted to the tree model, called the new inserted data point, and a single data point in the tree model is randomly removed from this tree model to maintain the amount of current data points, called the expired data point. The new inserted data point and the expired data point will adjust the tree maintaining the linear time complexity. We compared our algorithm with LOF algorithm and COF algorithm on the synthetic dataset and three UCI datasets. In the UCI datasets, a majority class is selected and other classes are randomly picked as the outliers. The results show that our algorithm outperformed when compared to LOF and COF using precision, recall, and F-measure.

Keywords Outlier detection · LOF · COF · Extreme-centroid tree

1 Introduction

Outlier detection method [1] is an algorithm to find the data points in a dataset which are different from the majority data points. It has been used in varieties of

P. Songwattanasiri(✉) · K. Sinapiromsaran
Department of Mathematics and Computer Science, Faculty of Science,
Chulalongkorn University, Bangkok 10330, Thailand
e-mail: panote.song@gmail.com, krung.s@chula.ac.th

K. Lavangnananda et al. (eds.), *Intelligent and Evolutionary Systems*,
Proceedings in Adaptation, Learning and Optimization 5,
DOI: 10.1007/978-3-319-27000-5_9

application domains such as intrusion detection in network security [2], fraud detection in a bank transaction [2]. Outlier detection can be categorized into three scenarios [3], 1) Supervised outlier detection which labels outlier instances, 2) Semi-supervised outlier detection algorithm which assumes the model from the majority data points and all data points conflicting with the model are outliers, 3) Unsupervised outlier detection which does not label the outlier class and no model assumption for a dataset.

In this paper, we propose the novel outlier detection algorithm for identifying the outliers based on an extreme-centroid tree. Our methodology is split into two steps. The first step is performed once by constructing the extreme-centroid tree from a subset of a dataset. Two extreme-centroid data points are extracted with a centroid of the data points in this subset. They are used for splitting current data points into two groups. Each group is split recursively until the maximum depth or there is a single data point in a group. The leaf node containing a single data point is classified as the suspected outlier. In each node of the tree, it kept the centroid, the extreme-centroids, the height of the node, and the number of the data points. The second step is performed iteratively until all data points are considered. It adjusted the tree when 1) the inserted data point is the new extreme-centroid, 2) the size of the highest node of the inserted data point is satisfying the splitting criterion, 3) the expired data point is an extreme-centroid, and 4) the expired data point is a single data point in its highest node.

The next section contains literature reviews related to the outlier detection. In section 3, definitions and notations are defined. Then, our algorithm is introduced and analyzed its time complexity. In section 4, the results of our algorithm are demonstrated. At the end, the conclusions and our future works are described.

2 Literature Review

Knox, E. M., and Ng, R.T. proposed the definition of the outliers with distance-based in 1998 [4]. They defined the outliers with 2 parameters, which are the radius (eps) and the minimum number of neighbor points (minpts). A data point is considered as an outlier when the number of its neighbors in radius eps is smaller than minpts. This is called Distance-Based (DB) outlier definition. However, this definition cannot identify the outliers in various densities of clusters.

In 2000, the most well-known technique, named Local Outlier Factor (LOF) [5], was created to remedy the problem of DB. LOF is a density-based outlier detection technique that uses a local cluster to identify the outliers. Instead of labeling outlier level, LOF computes the outlier degree of each data point, called outlier factors. High outlier factor means the corresponding data point should be denoted as the outliers.

In 2002, Connectivity Outlier Factor (COF) [6] was proposed by Tang, J., et al. They proposed this algorithm to identify the outliers in a low density cluster which is the weakness of the LOF. COF connects each data point with a path in a local cluster and uses it to compute an outlier factor. However, the running time of COF is slower than that of LOF in a large size of data points.

Angle-Based Outlier Detection (ABOD) [7] was proposed in 2010. This technique aims to detect the outlier in high dimensional dataset. To avoid the curse of dimensionality, ABOD uses the angle-based distance to compute the outlier factors. However, ABOD cannot identify the outliers in multi-cluster dataset.

Outlier Detection Score Based on Ordered Distance Difference (OOF) [8] was proposed in 2013. It computes the outlier factors with the difference of the ordered distance matrix of the entire data points. OOF is a parameter-free algorithm, which is an advantage of OOF. However, the running time of OOF is slower than that of LOF and COF due to the ordered distance matrix.

Histogram-based Outlier Score (HBOS) [1] was proposed by Markus Goldstein and Andreas Dengel. HBOS calculates the outlier factors for each dimension by using histogram. The outlier factors of each dimension are adjusted with the Naive Bayes probability.

In the next section, we proposed a novel outlier detection algorithm, Extreme-centroid Tree for Outlier detection.

3 Extreme-Centroid Tree for Outlier Detection

3.1 Definitions and Notations

The definitions and notations are described below.

Assume $p, q \in D \subseteq \mathbb{R}^n$, where $p = (p_1, p_2, \ldots, p_m)$, and $q = (q_1, q_2, \ldots, q_m)$, and $G \subseteq D$.

Definition 1 (Extreme-centroid poles)

Extreme-centroid poles are a pair of two data points which consist of the extreme-centroid pole and the pseudo extreme-centroid pole.

The first point is the extreme-centroid pole $e1_G$, which is a data point with the largest distance from a centroid c_G.

$$e1_G = \underset{p}{argmax}\{dist(p, c_G) | p \in G\}$$

The second point is the pseudo extreme-centroid pole $e2_G$, which is a data point with the largest distance from the extreme-centroid pole.

$$e2_G = \underset{p}{argmax}\{dist(p, e1_G) | p \in G\}$$

Definition 2 (Outlier)

An outlier is an isolated individual data point in a dataset D.

Let c be the centroid of G, where G is a subset of D, which is recursively split by the extreme-centroid tree.

Let o be an outlier.

$$dist(o, p) > max(dist(p_1, p_2)), \quad \forall p \in G \setminus \{o\}, \forall p_1 \forall p_2 \in G$$

3.2 *Extreme-Centroid Tree for Outlier Detection (ETO) Algorithm*

Extreme-centroid Tree for Outlier detection (ETO) is a distance-based outlier detection algorithm. It executes two steps. The first step is only performed once on the sampling dataset. It construct the extreme-centroid tree from the subset of the original dataset. The second step will refine the extreme-centroid tree based on the rest of data points.

1) Constructing the extreme-centroid tree based on extreme-centroid poles

The root initially contains all current data points in a subset. This subset is a set of data points that randomly collected from a dataset. Two extreme-centroids (Definition 1) are identified and used for splitting current data points into two groups. Each data point belongs to a group of the closest the extreme-centroid as Fig. 1 a) and b). The clusters that split from the root are described as the child nodes (Fig. 1 c)). Each node is recursively split until there is no need to partition such as there is only an individual data point, see Fig. 2 a), b), and c). A node with a single data point is denoted as a leaf node. The information for a node of the tree contains the centroid, the extreme-centroids, level of the tree, and the number of the data points. In this step, if a leaf node is identified, then it will be assigned as a suspected outlier.

From Fig. 3, we limits the tree size to be three. No leaf nodes is labeled as the suspected outliers because both of them are the sibling of each other. In Fig. 4, the leaf nodes are labeled as the suspected outliers because their siblings are not the leaf nodes.

The suspected outliers are trimmed from the tree model. The suspected outliers are returned to the rest of a dataset. So the tree will maintain only the majority data points.

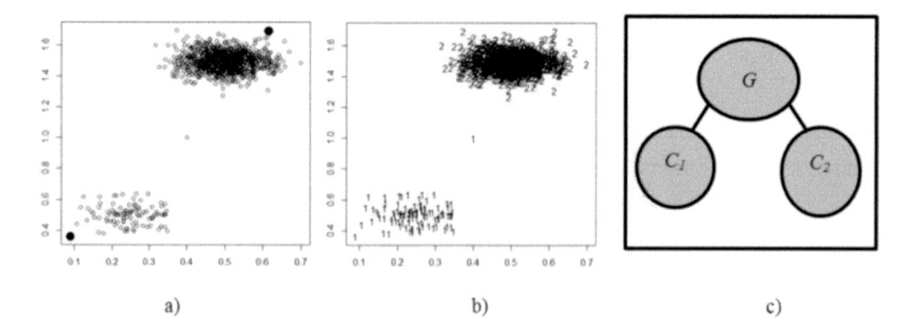

Fig. 1 a) A pair of the extreme-centroids of the entire data points b) Clusters that separated with each of the extreme-centroid c) The clusters are formed the tree structure.

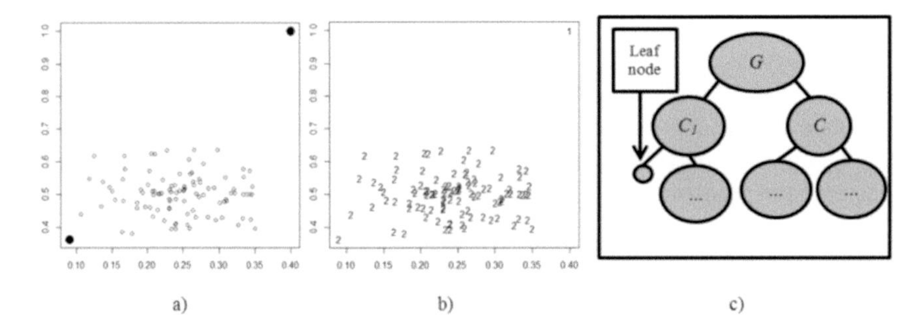

Fig. 2 a) A pair of the extreme-centroids of a cluster after the first splitting b) Candidate outlier is detected after clustering (data point "1" in a circle) c) Candidate outlier is collected as a leaf node.

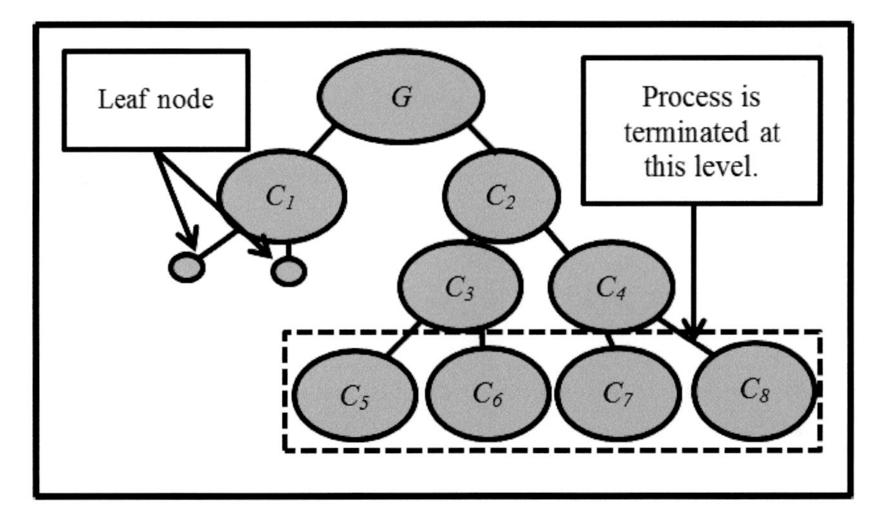

Fig. 3 The terminal criterion is set to have the height of the tree equal to 3, thus the algorithm has to terminate when it reaches that height.

2) Identifying outliers using the extreme-centroid tree

To identifying outliers using the extreme-centroid tree, it requires two subprocesses, 1) inserting the new data point and 2) removing the expired data point. The new inserted data point is a data point which randomly collected from the rest of a dataset and entered to this tree. In each level of the tree, the new inserted data point will adjust the tree.

1. The inserted data point is the new extreme-centroid in some node of the tree. In this case, we will reconstruct this node and its descendants with Algorithm 1.
2. An inserted data point reaches a leaf node and the level of this node is less than the terminal criterion. In this case, the extreme-centroids of this leaf node are identified and this node will be split into two new leaf nodes.

In this step, the new inserted data point is classified as an outlier, if it belongs to the leaf node and this leaf node reaches the highest level when compared to other leaf nodes. After the new inserted data points is classified as the outlier, a data point in extreme-centroid tree is randomly collected as the expired data point and it will be removed from this tree. The extreme-centroid tree structure is adjusted from removing the expired data point.

1. The expired data point is an extreme-centroid in some node of tree. In this case, we will reconstruct this node and its descendants with Algorithm 1.
2. The expired data point is a single data point in its highest node. In this case, this node is deleted.

Algorithm 1: Constructing the extreme-centroid tree based on extreme-centroid poles

Procedure: Constructing the extreme-centroid tree

Inputs: a subset G, where G is a random subset of the original dataset D.

Let: G be the root.

p be a data point in G.

r_D be the height of the current node that the data point p belongs to, where $p \in G$.

Outputs: the extreme-centroid tree model.

```
1    Calculate the centroid c of G
3    for each data point p in G,
4    r_D = r_D + 1
5    Identify the extreme poles e1_G and e2_G (Definition 1)
6    G1 = a set of data points that belong to group of e1_G.
7    G2 = a set of data points that belong to group of e2_G.
8    for each point p,
9      if dist(p, e1_G)< dist(p, e2_G) then
10       p belongs to group G1
11     else,
12       p belongs to group G2
13   if (r_0+1) == r_D then,
14     stop the process
15   if w1 > 1 then,
16     ETO(G1)
17   if w2 > 1 then,
18     ETO(G2)
```

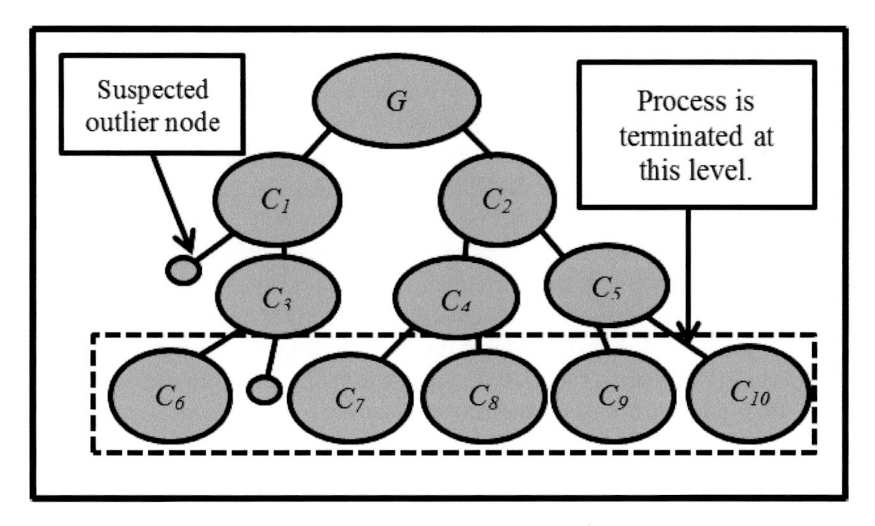

Fig. 4 An example of the leaf nodes which labeled as a suspected outlier in the initial step.

3.3 Time Complexity of the Outlier Detection Based on Extreme-Centroid Poles

In this section, we analyze the time complexity of the ETO. ETO composes of two-step processes (1) constructing the extreme-centroid tree, and (2) adjusting the extreme-centroid tree. In step (1), the process needs to find a) a centroid ($O(N)$), b) two extreme-centroids ($O(2 \cdot N)$), and c) split the data points into two groups ($O(N)$) for each node of tree until the criterion r_0 is met ($O(r_0 \cdot N)$), where r_0 is the height of the tree. Therefore, the time complexity of step (1) can be stated as $O(r_0 \cdot N) + O(3 \cdot N) = O(r_0 \cdot N)$. For the time complexity in step (2), we separate into 4 cases as we described in adjusting the extreme-centroid tree step. In case 1, the time complexity of reconstructing a node and its descendants is $O(r_0 \cdot N_c)$, where N_c is the number of the data points in the adjusted cluster. In case 2, the time complexity of splitting the new node with a single data point is $O(1)$. In case 3, the time complexity of reconstructing a node and its descendants is $O(r_0 \cdot N_c)$. In case 4, the time complexity of deleting an empty leaf node is $O(1)$. From all cases, the time complexity of this step is $O(r_0 \cdot N_c) + O(1) + O(1) + O(r_0 \cdot N_c) = O(r_0 \cdot N_c)$. From all steps, the time complexity of outlier detection with ETO is $O(r_0 \cdot N) + O(r_0 \cdot N_c) = O(r_0 \cdot N)$.

4 Experimental Results

In this paper, LOF and COF are compared to our proposed algorithm with the synthetic datasets and three real-world datasets.

For the synthetic datasets, we synthesize 1000 data points with the multivariate normal distribution. We set the number of the outliers to be 10 [9]. We use precision, recall, and F-measure as the performance measures for our experiments.

The K-nearest neighbors of LOF and COF are set to be 5, 10, and 20. The terminal criterion of our algorithm is set to be 7. We repeat the experiments 10 times. For LOF and COF, we use the optimal cut-off line to collect the outliers, which means the recall of LOF and COF are always 1.

For the real world datasets, we use the Breast Cancer Wisconsin (Diagnostic) dataset, the E. coli dataset, and the Satellite image dataset. For the Breast Cancer Wisconsin (Diagnostic) dataset, it consists of 357 benign and 212 malignant medical diagnosis data points with 30 dimensions. We randomly correct 10 malignant data points as the outliers and drop the rest malignant data points. For the E. coli dataset, we choose class "cp" as the majority class and randomly choose 10 data points from the rest as the outliers. For the Satellite image dataset, we choose class 1 as the majority class and randomly choose 10 data points from the rest as the outliers.

For each dataset, a size of the subset G is set to be 30% of the entire data points. We repeat the experiments 10 times for each dataset.

Table 1 Dataset descriptions

Information	# of data points	# of dimensions	# of outliers
Synthetic data points	**1000**	**2**	**10**
E. coli	**152**	**7**	**10**
Breast Cancer	**357**	**30**	**10**
Satimage	**1057**	**36**	**10**

Table 2 Precisions, Recalls, and F-measures of LOF, COF, and ETO, respectively

Dataset		Precision	Recall	F-measure
single cluster (multivariate normal)	**LOF**	**0.828**	**1**	**0.901**
	COF	**0.756**	**1**	**0.861**
	ETO	**0.905**	**1**	**0.944**
E. coli	**LOF**	**0.457**	**1**	**0.627**
	COF	**0.290**	**1**	**0.450**
	ETO	**0.625**	**1**	**0.769**
Breast Cancer	**LOF**	**0.291**	**1**	**0.412**
	COF	**0.113**	**1**	**0.190**
	ETO	**0.421**	**0.8**	**0.552**
Satimage	**LOF**	**0.127**	**1**	**0.26**
	COF	**0.046**	**1**	**0.09**
	ETO	**0.364**	**0.8**	**0.50**

From Table 2, LOF's performance is higher than COF's performance in the synthetic datasets and three real-world datasets, because COF is not suitable for a dataset with the high density distribution. However, our proposed technique outperforms when compare with both LOF and COF. Due to the outlier characteristics, LOF and COF may not identify the outliers from the majority data points. ETO constructs the model from a subset of the entire data points and adjusted the model when the new data point is inserted, therefore it can identify the outlier while it is inserted to the tree.

5 Conclusion

From the previous section, we compare our algorithm to LOF and COF on the datasets with various distributions. LOF is more sensitive to its parameter than COF's. When the parameter of LOF is changed the performance of LOF is also changed while the parameter of COF shows a little effect to the datasets. However, the performance of COF is slightly better when its parameter is increased. For our algorithm, the parameter does not affect the performance because it is used for controlling the height of the tree which reduces the running time. For the power of the detection, our algorithm shows the best performance in the synthetic datasets and the real-world datasets. In the future, we intend to adapt our algorithm to the temporal dataset such as a data stream.

References

1. Goldstein, M., Dengel, A.: Histogram-based outlier score (HBOS): A fast unsupervised anomaly detection algorithm. KI-2012 Poster Demo Track, 59–63 (2012)
2. Amer, M., Abdennadher, S.: Comparison of unsupervised anomaly detection techniques. Bachelor's Thesis (2011)
3. Chandola, V., Banerjee, A., Kumar, V.: Anomaly detection: A survey. ACM Comput. Surv. **41**, 15 (2009)
4. Knox, E.M., Ng, R.T.: Algorithms for mining distancebased outliers in large datasets. In: Proceedings of the International Conference on Very Large Data Bases, pp. 392–403. Citeseer (1998)
5. Breunig, M.M., Kriegel, H.-P., Ng, R.T., Sander, J.: LOF: identifying density-based local outliers. In: ACM Sigmod Record, pp. 93–104. ACM (2000)
6. Tang, J., Chen, Z., Fu, A.W.-C., Cheung, D.W.: Enhancing effectiveness of outlier detections for low density patterns. In: Advances in Knowledge Discovery and Data Mining, pp. 535–548. Springer (2002)
7. Kriegel, H.-P., Zimek, A.: Angle-based outlier detection in high-dimensional data. In: Proceedings of the 14th ACM SIGKDD International Conference on Knowledge Discovery and Data Mining, pp. 444–452. ACM (2008)
8. Buthong, N., Luangsodsai, A., Sinapiromsaran, K.: Outlier detection score based on ordered distance difference. In: 2013 International Computer Science and Engineering Conference (ICSEC), pp. 157–162. IEEE (2013)
9. Kriegel, H.-P., Kröger, P., Schubert, E., Zimek, A.: LoOP: local outlier probabilities. In: Proceedings of the 18th ACM Conference on Information and Knowledge Management, pp. 1649–1652. ACM (2009)

Gaussian Fuzzy Integral Based Classification

Wang Jinfeng and Wang Wenzhong

Abstract Fuzzy integral is a kind of effective fusion tool. Traditionally, fuzzy integral can project the data with n-dimension into one line, in which the projection is along with a group of linear lines. In reality, data distribution is not regular, so the straight line for projection is too limited. Gaussian function is applied to natural science widely. It is close to normal distribution and can cover more data. In this article, a new generalization of fuzzy integral is proposed. The Gaussian function is used as integrand. A new classifier is constructed based on Gaussian Fuzzy integral and applied into several benchmark data sets. The results show that the new version can improve the property of fuzzy integral and obtain the better performance.

Keywords Gaussian function · Fuzzy integral · Generalization · Classification

1 Introduction

In reality, nonlinear is ubiquitous. The classical linear method cannot deal with those nonlinear problems with the lower accuracy. Fuzzy measure was proposed to describe the contribution of each feature and combination of features for decision [1, 2]. It can show the interaction among features, and the contribution is nonadditive. Each fuzzy measure represents one subset of the feature universal set.

Traditionally, fuzzy integral can classify the n-dimensional data by projecting into straight line and using linear classifier on virtual space [3, 4]. But, there are

W. Jinfeng
College of Mathematics and Informatics, South China Agricultural University, Guangzhou 510642, China
e-mail: wangphoenix@163.com

W. Wenzhong(✉)
College of Economics and Management, South China Agricultural University, Guangzhou 510642, China
e-mail: wangwenzhong@163.com

K. Lavangnananda et al. (eds.), *Intelligent and Evolutionary Systems*, Proceedings in Adaptation, Learning and Optimization 5,
DOI: 10.1007/978-3-319-27000-5_10

many data will be lost in the projecting process. In actual problems, dataset is not distributed linearly. For example, some data may be surrounded by other data belonged to different classes; or there are overlapping situations among multiple classes. As so far, research on Fuzzy integral mainly focuses on two points. One is about the application of classical Fuzzy Integral; the other is fundamental theory of Fuzzy Measure [5, 6]. There are few studies to consider the data distribution. When modeling these problems, Gaussian function is used as a distribution function usually. So, a new generalized fuzzy integral---Gaussian Fuzzy Integral is proposed, which adopts Gaussian function as integrand. All data in same class can be projected along same Gaussian curves. The classifier can be constructed based on Gaussian Fuzzy Integral. Experiments show that this new method can cover most datasets and improved the classifying accuracy greatly.

This article is arranged as follows. Section 1 gave out the introduction about this research. The classical Fuzzy Integral and Fuzzy Measure are presented in Section 2. Gaussian Fuzzy measure and classifier are proposed in Section 3. The next section shows the experimental results and related analysis. The last section gives the conclusion.

2 The Fundamental Materials

The signed fuzzy measure [9] was used in previous research, which can break the limitation of traditional fuzzy measure and be extended to more fields. So, in this article, we still adopt the signed fuzzy measure to our new model. Then, the signed fuzzy measure and fuzzy integral. The fundamental materials will be introduced as follows.

2.1 Signed Fuzzy Measure

Denote by X and call it the universal set. Then $(X, P(X))$ is a measurable space, where $P(X)$ is the power set of X. In almost all real problems, the universal set is finite. For example, in any database, the number of features is always finite. Thus, throughout this paper we assume that $X = \{x_1, x_2, ..., x_n\}$, where each x_i, $i = 1, 2, \ldots, n$, is a feature. In a multi-regression problem, $x_1, x_2, ..., x_n$ are called predictive features. They are usually numerical. There is another numerical feature Y called the objective feature in the database as a fusion target. The observation value of Y is denoted by y generally.

Definition 1. A Fuzzy Measure [10], μ is a mapping from $P(X)$ to the finite space $[0, \infty)$ satisfying the following conditions:

1) $\mu(\phi) = 0$;
2) $A \subset B \Rightarrow \mu(A) \leq \mu(B), \forall A, B \in P(X)$.

Set function μ is nonadditive in general. If $\mu(X)=1$, then μ is said to be regular.

To further understand the practical meaning of Fuzzy Measure, let us consider the elements in a universal set X as a set of predictive features to predict a certain objective. Then, for each individual predictive feature as well as each possible combination of the predictive features, a distinct value of a Fuzzy Measure is assigned to describe its influence to the objective. Due to the nonadditivity of the Fuzzy Measure, the influences of the predictive features to the objective are dependent such that the global contribution of them to the objective is not just the simple sum of their individual contributions. However, the monotonicity and nonnegativity of Fuzzy Measure are too restrictive for real applications. The Signed Fuzzy Measure was proposed.

Definition 2. Any set function, $\mu : P(X) \rightarrow (-\infty, \infty)$, is called a Signed Fuzzy Measure if $\mu(\varnothing)=0$, where $\varnothing$ is the empty set. Any nonnegative Signed Fuzzy Measure is called Fuzzy Measure. The Signed Fuzzy Measure is also called Generalized Fuzzy Measure [7, 8, 9], [11].

The following example explains that a Signed Fuzzy Measure can be used for describing the interaction among the contributions from the information sources towards a certain target. A similar example of workers appeared in [7] first. Worker efficiency corresponds to the Signed Fuzzy Measure.

Example 1. Let $X = \{x_1, x_2, x_3\}$ be the set of three workers. They are hired for manufacturing a certain type of products. Their individual and joint efficiency μ: $P(X) \rightarrow (0\ ,\infty]$ is given as follows:

$$\mu(E) = \begin{cases} 0 & \text{if } E = \varnothing \\ 5 & \text{if } E = \{x_1\} \\ 3 & \text{if } E = \{x_2\} \\ 10 & \text{if } E = \{x_1, x_2\} \\ 4 & \text{if } E = \{x_3\} \\ 4 & \text{if } E = \{x_1, x_3\} \\ 6 & \text{if } E = \{x_2, x_3\} \\ 9 & \text{if } E = X \end{cases} .$$

Set function μ is a Fuzzy Measure that describes the interaction among the contributions from individual workers towards the target, the total number of toys manufactured by these workers. In this example, $\mu(\{x_1, x_2\}) > \mu(\{x_1\}) + \mu(\{x_2\})$ shows that workers x_1 and x_2 cooperate well, therefore, the interaction between their contributions is mutually promoting. While $\mu(\{x_1, x_3\}) < \mu(\{x_1\}) + \mu(\{x_3\})$, even $\mu(\{x_1, x_3\}) < \mu(\{x_1\})$, shows that workers x_1 and x_3 cooperate very badly and the interaction between their contributions is mutually inhibitive. This set function is not monotonic.

2.2 Nonlinear Integral with Respect to the Signed Fuzzy Measure

Let $X = \{x_1, x_2, ..., x_n\}$ be Signed Fuzzy Measures, and $f : X \rightarrow [0, \infty)$ be non-negative functions.

To calculate the value of the Nonlinear Integral of a given real-valued function f, usually the values of f, i.e., $f(x_1), f(x_2), \cdots, f(x_n)$, should be sorted in a non-decreasing order so that $f(x') \leq f(x') \leq \ldots \leq f(x')$, where $(x_1', x_2', \ldots x_n')$ is a certain permutation of $(x_1, x_2, \ldots x_n)$. So the value of Nonlinear Integral can be obtained by

$$\int f d\mu = \sum_{i=1}^{n} [f(x'_i) - f(x'_{i-1})] \mu(\{x'_i, x'_{i+1}, \ldots\ldots x'_n\}),$$

where $f(x_0') = 0$.

The Nonlinear Integral is based on linear operators to deal with nonlinear space. When the linear space cannot provide the accurate classification, we need to extend the projection to 2-dimensional space in order to stretch and classify the confused data. So the second Fuzzy Measure ν is introduced to finish the second projection.

Fuzzy integral can deal with nonlinear problem with linear operator. Traditional fuzzy integral make the original information by projecting data into one dimensional space as figure 1. Projection lines are determined by integrand which is linear function in classical fuzzy integral. The direction of projecting relies on the sign of fuzzy measures. But, linear projection cannot reflect the data distribution leading to the crossover among classes. We need a kind of projection model close to the real distribution. So, a new fuzzy integral based on Gaussian function is proposed to cover more data and obtain better performance.

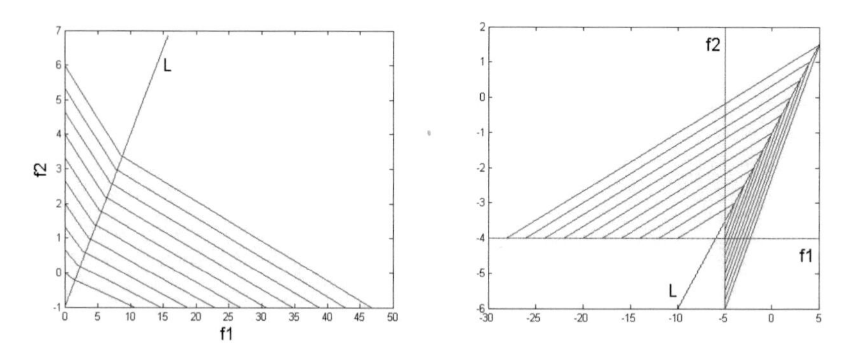

Fig. 1 Projection by classical fuzzy integral

3 Gaussian Fuzzy Integral Based Classifier

The standard Gaussian Function can be defined as

$$\mathrm{G}(x) = ae^{-(x-b)^2/2c^2}$$

In which a, b and c are real constant data, and $a>0$.

Gaussian function can describe the distribution of most datasets. Different parameters infect function figure. Gaussian function is applied to describe normal distribution in statistics, to deal with signal by defining Gaussian filter and image processing by Gaussian fuzzification. For simplicity, let $a=1$, $b=0$, $c=1$ usually.

3.1 *Fuzzy Integral Based on Gaussian Function*

The fundamental setup still keeps the classical fuzzy integral format. Gaussian function based fuzzy integral can be defined as follows.

$$\int e^{-\frac{(f(x)-b)^2}{2c^2}} \mathrm{d}\mu = \sum_{i=1}^{n}\left[e^{-\frac{(f'(x_i)-b_i)^2}{2c_i^2}} - e^{-\frac{(f'(x_{i-1})-b_{i-1})^2}{2c_{i-1}^2}} \right] \mu(\{x'_i, x'_{i+1}, \cdots, x'_n\})$$

in which $f(x)$ is the feature vector, b_i is the mean value of $f(x)$, and c_i^2 is the variance. The formulas are defined as follows.

$$b_i = \frac{1}{l}\sum_{j=1}^{l} f(x_{ij}) \quad ; \quad c_i^2 = \frac{1}{l}\sum_{j=1}^{l}\left(f(x_{ij}) - b_i\right)$$

Assumed that the feature number n=2, and $\mu_1 = 0.4; \mu_2 = 0.5; \mu_{12} = 0.6$, the projection is shown in Fig. 2(a). The statues of projection curves rely on the fuzzy measure. When $\mu_1 = 0.1; \mu_2 = 0.5; \mu_{12} = 0.9$, the corresponding projection is presented in Fig. 2(b). Gaussian fuzzy integral can cover most data distribution by changing fuzzy measure. Data is projected into 1-D space by Gaussian fuzzy measure and classified with the higher accuracy.

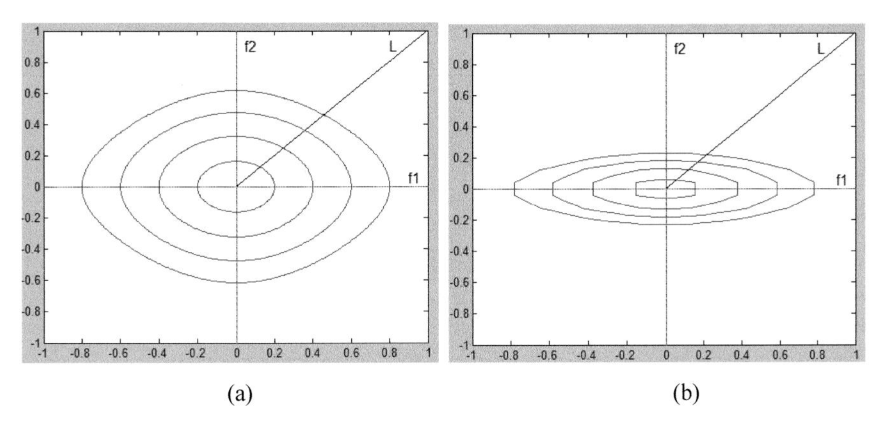

(a) (b)

Fig. 2 The projection of Gaussian Fuzzy Integral

3.2 Classifier Based on Gaussian Fuzzy Integral

In classification problems, classifier based on Gaussian fuzzy integral is constructed by projecting multiple dimensional data to one dimension space along the Gaussian curves and classifying the final virtual data. If the classification performance is not satisfying, fuzzy measures must be learned again, in which parameters a, b and u can be learned with fuzzy measures using GA. The concrete technical route is presented as figure 3.

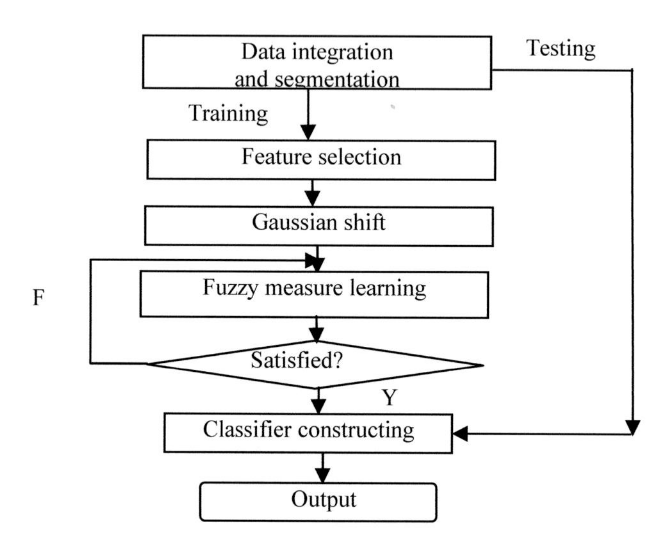

Fig. 3 The technical route for constructing classifier

4 Experiment and Analysis

In this article, the new model was constructed by Matlab2012. The validation was finished in two parts of experiments. Firstly, several classical data sets were selected from UCI data warehouse for simulating[10]. Secondly, a group of HBV Gene data was adopted for testing compared with several classical algorithms.

4.1 Benchmark Data

The description about benchmark data are listed in table 1. The cross-validate method was adopted for reducing overfitting. All results are shown in table 2 which includes classical Fuzzy Integral and Gaussian Fuzzy Integral. The best values on accuracy are marked in bold. We can see that the Gaussian Fuzzy Integral is superior to the classical one.

Table 1 Data Description

Data sets	cases	feature	class
Monk1	556	6	2
Monk2	601	6	2
Monk3	554	6	2

Table 2 Classification Results

Data sets		FI	Gaussian-FI
Monk1	Training accuracy	0.867	**0.950**
	Testing accuracy	0.789	**0.941**
Monk2	Training accuracy	0.720	**0.983**
	Testing accuracy	0.677	**0.963**
Monk3	Training accuracy	**0.954**	0.942
	Testing accuracy	**0.950**	0.877

4.2 HBV Diagnosis

HBV data was collected from Wilson Hospital which contains 98 patients without cancer and 100 positive patients. All DNA sequences of HBV cases were selected carefully by biologists for minimizing the statistical bias. Database was clustered into four data sets which are listed in table 3. All data were processed by molecular analysis, clustering and feature selection [11, 12].

Table 3 HBV Database Description

Data sets	Negative	Positive	Total
B	51	37	88
C1	10	16	26
C2	18	22	40
C3	19	25	44
Total	98	100	198

In classifying an unknown case, depending on the class predicted by the classifier and the true class of the patient, four possible types of results can be observed for the prediction as follows:

(1) True positive (TP) - The result of the patient has been predicted as positive (Cancer) and the patient has cancer.
(2) False positive (FP) - The result of the patient has been predicted as positive (Cancer) but the patient does not have cancer.
(3) True negative (TN) - The result of the patient has been predicted as negative (Control), and indeed the patient does not have cancer.
(4) False negative (FN) - The result of the patient has been predicted as negative (Control) but the patient has cancer.

For each learning and evaluation experiment, *Accuracy*, *Sensitivity* and *Specificity* defined below are used as the fitness or performance indicators of the classification.

$$Accuracy = (TP + TN)/(TP + TN + FP + FN),$$

$$Sensitivity = TP/(TP + FN),$$

$$Specificity = TN/(TN + FP).$$

Table 4 All Comparison Results

Data sets \ Algorithms		FI	Gaussian-FI	SVM	DT	NB
B	Training Accuracy	0.682	0.773	0.674	0.682	0.689
	Training Sensitivity	0.811	0.902	0.794	0.811	0.790
	Training Specificity	0.588	0.595	0.589	0.588	0.617
	Testing Accuracy	0.674	0.753	0.680	0.681	0.650
	Testing Sensitivity	0.813	0.900	0.795	0.812	0.758
	Testing Specificity	0.574	0.550	0.597	0.571	0.573
C1	Training Accuracy	0.961	0.961	0.897	0.937	0.894
	Training Sensitivity	1.000	0.9	0.965	1.000	0.722
	Training Specificity	0.899	1.000	0.810	0.836	1.000
	Testing Accuracy	0.847	0.780	0.961	0.717	0.650
	Testing Sensitivity	0.980	0.600	1.000	1.000	0.300
	Testing Specificity	0.640	0.950	0.899	0.280	0.850
C2	Training Accuracy	0.918	0.881	0.725	0.839	0.773
	Training Sensitivity	0.903	0.888	0.665	0.749	0.993
	Training Specificity	0.937	0.883	0.785	0.953	0.589
	Testing Accuracy	0.817	0.700	0.848	0.728	0.727
	Testing Sensitivity	0.788	0.800	0.789	0.615	0.897
	Testing Specificity	0.860	0.6	0.907	0.880	0.592
C3	Training Accuracy	0.731	0.755	0.604	0.684	0.697
	Training Sensitivity	0.913	1.000	0.475	0.504	0.688
	Training Specificity	0.491	0.569	0.780	0.905	0.702
	Testing Accuracy	0.600	0.695	0.753	0.645	0.587
	Testing Sensitivity	0.738	0.950	0.663	0.442	0.600
	Testing Specificity	0.410	0.483	0.871	0.920	0.567
AVE.	Training Accuracy	0.777	**0.843**	0.698	0.748	0.735
	Training Sensitivity	0.877	**0.923**	0.720	0.755	0.799
	Training Specificity	0.678	0.762	0.700	**0.765**	0.680
	Testing Accuracy	0.709	0.732	**0.767**	0.687	0.652
	Testing Sensitivity	0.813	**0.813**	0.791	0.715	0.691
	Testing Specificity	0.604	0.646	**0.760**	0.673	0.612

Medicine Specialist prefer to the higher sensitivity with lower acceptable accuracy and specificity. We rather confirm more patients with cancer than miss those true patients. In these data sets, all features are symbolic types. Each feature contains four values A, C, G and T. For using our nonlinear model, we use 0, 1, 2 and 3 to represent the feature as initial values.

We applied the Gaussian Fuzzy Integral(Gaussian-FI) to HBV Gene Data Sets and compared with the results in previous research [13] as shown in Table 4, which includes Neural Network(NN)[14], Decision Tree(DT)[15], Naive Bayesian(NB)[16], Support Vector Machine(SVM)[17] and classical Fuzzy Integral(FI). The average is used to measure performance and the best values are highlighted in bold. The results show that SVM has the best accuracy and the poor sensitivity. On the contrary, Gaussian FI has the best testing sensitivity in favor of shifting test. We hope not to miss true patients. So Gaussian FI is the best choice as the diagnosis assisting tool.

5 Conclusion

Fuzzy Integral is a good tool of information fusion. But it has many limits and defects. In this article, a new generalized fuzzy integral was proposed based on Gaussian Function---Gaussian Fuzzy Integral. It extended the classical Fuzzy Integral by projecting data along Gaussian curve into virtual space. Linear classification was implemented on the virtual values. Experimental results showed that the performance of Gaussian FI is better than classical ones on classifying, especially for cancer diagnosis. Gaussian FI has the best training and testing sensitivity which is important for medicine specialist and doctors. The research on generalized Fuzzy Integral is worthy digging, such as adjusting parameters in Gaussian function, reducing the complexity of learning parameters. We will continue to develop the related work in the future.

Acknowledgements This research is supported by the National Natural Science Foundation of China (No. 61202295), and the National Social Science Foundation of China (Projects No. 10CJY024).

References

1. Sugeno, M.: Theory of fuzzy integrals and its applications. Ph.D. dissertation. Tokyo Institute of Technology (1972)
2. Sugeno, M.: Fuzzy measures and fuzzy integrals: a survey. In: Gupta, Saridis, Gaines (eds.), pp. 89–102 (1977)
3. Grabisch, M., Nicolas, J.M.: Classification by fuzzy integral: Performance and tests. Fuzzy Sets and Systems **65**, 255–271 (1994)
4. Keller, J.M., Yan, B.: Possibility expectation and its decision making algorithm. In: 1st IEEE Int. Conf. On Fuzzy Systems, San Diago, pp. 661–668 (1992)
5. Chen, B., Chen, S., Feng, J.: A study of multisensor information fusion in welding process by using fuzzy integral method. Int. J. Adv. Manuf. Technol. **74**, 413–422 (2014)

6. Cavrini, F., Rita, L., Bianchi, Q.L., Saggio, G.: Combination of Classifiers Using the Fuzzy Integral for Uncertainty Identification and Subject Specific Optimization: Application to Brain-Computer Interface (2015). doi:10.5220/0005035900140024
7. Mikenina, L., Zimmermann, H.J.: Improved feature selection and classification by the 2-additive fuzzy measure. Fuzzy Sets and Systems **107**, 197–218 (1999)
8. Xu, K.B., Wang, Z.Y., Heng, P.A., et al.: Classification by Nonlinear Integral projections. IEEE Transactions on Fuzzy System **11**(2), 187–201 (2003)
9. Wang, J. F, Lee, K. H, Leung, K. S, Wang, Z. Y.: Projection with Double Nonlinear Integrals for Classification. Book of Advances in Data Mining, 5077, 142-152 (2008)
10. Asuncion, A, Newman, D. J. UCI Machine Learning Repository, Irvine, CA, University of California, Department of Information and Computer Science (2007). http://www.ics.uci.edu/~mlearn/MLRepository.html
11. Kumar, S., Tamura, K., Nei, M.: MEGA3: Integrated Software for Molecular Evolutionary Genetics Analysis and Sequence Alignment. Brief. Bioinformatics **5**, 150–163 (2004)
12. Sugauchi, F., Kumada, H., Sakugawa, H., Komatsu, M., Niitsuma, H., Watanabe, H., Akahane, Y., Tokita, H., Kato, T., Tanaka, Y., Orito, E., Ueda, R., Miyakawa, Y., Mizokami, M.: Two Subtypes of Genotype B (Ba and Bj) of Hepatitis B Virus in Japan. Clinical Infectious Diseases **38**, 1222–1228 (2004)
13. Leung, K.S., Lee, K.H., Wang, J.F., et al.: Data Mining on DNA Sequences of Hepatitis B Virus. IEEE/ACM Transactions on Computational Biology and Bioinformatics **8**(2), 428–440 (2011)
14. SAS1 Enterprise Miner (EM) (2009). http://www.sas.com/technologies/analytics/datamining/miner/
15. Data Mining Tools See5 and C5.0: Software (2006). http://www.rulequest.com/see5-info.html
16. Borgelt, C.: Bayes Classifier Induction. Software (2009). http://fuzzy.cs.uni-magdeburg.de/~borgelt/bayes.html
17. Chang, C.C., Lin, C.J.: LIBSVM: A Library for Support Vector Machines (2001). http://www.csie.ntu.edu.tw/~cjlin/libsvm

Predicting Duration of CKD Progression in Patients with Hypertension and Diabetes

Warangkana Khannara, Natthakan Iam-On and Tossapon Boongoen

Abstract Renal failure is one of major medical diseases that is recently on the rise, especially in Thailand. In general, patients with hypertension and diabetes are at high risk of encountering this disorder. The medical cost for a large group of chronic-disease patients has been the burden not only to the local hospitals, but also the country as a whole. Without forward planning, the allocated budget may not cover the expense of increasing cases. This research aims to develop an intelligent model to predict the duration to progress kidney disease in those patients with hypertension and diabetes. As such, the predictive model can help physicians to acknowledge patients' risk and set up a plan to prolong the progression duration, perhaps by modifying their behaviors. The methodology of data mining is employed for such cause, with records of 360 patients from Phan hospital's database in Chiang Rai province between 2004 and 2014. Prior model generation, the underlying data has gone through conventional steps of data cleaning and preparation, such that the problems of incomplete and biased data are resolved. To explore the baseline of prediction performance, four classical classification techniques are exploited to create the desired model. These include decision tree, K-nearest neighbor, Naive Bayes, and Artificial Neural Networks. Based on 10-fold cross validation, the overall accuracy obtained with the aforementioned techniques is around 70% to 80%, with the highest of 86.7% being achieved by Artificial Neural Networks.

Keywords CKD Progression · Diabetes · Hypertension · Data mining · Prediction

W. Khannara(✉) · N. Iam-On
School of Information Technology, Mae Fah Luang University, Chiang Rai 57100, Thailand
e-mail: tao-narak@hotmail.com, nt.iamon@gmail.com

T. Boongoen
Department of Mathematics and Computer Science,
Royal Thai Air Force Academy, Bangkok 10220, Thailand
e-mail: tossapon_b@rtaf.mi.th

K. Lavangnananda et al. (eds.), *Intelligent and Evolutionary Systems*,
Proceedings in Adaptation, Learning and Optimization 5,
DOI: 10.1007/978-3-319-27000-5_11

1 Introduction

Chronic Kidney Disease (CKD) has long been one of important issues in public health worldwide. With this disease, many patients inevitably encounter pain and high cost of medical treatment. Specific to Thailand, patients with CKD were approximately 17.5 percent of the population in 2010 [1]. In addition, the cost of treating patients with CKD on Renal Replacement Therapy (RRT) has inclined over the previous years. In 2010, there were approximately 40,000 new cases of RRT and the yearly increase was about 12,000 cases [2]. Given this observation, governmental hospitals as well as other medical service agencies may face a financial difficulty as to cope with CKD cases in the near future.

In particular, the association between CKD with diabetes and high blood pressure (i.e., hypertension) has been widely identified in the medical literature. With respect to the report of Nephrology Society of Thailand in 2013, 18.7 to 43.5 percent of patients with diagnosis of CKD were caused by diabetes, while 6.1 percent cased by high blood pressure [3, 4, 5]. According to the investigation of [6], diabetes appears to be the main cause of kidney disease in patients with end-stage renal disease, at the rate around 30.1 percent.

There are studies on the prognosis using statistical methods such as Cox model [7, 8] and Markov model [9]. To determine the duration of the disease, the exploitation of risk factor of disease progression were analyzed. The model used in the study was to discover the factors that affect the progression of the disease and help to predict the likelihood of disease. The cross-sectional study was conducted to determine the prevalence of kidney disease [10]. There are other related works that explore the literature to identify risks expected to result in kidney disease [11]. The study makes use of data mining techniques that help to determine the decision to treat the disease [12, 13, 14], in addition to the typical calculation of the GFR (Glomerular Filtration Rate). But, there has been no study focusing on predicting progression of CKD in patients with diabetes or hypertension. This provides an opportunity to explore the use of data mining techniques to classify the progression interval, which can be used to determine the risk faced by each individual patient and the demand of medical care upon the hospital.

The study is based on the specific population of patients with prior diagnosis of diabetes or high blood pressure, who have been receiving medical treatment at Phan hospital. A set of potential patient records have been specified in Phan database during the first quarter of 2015. These are extracted with the targeted attributes corresponding to those commonly identified in the literature and others suggested by medical experts. A number of clinical variables can be retrieved and employed in the empirical investigation. These include patients age, sex, Body Mass Index (BMI), and the results of diagnostic laboratory, e.g., High-Density Lipoprotein (HDL) and Blood Urea Nitrogen (BUN). In addition, patients medical history, especially those associative diseases to CKD such as gallstones and gout, are also examined. As the research has been shaped as a classification problem, a number of conventional or basic classifiers like decision tree (J48), Naive Bayes (NB), K-nearest nieghbor (KNN) and Artificial Neural Networks (ANN) are initially exploited to present the baseline for accuracy level with the data set. Given such background, more advance techniques can be formulated to maximize the potential of this research work and its applications.

The rest of this paper is organized as follows. Section 2 introduces background of CKD and related works as to set the scene for the following sections. The data-mining approach to progression prediction including details of data specification, extraction, cleaning and transformation are presented in Section 3. To tackle the problem of unbalanced data that is typical for this classification task, the proposed framework has integrated synthetic data objects into the learning process. As a result, a conventional classifier can be applied to create the desired predictive model. Section 4 provides experimental design and the corresponding results with standard classifiers. The paper is concluded in Section 5 with possible future work.

2 Background and Related Work

The definition of CKD is currently under the standard of the Kidney Disease Improving Global Outcome (KDIGO) 2012 [15]. To be precise, this term means that a person's kidneys are abnormal in structure or working on disorder conditions for more than 3 months. Fig. 1 presents the conceptual model of the course of CKD. According to this figure, the shaded ellipses represent different stages of CKD, while the unshaded ones show potential antecedents or impact of CKD. The thick arrows between ellipses represent risk factors associated with the initiation and progression of the disease that can be affected or detected by interventions: susceptibility factors (black), initiation factors (dark gray), progression factors (light gray), and end-stage factors (white). See Table 1 for more details. Interventions of each stage are given beneath the stage. In particular, a person who appears to be normal should be screened for CKD risk factors. Others with increasing risk of CKD should be screened for CKD. Note that the term 'Complications' refers to all complications of CKD and treatment, including complications of GFR (e.g., hypertension, anemia, malnutrition, bone disease, neuropathy, and decreased quality of life) and cardiovas-

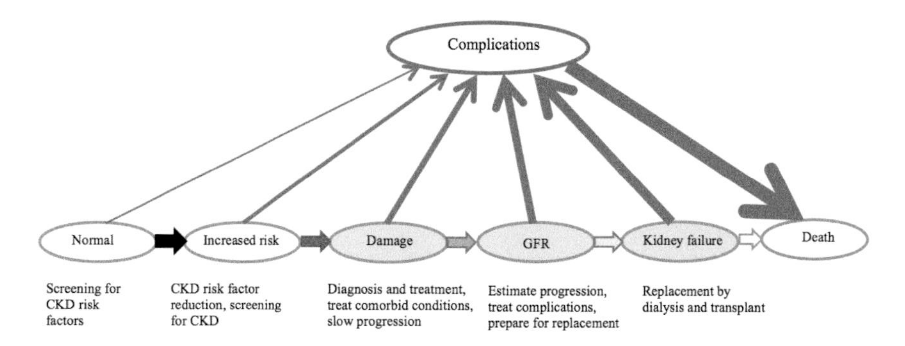

Fig. 1 Conceptual model of the course of chronic kidney disease and the corresponding therapeutic strategies [15].

Table 1 Risk factors for chronic kidney disease (CKD) and its outcomes [15].

Risk factor	Definition	Examples
Susceptibility	Increase susceptibility to kidney damage	Older age, family history of chronic kidney disease, reduction in kidney mass, low birthweight, U.S. racial or ethnic minority status, low income or education
Initiation	Directly initiate kidney damage	Diabetes, high blood pressure, autoimmune diseases, systemic infections, urinary tract infections, urinary stones, lower urinary tract obstruction, drug toxicity
Progression	Cause worsening kidney damage and faster decline in kidney function after initiation of kidney damage	Higher level of proteinuria, higher blood pressure, poor glycemic control in diabetes, smoking
End-stage	Increase morbidity and mortality in kidney failure	Lower dialysis dose, temporary vascular access, anemia, low serum albumin, high serum phosphorus and late referral

cular disease (CVD). The increased thickness of arrows connecting later stages to complications represents the increased risk of complications as kidney disease.

The value of abnormal renal function can be measured by the GFR, which is divided into the following five stages of kidney disease [16]. See Table 2 details. A number of recent studies have commonly point out blood sugar levels of diabetic patients, hypertension and cardiovascular disease as important factors, which commonly cause kidney disease [17, 18, 19]. This leads to the focused population of examined patients, for which the present research aims to predict CKD progression patterns. To accomplish this objective, the methodology of data mining [20] is identified as an effective approach to extract a useful and accurate predictive model. Data mining is able to disclose patterns hidden in the data under examination, which help to improve the decision making process. It typically consists of various steps: (1) collecting and selecting data to be used, (2) cleaning and preparing data such that it becomes suitable for the following tasks, (3) transforming and consolidating data types and domains, (4) creating the desired model using an appropriate learning method (e.g., classification and clustering), and (5) evaluating the resulting model with the preparation to deploy.

Table 2 Classification of CKD based on GFR.

Stage	State of kidney function	Classification of severity by GFR
1	Kidney damage with normal or increasing in GFR	GFR $\geq$ 90
2	Kidney damage with mild decreasing in GFR	GFR between 60-89
3	Moderate with decreasing in GFR	GFR between 30-59
4	Severe with decreasing in GFR	GFR between 15-29
5	Kidney failure	GFR $<$ 15 (or dialysis)

In recent years, there have been many researches that make use of data mining techniques to healthcare applications. These include the search for factors and risks in diseases occurrence [21]. Also, there are studies that aim to find the duration of diseases occurrence by data mining method. Specific to kidney disease, there are researchers attempts to generate a decision to identify the risk factors of early arteriovenous fistula failure in hemodialysis patients [22]. Similarly, another work by [23] predicts survival of kidney dialysis patients using a decision tree model.

3 Method

3.1 Overview

Fig. 2 illustrates the overview of this research study, which initially starts with the selection of important variables or attributes to predict the progression of kidney disease. These can be identified based on previous works found in the literature and suggestions of medical professionals. Following that, the aforementioned collection of attributes is preprocessed such that any error found can be corrected. This is to ensure the quality of the resulting prediction. Having done this, several classification techniques are applied to the selected data set. These include standard classifiers like J48, KNN, NB and ANN. Note that the underlying classification process aims to categorize a new patient into classes of time interval between diagnosis of diabetes/hypertension and diagnosis of CKD (i.e., Class1 and Class2 that are shown in Fig. 2 and will be discussed in the following section). After formulating an accurate and robust predictive model, it will be used to support decision-making aids. For instance, it can report the risk level of each patient that may lead to modification on medical assessment and plan for treatment.

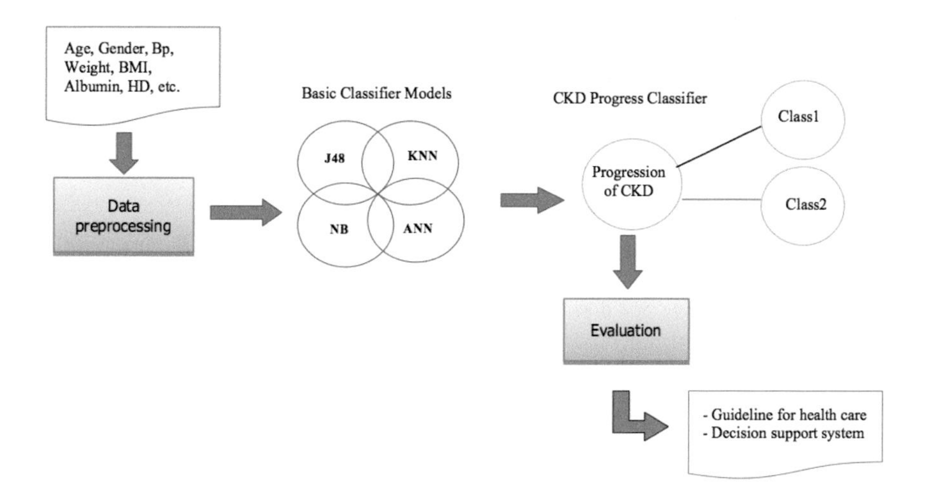

Fig. 2 The overview of research study.

Moreover, the hospital can use it to better evaluate the demand of resources arising from new CKD cases likely to occur in the future.

3.2 Data Collection and Preparation

Fig. 3 summarizes the process of data collection and preparation, such that the well-formed data can be achieved for the following learning stage. Within the database at Phan hospital, there are totally 364 patient records with the set of desired attributes. These patients have been diagnosed with diabetes or hypertension before the progression to kidney disease (i.e., CKD stage3) is validated. Note that the duration between the aforementioned key observations found in this set of population is between one to ten years. Based on the consultation with CKD experts and the result of applying a standard discretization (i.e., binning) technique, two categories of progress duration are appropriate for further medical and budget planning. One is the duration of less than 5.5 years, the other is more than or equal to 5.5 years, which are denoted as Class1 and Class2, respectively.

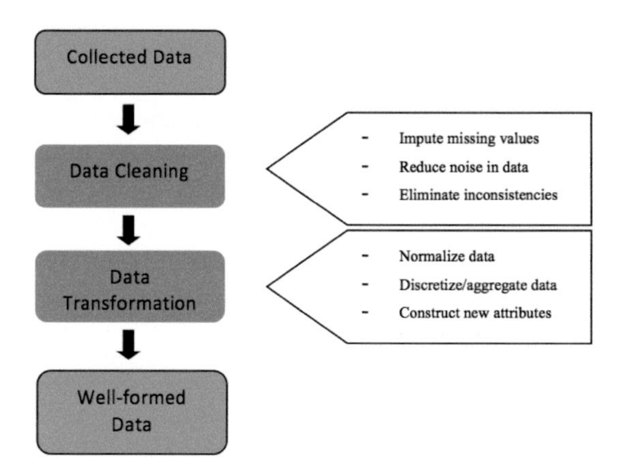

Fig. 3 Data preparation process.

Specific to this population, the following original attributes are extracted.

- General information: sex and age being diagnosed with hypertension or diabetes (Age HT/DM).
- Laboratory results:
  - total number of times with blood-sugar result (x_2), number of times with high-sugar value (x_1), and number of times with normal-sugar value (x_3).
  - total number of times with blood-fat result (y_2), number of times with high-fat value (y_1), and number of times with normal-fat value (y_3).

- total number of times with blood-waste result (z_2), number of times with high-waste value (z_1), number of times with low-waste value (z_3) and number of times with normal-waste value(z_4).

– Clinical data:

- body mass index higher than standard (BMIHigh).
- prescription of pain killer with power to damage kidney consecutively more than 6 months (NSAID).
- ability to control blood pressure (ControlBP).
- having congenital disease of diabetes (DM).
- having congenital disease of hypertension (HT).
- having record of stone (N20).
- having record of Ischemic heart disease (I200).
- having record of disease related to heart failure (I500).
- having record of Gout or Rheumatoid (M109).
- having record of Urinary disease (N30).
- having record of protein in the urine microalbuminuria (Albumin).

Having obtained the initial collection of data, it is to proceed to the step of data cleaning in Fig. 3. Firstly, the problem with missing value is resolved, using the basic norm of leaving those records with a missing value(s) out of the target data set. As a result, this has reduced the size of the initial population to 360 patient records, with 223 records belonging to Class1 and the other 137 are of Class2. In addition, problems with inconsistent data types and incorrecct data values are taken care of in this processing phase.

In the next preparation step, both data transformation and normalization are conducted. In particular to different laboratory results aggregated in the data set, the value ranges are different and may raise an uneven preference of one attribute to another in the learning process. Therefore, some of initial attributes related to the laboratory results are transformed to the following seven new ones, whose values are in the range of [0, 1].

FBS-high: Calculated from number of times with high-sugar value in blood (x_1) and total number of times with blood-sugar laboratory result (x_2):

$$FBS - high = \frac{x_1}{x_2} \tag{1}$$

FBS-normal: Calculated from number of times with normal-sugar value in blood (x_3) and total number of times with blood-sugar laboratory result (x_2):

$$FBS - normal = \frac{x_3}{x_2} \tag{2}$$

HDL-high: Calculated from number of times with high-fat value in blood (y_1) and total number of times with blood-fat laboratory result (y_2):

$$HDL - high = \frac{y_1}{y_2} \tag{3}$$

HDL-normal: Calculated from number of times with normal-fat value in blood (y_3) and total number of times with blood-fat laboratory result (y_2):

$$HDL - normal = \frac{y_3}{y_2} \tag{4}$$

BUN-high: Calculated from number of times with high-waste value in blood (z_1) and total number of times with blood-waste laboratory result (z_2):

$$BUN - high = \frac{z_1}{z_2} \tag{5}$$

BUN-low: Calculated from number of times with low-waste value in blood (z_3) and total number of times with blood-waste laboratory result (z_2):

$$BUN - low = \frac{z_3}{z_2} \tag{6}$$

BUN-normal: Calculated from number of times with normal-waste value in blood (z_4) and total number of times with blood-waste laboratory result (z_2):

$$BUN - normal = \frac{z_4}{z_2} \tag{7}$$

Table 3 presents the final 20 attributes employed in the empirical study, which have been modified from the original group of 23 attributes. Note that the attribute Age HT/DM $\in [0, 1]$ has gone through a normalization as to overcome the bias with other numeric attributes.

$$AgeHT/DM = \frac{Age - Age_{min}}{Age_{max} - Age_{min}}, \tag{8}$$

where Age denotes the patient's age in the original value range, while Age_{min} and Age_{max} are the minimum and the maximum age values among 360 patients.

4 Results

This research makes use of the tool of Weka 3.7.1 program[1] for the experiments. The classification techniques investigated are decision tree (J48), K-nearest neighbor (KNN), Artificial neural network (ANN), and Naive Bayes (NB). The framework of 10-fold cross validation is used to assess the performance of these classifiers with the current prediction problem. In particular, the aforementioned well-formed dataset is

[1] Available at www.cs.waikato.ac.nz/ml/weka/

Table 3 Details of the attributes used in this research study.

Attribute	Data Type	Value Domain
Sex	Nominal	{Male, Female}
Age HT/DM	Numeric	[0, 1]
FBS-high	Numeric	[0, 1]
FBS-normal	Numeric	[0, 1]
HDL-high	Numeric	[0, 1]
HDL-normal	Numeric	[0, 1]
BUN-high	Numeric	[0, 1]
BUN-low	Numeric	[0, 1]
BUN-normal	Numeric	[0, 1]
BMIHigh	Nominal	{Yes, No}
NSAID	Nominal	{Yes, No}
ControlBP	Nominal	{Yes, No}
DM	Nominal	{Yes, No}
HT	Nominal	{Yes, No}
N20	Nominal	{Yes, No}
I200	Nominal	{Yes, No}
I500	Nominal	{Yes, No}
M109	Nominal	{Yes, No}
N30	Nominal	{Yes, No}
Albumin	Nominal	{Yes, No}

repeatedly split into training and test sets, with the ratio of 70% to 30%. For each of ten repetitions, the training set is fed to the desired classification algorithm, with the resulting classifier model is assessed with the corresponding test set. The evaluation is based on the confusion matrix of binary classes shown in Fig. 4, from which the metric Accuracy $\in$ [0, 100] can be estimated. Note that, Class1 and Class2 of the present problem can be regarded as positive and negative cases of the general binary classification.

$$Accuracy = \frac{TP + TN}{TP + FP + TN + FN} \times 100 \tag{9}$$

		Predicted Label	
		positive	negative
Known Label	positive	TP	FN
	negative	FP	TN

Fig. 4 Illustration of confusion matrix of binary classes, where TP = True Positive, TN = True Negative, FP = False Positive, and FN = False Negative.

Based on this evaluation design, Table 4 shows the accuracies obtained by four different classification methods, with the corresponding standard deviation values. The results suggest that ANN is the most accurate, with NB giving the lowest accuracy value. It is note worthy that the result given for KNN is acquired with the setting of K = 1. As for ANN, the back-propagation algorithm is used to train the networks, with weights/bias values being initialized to small random values (between 0.001 and 0.0001) and the hidden layer size of 10. The learning rate and momentum are set to 0.05 and 0.01, respectively. The training process stops after 2,000 iterations, or when a mean squared error term drops below 0.001.

Table 4 Experimental results - classification accuracy and the corresponding standard deviation (SD) obtained from 10-fold cross validation.

Algorithm	Accuracy	SD
J48	80.242	1.925
KNN (K=1)	76.835	0.773
NB	72.565	0.911
ANN	86.750	1.660

It is shown in the previous table that KNN is not as effective as ANN, with the accuracy around 80%, 6% lower than the other. Therefore, it may be useful to investigate the KNN model further, with respect to its perfoomance and the neighbor sizes. To this end, Fig. 5 depicts the decrease of classification accuracy as the size of nearest neighbors gets larger. This has dropped from 76% to 71%, with the increase of K from 1 to 5. Intuitively, this indicates that attributes are not equally important for the prediction task under examination. This is reinforced by the results of NB, which makes use of all given attributes without any bias or preference. Fig. 6 provides

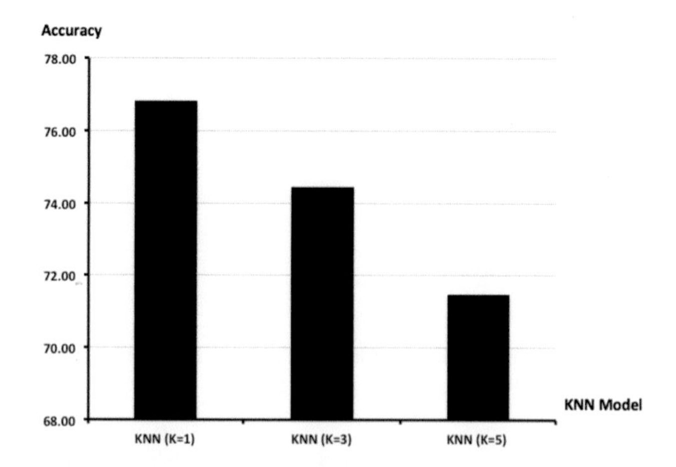

Fig. 5 Performance of KNN classifier with different sizes of nearest neighbors, i.e. $K \in \{1, 3, 5\}$.

```
N20 = N
|   BMIHigh = Y
|   |   NSAID = N
|   |   |   I200 = N
|   |   |   |   BunHigh = '(-inf-0.5]': '(5.5-inf)' (6.0/2.0)
|   |   |   |   BunHigh = '(0.5-inf)'
|   |   |   |   |   dm = Y
|   |   |   |   |   |   I500 = N
|   |   |   |   |   |   |   ageHT/DM = '(-inf-53.666667]': '(-inf-5.5]' (11.0/2.0)
|   |   |   |   |   |   |   ageHT/DM = '(53.666667-71.333333]'
|   |   |   |   |   |   |   |   sex = F
|   |   |   |   |   |   |   |   |   FBShigh = '(-inf-0.5]': '(5.5-inf)' (4.0)
|   |   |   |   |   |   |   |   |   FBShigh = '(0.5-inf)': '(-inf-5.5]' (3.0/1.0)
|   |   |   |   |   |   |   |   sex = M: '(-inf-5.5]' (5.0/1.0)
|   |   |   |   |   |   |   ageHT/DM = '(71.333333-inf)': '(5.5-inf)' (3.37/0.37)
|   |   |   |   |   |   I500 = Y: '(-inf-5.5]' (3.0)
|   |   |   |   |   dm = N: '(-inf-5.5]' (20.0)
|   |   |   I200 = Y: '(5.5-inf)' (4.0/1.0)
|   |   NSAID = Y: '(-inf-5.5]' (33.0)
|   BMIHigh = N
|   |   ControlBP = N
|   |   |   I500 = N
|   |   |   |   sex = F: '(-inf-5.5]' (33.63/6.0)
|   |   |   |   sex = M
|   |   |   |   |   HDLnormal = '(-inf-0.5]'
|   |   |   |   |   |   NSAID = N: '(5.5-inf)' (9.0/1.0)
|   |   |   |   |   |   NSAID = Y
|   |   |   |   |   |   |   HDLHigh = '(-inf-0.5]': '(5.5-inf)' (3.0/1.0)
|   |   |   |   |   |   |   HDLHigh = '(0.5-inf)': '(-inf-5.5]' (6.0)
|   |   |   |   |   HDLnormal = '(0.5-inf)': '(-inf-5.5]' (8.0)
|   |   |   I500 = Y: '(5.5-inf)' (6.0/2.0)
|   |   ControlBP = Y
|   |   |   BunNormal = '(-inf-0.46]'
|   |   |   |   FBSnormal = '(-inf-0.5]'
```

Fig. 6 Example of rules obtained by J48 classifier.

example of rules acquired by J48 classification model. It is shown that only some attributes are included in the decision tree, with different levels of signficance. Based on this, N20 (i.e., patient has a record of stone) is the most discriminative, and more important than information related to age and gender. Given this insight, a model that assigns a set of weights to attributes such as ANN may perform well. This also raises the further investigation of using the techniques of feature selection and transformation.

5 Conclusion

This paper has presented the data mining approach to predicting the progression of kidney disease occurrence in the patients having hypertension and diabetes. It is the first and useful step towards the provision of a computerized model for identifying patient groups, hence a set of appropriate medical care and behavioral guidance. This is not only effective to prolong the disease progress, but also to estimate hospitals' budget in the coming years. The empirical study is based on a collection of patients at Phan hospital, Chiang Rai province, Thailand. The research follows the conventional data mining steps of data cleaning and transformation, as to prepare for the classification model development. Having obtained a set of well-formed data, four benchmark classifiers: J48, KNN, NB and ANN; are exploited to create the desired predictive model. Based on the framework of 10-fold cross validation, J48 and ANN are more accurate than the others, with ANN achieving the highest accuracy of 86.7%. Intuitively, this difference is due to the fact that attributes might not be equally significant to the classification task, as suggested by the worse results of KNN and NB. To verify this, the future work is to expand the population such that the corresponding outcome can consolidate the aforementioned assumption. Another is to make use of feature selection and transformation techniques to improve the performance of those conventional models.

Acknowledgments This research study has been sponsored by Phan Hospital and Mae Fah Luang University, Chiang Rai, Thailand.

References

1. Ingsathit, A., Thakkinstian, A., Sangthawan, P., Kiattisunthorn, K.: Prevalence and risk factors of chronic kidney disease in the Thai adult population: Thai SEEK study. Nephrol Dial Transplant **25**(5), 16–23 (2010)
2. Praditpornsilpa, K.: Annual Report. Thailand renal replacement therapy (2010)
3. Suwanwalaikorn, S.: Vascular complications in type 2 diabetics in Thailand. A preliminary report from a multicenter research study group (2013)
4. Suwanwalaikorn, S.: High prevalence of microalbuminuria in Thai type 2 diabetes patients: Results form DEMAND Campaign (A collaborative multicenter DEMAND Study). Annual Meeting of the Royal College of Physicians of Thailand (2004)
5. Thai Multicenter Research Group on Diabetes Mellitus: Vascular complications in non-insulin dependent diabetes in thailand. Diabetes Res Clin Pract **25**, 61–69 (1994)
6. Thailand renal replacement therapy registry (trt registry) 1997-2000. Journal of the Nephrology Society of Thailand **8** (2002)
7. Tangri, N., Stevens, L.A., Griffith, J., Tighiouart, H., Djurdjev, O., Naimark, D., Levin, A., Levey, A.S.: A predictive model for progression of chronic kidney disease to kidney failure. The Journal of the American Medical Association **305**(15), 1553–1559 (2011)
8. Hippisley-Cox, J., Coupland, C.: Predicting the risk of chronic kidney disease in men and women in England and Wales: prospective derivation and external validation of the QKidney Scores. BMC Family Practice **11**, 49 (2010)
9. Xun, L., Linsheng, L., Li, L., Tanqi, L.: A markov model study on the hierarchical prognosis and risk factors in patients with chronic kidney disease. In: Proceedings of International Conference on Computer Science and Electronics Engineering, pp. 334–338 (2012)
10. Kearns, B., Gallagher, H., de Lusignan, S.: Predicting the prevalence of chronic kidney disease in the English population : a cross-sectional study. BMC Nephrology **14**, 49 (2013)
11. Echouffo-Tcheugui, J.B., Kengne, A.P.: Risk models to predict chronic kidney disease and its progression: A systematic review. PLoS Med **9**(11), e1001344 (2012)

12. Xun, L., Xiaoming, W., Ningshan, L., Tanqi, L.: Application of radial basis function neural network to estimate glomerular filtration rate in chinese patients with chronic kidney disease. In: Proceedings of International Conference on Computer Application and System Modeling, pp. V15–332–V15–335 (2010)
13. Al-Hyari, A.Y., Al-Taee, A.M., Al-Taee, M.A.: Clinical decision support system for diagnosis and management of chronic renal failure. In: Proceedings of IEEE Jordan Conference on Applied Electrical Engineering and Computing Technologies, pp. 1–6 (2013)
14. Chiu, R.K., Chen, R.Y., Wang, S.A., Jian, S.J.: Intelligent systems on the cloud for the early detection of chronic kidney disease, pp. 1737–1742 (2012)
15. National Kidney Foundation: KDOQI clinical practice guidelines for chronic kidney disease: evaluation, classification and stratification. American Journal of Kidney Diseases **39**, S1–S266 (2002)
16. Praditpornsilpa, K., Townamchai, N., Chaiwatanarat, T., Tiranathanagul, K., Katawatin, P., Susantitaphong, P.: The need for robust validation for MDRD-base glomerular filtration rate estimation in various CKD populations. Nephrol Dial Transplant **26**, 2780–2785 (2011)
17. Levey, A., Atkins, R., Coresh, J., Cohen, E.P.: Chronic kidney disease as a global public health problem: approaches and initiatives - a position statement from kidney disease improving global outcomes. Kidney International **72**(3), 247–259 (2007)
18. International Diabetes Federation, International Society of Nephrology: Diabetes and kidney disease: time to act. International Diabetes Federation and International Society of Nephrology, Brussels (2003)
19. Association, A.D.: Diabetic nephropathy (position statement). Diabetes Care **27**(Suppl 1), S79–83 (2004)
20. Han, J., Kamber, M.: Data Mining: Concepts and Techniques, 1st edn. Morgan Kaufmann (2000)
21. Kharya, S.: Using data mining techniques for diagnosis and prognosis of cancer disease. International Journal of Computer Science, Engineering and Information Technology **2**(2), 55–66 (2012)
22. Parthiban, G., Rajesh, A., Srivatsa, S.K.: Diagnosis of heart disease for diabetic patients using Naive Bayes method. International Journal of Computer Applications **24**, 7–11 (2011)
23. Ernandez, T., Saudan, P., Berney, T., Merminod, T., Bednarkiewicz, M., Martin, P.Y.: Risk factors for early failure of Native Arteriovenous Fistulas. Nephron Clin Pract **101**, c39–c44 (2005)

Prediction of Student Dropout Using Personal Profile and Data Mining Approach

Phanupong Meedech, Natthakan Iam-On and Tossapon Boongoen

Abstract The problem of student dropout has steadily increased in many universities in Thailand. The main purpose of this research is to develop a model for predicting dropout occurences with the first-year students and determine the factors behind these cases. Despite several classification techniques being made available in the literature, the current study focuses on using decision trees and rule induction models to discover knowledge from data of students at Mae Fah Luang University. The resulting classifiers that is interpretable and analyzed by those involved in the assistant and consultation aid, are built from the collection of different attributes. These include student's academic performance in the first semester, student social behavior, personal background and education background. With respect to the experiments with various classifiers and the application of data rebalancing algorithm, the results indicate a promising accuracy, hence the reliability of this study as a decision support tool.

Keywords Student dropout · Personal profile · Data mining · Classification · Unbalanced data

1 Introduction

Within more competitive environment, many modern universities and higher education institutes turn to performance analysis, as the basis to setting up a strategic

P. Meedech(✉) · N. Iam-On
School of Information Technology, Mae Fah Luang University, Chiang Rai 57100, Thailand
e-mail: {phanupongm,nt.iamon}@gmail.com

T. Boongoen
Department of Mathematics and Computer Science, Royal Thai Air Force Academy, Bangkok 10220, Thailand
e-mail: tossapon_b@rtaf.mi.th

K. Lavangnananda et al. (eds.), *Intelligent and Evolutionary Systems*,
Proceedings in Adaptation, Learning and Optimization 5,
DOI: 10.1007/978-3-319-27000-5_12

plan [1]. Increasing amount of information regarding student details and course management provides a goldmine that can be explored to understand students learning behavior, preference and performance [2]. Given this, the application of data mining (DM) in education, i.e., educational data mining (EDM), has been a fast-growing area of research [3]. Disclosed knowledge can be useful to understand how students learn, and capture effects of different settings to their achievement.

Recent EDM research works have focused on student categories and targeted marketing [4]. Example of these include predictive modeling for maximizing student retention [5], enrollment prediction models based on admission data [6], prediction of student performance [7]. With an accurate prediction model, it can be used to gain insights of success and risk factors with respect to the course curriculum [8]. Awareness of these issues by educational staffs and management will help identifying the risk group and determining the appropriate course of measures.

The problem of student dropout has steadily increased in many universities and higher education institutes [9]. This causes negative impacts to both student and university ends. On one hand, students have wasted a great deal of time, financial support and morale to learning over the years. On the other, the aforementioned dilemma may dramatically damage the university reputation and budget, as less tuition fee and governmental funding are probably obtained. Revealed by previous works [10], there are many factors accounting for student dropout. At first, students might not be interested in studying. In addition to this, they may disapprove of their branch of learning. Another important issue is due to environmental issues such as friends, family, social setting and adaptation to university life.

To deal with student dropout better, there have been many international studies focusing specifically on the analysis of student performance and marketing-led student categorization [11]. Moreover, some others direct their investigations towards the discovery of dropout profile, from which an effective plan of enrollment can be derived [12]. Particular to cases in Thailand, the problem of student dropout can be observed in most of the universities, including those in governmental and private sectors. The research of student dropout in Thailand is still in its early stage as compared to the projects conducted in Europe and North America. A few social and psychological studies are found in the literature with an objective of determining factors of dropping out of Thai students [13]. As such, they suggest that first-year students are at high risk of leaving a university. The main reasons behind this are the financial difficulty and unsatisfactory with selected major of enrollment. Yet, the data-mining model development is limited in number, whilst restricted to the coupling of a particular learning model and exclusive dataset [14], in terms of type and interpretation.

Following the success of data mining approach to dropout prediction found in the literature [15], this research aims to investigate the use of different classification methods with a collection of student data specific to Mae Fah Luang University (MFU), Thailand, between the academic years of 2012-2013. The work focuses on the dropout problem observed with first-year students who have commonly encountered difficulty of staying in the courses. This predictive modelling exploits both prior and at-university academic results to disclose meaning patterns. In addition, information related to demographic and event-participation is included to refine the learning process. The underlying set of data that is extracted from MFU Management Information System (MIS) is highly similar to those of other Thai universities. Hence, the resulting analytical framework is generalized and can be effectively employed for such a problem. With the focus on human-interpretable classification models like decision tree and rule-based techniques, the discovered knowledge can be used not only to predict the dropout, but also to explain the rationale of this conclusion [16]. Besides, the present research provides an important set of empirical findings, from which more sophiticated and accurate classification models can be built.

The rest of this paper is organized as follows. Section 2 introduces related works as to set the scene for the following sections. The data mining approach to student dropout including details of data specification, extraction, cleaning and transformation are presented in Section 3. To tackle the problem of unbalanced data that is typical for this classification task, the proposed framework has integrated synthetic data objects into the learning process. As a result, a conventional classifier can be applied to create the desired predictive model. Section 4 provides experimental design and the corresponding results with various classifiers such as decistion tree, rule induction and K-nearest neghbor algorithms. The paper is concluded in Section 5 with possible future work.

2 Related Work

With respect to the investigation of [9], student dropout appears to be a significant problem in higher education such that about one fourth of students leaving academic institutes during the first year. In particular, the determinants to course completion revealed by this study include family background, personality, social involvement, and prior-college academic results. Dropout is likely to take place when any student achieves a grade less than a threshold set by the university. This unfortunate outcome is due to many factors; for instance, a habit of skipping classes, attitude towards the degree or course, social participation, and personal adjustment to university lifestyle and friends. In general, a university makes use of students overall academic evaluation (or GPAX) in each academic year to justify the termination. That means a student will be retired if he or she obtains the GPAX below the university standard. At Mae Fah Luang University, the aforementioned threshold value on semester-specific GPAX is 1.5. In fact, dropout may occur at any stages during a degree course, but it is reported that first-year students are at the highest risk of encountering one [17].

Following this finding, a line of research has been dedicated to exploiting EDM to disclose dropout patterns and relations to students characteristics and learning conditions [11, 12, 18–24]. These are subjected to a common assumption that early identification of vulnerable students may lead to an effective retention strategy. Ning and Jingui [12] conduct an empirical study, which aims to find the best data mining model for student performance prediction. The investigated models include decision tree and linear regression, with the former providing better accuracy than the other. Similarly, Ahmed et al. [18] apply classification techniques to predict student performance. In particular, nine factors are exploited in the classification process, which is designed to predict the final grade. These are department, high school degree, mid-term mark, lab test grade, seminar performance, assignment, measure of student participation, attendance and homework. It is reported that the midterm mark has the highest impact on student performance.

In addition, Kumar and Singh [20] also use ID3 decision tree technique for predicting performance of students in Information Technology (IT) faculty. The result suggests that those in CS major have better grade than others belonging in different academic branches. Pumpuang et al. [21] compare Bayesian Network, C4.5, Decision Forest and NBTree classification techniques for the task of course registration planning that helps to lessen the dropout probability. Based on WEKA[1] tool, it is found that NBTree exhibits the best accuracy level. In addition, Bunkar et al. [22] examine the performance of C4.5 (i.e., referred to as J48 in Weka), CART and ID3 decision tree algorithms for predicting the performance of first-year students. In this research, ID3 has been identified as the most effective among the models under investigation. The work of Siddiqui and Gemalel-Din [23] concentrates on course planning that can be useful for student enrollment. As a result of this study where regression tree and support vector regression techniques are used, it is recommended that an advisor can use both models to advise a student, as for his or her study plan.

Mishra et al. [11] tackle the problem of student dropout by applying the collection of attributes specific to the field of learning and emotional skills. This is based on the classification algorithms of J48 and random tree, with the latter showing more accurate prediction results than the other. Carlos Marquez-Vera et al. [24] introduce the prediction of school failure and dropout by applying the SMOTE algorithm [25], which is able to solve the problem of unbalanced data. Moreover, the built-in WEKA function for attribute selection is employed to get rid of some attributes that possess small or none correlation with the target classes. Having accomplished that, the selected attributes have been used with different classifiers. It is also worth looking up other data mining studies that make use of different learning techniques, such as k-means clustering technique [26] and decision tree model [15].

3 Method

This section describes the data mining framework specifically designed and exploited in the present research. Fig. 1 illustrates processing steps enapsulated within this approach, including data collection, data preparation, data analysis and model generation. There are many factors of student dropout such as social factors, family factors and education background including a major that students selected. In Thai education system, every high school normally gives advice to students on the selection of major in a university, which complies with student's interest and academic performance. This appears to be suboptimal to identify a possible dropout, as there are several other university-level issues worth taking into account. Hence, the targeted data that is used to developed a predictive model includes the followings.

[1] Available at www.cs.waikato.ac.nz/ml/weka/

- Students university performance: academic performance in the first semester only.
- Students social behavior: participation in different university events.
- Students personal background: demographic details.
- Students education background: academic performance in high school.

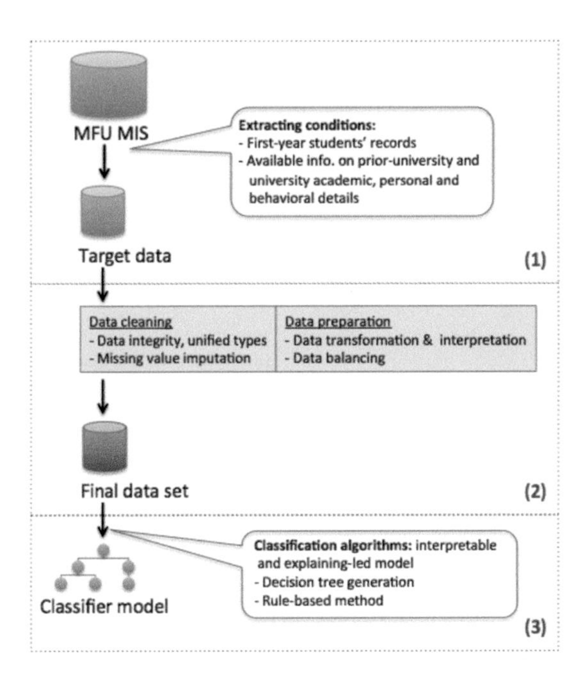

Fig. 1 The data mining approach to student dropout prediction, with respect to three phases of (1) data collection and extraction from MFU MIS, (2) data preparation inclduing cleaning and transformation processes, and (3) data analysis and the generation of classifier model, respectively.

3.1 Data Collection and Preparation

To demonstrate the proposed data mining approach to this problem at MFU, the current research picks up students in some related programs as a case study, where the incidence of dropout has been the largest in magnitude among academic schools and continued increasing in recent years. Specific to the academic years of 2012 to 2013, the trageted set of data is extracted from MFU MIS, where the underlying database relations are maintained by admission, registration and student affair departments. Table 1 summarizes 19 attributes used in this study, in accordance with the aforementioned categories.

Before coupling the extracted student records with any classification model, it is necessary to prepare the data such that the resulting set is complete, with uni-

Table 1 Description of investigated attributes, where Dropout attribute values are used as the target classes.

Category	Attribute	Type	Details
Student's university performance	GPAX1	Numeric	GPAX at the end of first semester
	Grade-A	Numeric	Amount of Grade A in first semester
	Grade-B+	Numeric	Amount of Grade B+ in first semester
	Grade-B	Numeric	Amount of Grade B in first semester
	Grade-C+	Numeric	Amount of Grade C+ in first semester
	Grade-C	Numeric	Amount of Grade C in first semester
	Grade-D+	Numeric	Amount of Grade D+ in first semester
	Grade-D	Numeric	Amount of Grade D in first semester
	Grade-F	Numeric	Amount of Grade F in first semester
	English	Nominal	Pass or fail English course in first semester
Student's social behavior	Event-Participation	Numeric	Number of university events a student participates in first semester
Student's personal background	Gender	Nominal	Male or female
	Region	Nominal	Region of student's famaily residence
Student's educational background	Entry-GPA	Numeric	High school GPA
	Entry-Type	Nominal	Type of entry admission to MFU: Quota, Direct admission, and Others
	English-GPA	Numeric	High school GPA of English subjects
	Science-GPA	Numeric	High school GPA of Science subjects
	Social-GPA	Numeric	High school GPA of Social subjects
	Maths-GPA	Numeric	High school GPA of Mathematics subjects
Student's status	Dropout	Nominal	Status at the end of first year: Yes (Dropout) or No (Not dropout)

fied representation. This is crucial to the quality of disclosed knowledge, hence the prediction accuracy. At first, relations aggregated from several relational MIS tables may encounter a missing value problem, i.e. unknown or incorrect attribute values. Specific to this research, any record with this problem is excluded from the final data collection, which consists of 509 records. For numerical attributes, data normalization as in the common range of [0, 1] is exploited to avoid bias in the learning process. With this, each $c \in \{GPAX1, Entry{-}GPA, English{-}GPA, Science{-}GPA, Social - GPA, Maths - GPA\}$ is normalized to c'as follows:

$$c' = \frac{c}{4} \tag{1}$$

where $c \in [0, 4]$.

For each of the grade frequencies obtained in the first MFU semester, it is standardized by the following example that is specific to the attribute Grade-A. Let *Grade-A** $\in [0, 1]$ be the normalized value of the count *Grade-A*,

$$Grade - A^* = \frac{Grade - A}{\Theta}, \tag{2}$$

provided that

$$\Theta = \sum_{\forall x \in A, B+, B, C+, C, D+, D, F} Grade - x \quad (3)$$

3.2 Model Generation

With the aim to create a classifier model that can be interpretable and analyzed by a user, a number of decision tree and rule-based methods are investigated. These include eight classical classification algorithms available in the WEKA data mining tool.

- Rule induction algorithms: JRip, OneR and Ridor
- Decision tree algorithms: J48, SimpleCart, ADTree, RandomTree and REPTree

To investigate the predictive performance and details of the resulting knowledge, the evaluation methodology of n-fold cross validation is employed. For this, the data set of 509 records are randomly partitioned into training and test sets, with the ratio of around 9 to 1 (i.e., 449 and 60 records). It is repeated for ten times as to set the assessment context of 10-fold cross valiation, i.e. $n = 10$. For each of the ten settings, the training set is fed to the learning process with a specific classification technique. After that, the corresponding test set is used to evaluate the accuracy (within the range of [0, 100]) of that classifier model. By iterating through ten data subsets, the overall classification perfromance can be identified as the average accuracy across those different runs.

However, the problem of imbalance data that appears in this data set has not been taken care of. In particular, the size of dropout class and that of not-dropout one is not even, with the number of records being 108 and 401, respectively. The problem with this imbalance data might occur because a learning algorithm tends to overlook minority class, while pay more attention to the other. So, the result obtained may not be accurate. The way to solve this problem is using oversampling technique, which is capable of balancing the sizes of different classes. In the present work, SMOTE (Synthetic Minority Over-sampling TEchnique [25]) technique that is provided in

Table 2 Results with original data set, where the presented accuracy is the average across 10-fold cross validation.

Algorithm	Accuracy	Standard deviation
JRip	79.2	4.354
OneR	78.0	2.459
Ridor	75.3	8.670
ADTree	75.0	6.382
J48	75.4	4.516
RandomTree	74.3	3.063
REPTree	74.9	8.190
SimpleCart	76.4	2.415

Table 3 Results with balanced data set (applying SMOTE), where the presented accuracy is the average across 10-fold cross validation.

Algorithm	Accuracy	Standard deviation
JRip	80.0	4.766
OneR	83.3	3.333
Ridor	79.0	5.454
ADTree	83.4	2.986
J48	82.3	2.744
RandomTree	75.7	2.851
REPTree	77.9	5.431
SimpleCart	80.0	5.665

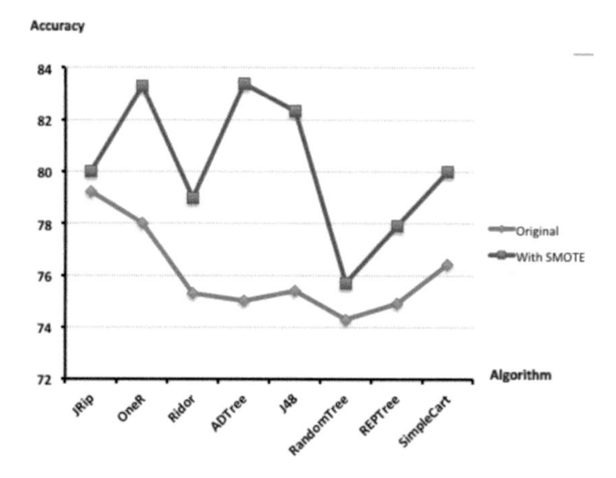

Fig. 2 The algorithm-specific comparison of the results with two different data settings: original data and balanced data using SMOTE.

WEKA, is exploited to solve the problem. Note that this algorithm will increase the size of the minority class, with simulated records created from the basis of k-nearest neighbors. As for the results shown in the next section, the value of k is set to 3, with similar results being observed when $k = 1$ and $k = 2$.

4 Results

In the first experiment, all specified classification algorithms are used to generate classifier models from the original training sets (i.e., wihtout the application of SMOTE). This is conducted in accordance with the predefined 10-fold setting. The corresponding results are shown in Table 2, where rule-based models like JRip and OneR methods appear to be more accurate than the category of decision tree techniques. The lowest accuracy has been observed with RandomTree model. This is

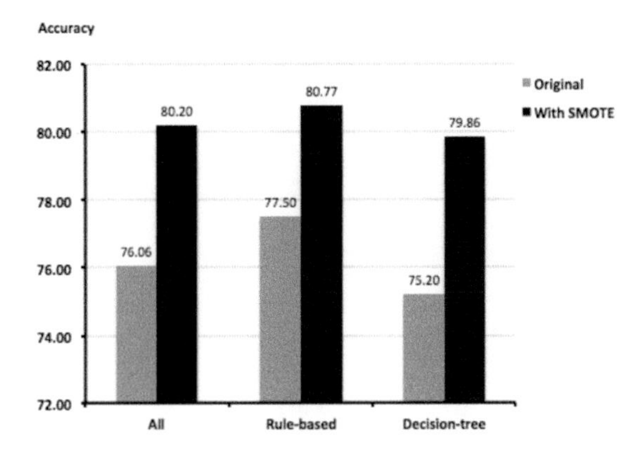

Fig. 3 The improvement over original data using SMOTE algorithm, as the averages across all, and two categories of classification models.

```
: 0
|  (1)GPAX/1 < 2.594: -0.603
|  |  (7)GPAX/1 < 1.347: -1.613
|  |  (7)GPAX/1 >= 1.347: 0.223
|  |  (9)GPA4_Social < 3.149: -0.261
|  |  (9)GPA4_Social >= 3.149: 0.434
|  (1)GPAX/1 >= 2.594: 1.25
|  |  (6)GPAX/1 < 2.899: -0.848
|  |  (6)GPAX/1 >= 2.899: 0.365
|  |  |  (8)SEX = M: -0.537
|  |  |  (8)SEX = F: 1.414
|  (2)F < 0.006: 0.308
|  |  (10)GPA2_Math < 2.025: 0.631
|  |  (10)GPA2_Math >= 2.025: -0.207
|  (2)F >= 0.006: -0.644
|  |  (5)F < 0.332: -1.74
|  |  (5)F >= 0.332: 0.326
|  (3)C < 0: 0.385
|  (3)C >= 0: -0.526
|  |  (4)C < 0.333: -1.935
|  |  (4)C >= 0.333: 0.346
Legend: -ve = Drop Out, +ve = Normal
```

Fig. 4 The set of rules obtained by ADTree algorithm.

rational as the selection of attribute for tree generation is random and not as intuitive as other decision tree models.

In the second experiment, the previous formulation of 10-fold cross validation is re-used, with each training set being balanced using SMOTE. The resulting training set contains 802 records after rebalanced (401 Dropout and 401 Not-dropout). Table 3 presents the results of this assessment, where all classification algorithms perform better than the case of unbalance data. Fig. 2 gives the illustration that supports this conclusion. The classifier models that obtain the high accuracy rates are J48, ADTree

```
GPAX/1 <= 2.587377
|   F <= 0
|   |   C <= 0
|   |   |   GPA2_Math <= 2.02: Normal (25.0/2.0)
|   |   |   GPA2_Math > 2.02
|   |   |   |   ENTRYTYPET = Direct
|   |   |   |   |   GPA4_Social <= 2.66: Normal (6.0)
|   |   |   |   |   GPA4_Social > 2.66
|   |   |   |   |   |   GPA4_Social <= 3.15: Drop Out (20.0/1.0)
|   |   |   |   |   |   GPA4_Social > 3.15
|   |   |   |   |   |   |   CountOfACTIVITYID <= 22.465123: Drop Out (7.0/2.0)
|   |   |   |   |   |   |   CountOfACTIVITYID > 22.465123: Normal (10.0)
|   |   |   |   ENTRYTYPET = Quota
|   |   |   |   |   B <= 0.332519
|   |   |   |   |   |   CountOfACTIVITYID <= 20: Normal (4.0/1.0)
|   |   |   |   |   |   CountOfACTIVITYID > 20: Drop Out (7.0)
|   |   |   |   |   B > 0.332519: Normal (7.0)
|   |   |   |   ENTRYTYPET = Other: Normal (5.0)
|   |   C > 0
|   |   |   ENTRYTYPET = Direct
|   |   |   |   CountOfACTIVITYID <= 21
|   |   |   |   |   C <= 0.332726: Drop Out (14.0)
|   |   |   |   |   C > 0.332726: Normal (20.0/6.0)
|   |   |   |   CountOfACTIVITYID > 21: Drop Out (57.0/2.0)
|   |   |   ENTRYTYPET = Quota
|   |   |   |   SEX = M: Drop Out (6.0)
|   |   |   |   SEX = F: Normal (7.0/1.0)
|   |   |   ENTRYTYPET = Other: Normal (4.0)
|   F > 0
|   |   ENTRYTYPET = Direct: Drop Out (205.0/11.0)
|   |   ENTRYTYPET = Quota: Drop Out (45.0/6.0)
|   |   ENTRYTYPET = Other: Normal (5.0/1.0)
GPAX/1 > 2.587377
|   GPAX/1 <= 3
|   |   SEX = M
|   |   |   B <= 0.4912
|   |   |   |   ENTRYTYPET = Direct: Normal (20.0/1.0)
|   |   |   |   ENTRYTYPET = Quota
|   |   |   |   |   B+ <= 0.010599: Normal (4.0)
|   |   |   |   |   B+ > 0.010599: Drop Out (6.0)
|   |   |   |   ENTRYTYPET = Other: Normal (3.0)
|   |   |   B > 0.4912: Drop Out (7.0/1.0)
|   |   SEX = F: Normal (65.0/6.0)
|   GPAX/1 > 3: Normal (183.0/2.0)
```

Fig. 5 The set of rules obtained by J48 algorithm.

and OneR. These findings suggest that the unbalance problem has a greater effect on decision tree category than the other, see Fig. 3.

Based on the experimental results discussed so far, the prediction of student dropout is practical with the rate of accuracy around 80%. This may be enhanced using a more advanced classification technique such as SVM (Support Vector Machine). However, the resulting model might not be understandable by a general user. This has turned to be the advantage of this work where rule induction methods are employed. These rules show the relationships between attributes and the dropout outcome. In Fig. 4 and 5, the rules obatined from ADTree and J48 algorithms are shown, respectively. According to these, some attributes are commonly regarded as effective indicators of dropout, GPAX1, Grade-F and Grade-C, for instance.

It is also worth noting that the size of rule set created by ADTree is smaller than that of J48. Of course, the requirements of time and space to use the former is smaller than the latter. Despite this, it is interesting to see that some attributes appearing in the rule set of J48 are not in that of the others. Hence, there is a tradeof between model efficiency and robustness. Nonetheless, these rule sets provide a great oppotunity to understand the critical factors to dropout, at least to this specific group of students. This allows a preventive strategy to be carefully drawn and executed, to identify and help those at risk of leaving the university prematurely.

5 Conclusion

This paper has presented a research work on the development of analytical framework for student dropout prediction. The proposed approach follows the basis of data mining to collect, prepare and generate the desired classifier model. At first, different groups of data have been identified and combined to form the target data set. These include academic perfromance both at university and from previous high-school level, personal characteristics, admission type, and participation in social events. As for the case of Thai students, this research provides a unique study where those diversed data types are blended as a base for predictive modelling. For the phase of data preparation, conventional techniques for solving missing values and normalization are employed to avoid bias and errors found in the data. Lastly, different rule induction models are explored for the classification purpose, such that the resulting knowledge is interpretable. To this end, an IF-THEN rule is a native form for decision making because it is easy to use and efficient. The accuracy obtained is around 80% with the use of SMOTE to solve the problem of imbalance data. Several findings worth noting include the fact that GPAX1 and Grade-F are the most discriminative attributes, with Entry-Type being used to refine the decision by some models. These can be useful for those trying not only to identify at-risk students, but also to understand the dropout circumstances. Besides the results shown in this paper, there are future works that can further improve the prediction quality. One is to include other types of student data, which can be informative to refine a classification process. Another involves the use of more advanced techniques that can be highly accurate, for the rule generation.

Acknowledgments The research publication is partly funded by Mae Fah Luang University, Chiang Rai, Thailand.

References

1. Bala, M., Ojha, D.B.: Study of applications of data mining techniques in education. International Journal of Research in Science and Technology **1**, 1–10 (2012)
2. Mostow, J., Beck, J.: Some useful tactics to modify, map and mine data from intelligent tutors. Natural Language Engineering **12**, 195–208 (2006)
3. Romero, C., Ventura, S.: Educational data mining: a review of the state-of-the-art. IEEE Transactions on Systems Man and Cybernetics, Part C **40**, 601–618 (2010)
4. Romero, C., Ventura, S.: Data mining in education. Wiley Interdisciplinary Reviews: Data Mining and Knowledge Discovery **3**(1), 12–27 (2013)
5. Yu, C., Gangi, S.D., Jannasch-Pennell, A., Kaprolet, C.: A data mining approach for identifying predictors of student retention from sophomore to junior year. Journal of Data Science **8**, 307–325 (2010)
6. Yadav, S.K., Pal, S.: Data mining application in enrollment management: A case study. International Journal of Computer Applications **41**(5), 1–6 (2012)
7. Ramaswami, M., Bhaskaran, R.: A CHAID based performance prediction model in educational data mining. International Journal of Computer Science **7**(1), 10–18 (2010)
8. Vandamme, J., Meskens, N., Superby, J.: Predicting academic performance by data mining methods. Education Economics **15**(4), 405–419 (2007)
9. Tinto, V.: Research and practice of student retention: What next? Journal of College Student Retention: Research, Theory and Practice **8**(1), 1–20 (2006)
10. Strayhorn, T.L.: An examination of the impact of first-year seminars on correlates of college student retention. Journal of the First-Year Experience and Students in Transition **21**(1), 9–27 (2009)
11. Mishra, T., Kumar, D., Gupta, S.: Mining students'data for performance prediction. In: Proceedings of International Conference on Advanced Computing & Communication Technologies, pp. 255–263 (2014)
12. Ning, F., Jingui, L.: Work in progress - a decision tree approach to predicting student performance in a high-enrollment, high-impact, and core engineering course. In: Proceedings of IEEE International Conference on Frontiers in Education Conference, pp. 1–3 (2009)
13. Gulati, H.: Predictive analytics using data mining technique. In: Proceedings of International Conference on Computing for Sustainable Global Development, pp. 713–716 (2015)
14. Kongsakun, K., Fung, C.C.: Neural network modeling for an intelligent recommendation system supporting srm for universities in thailand. WSEAS Transactions on Computers **11**(2), 34–44 (2012)
15. Kabra, R.R., Bichkar, R.S.: Performance prediction of engineering students using decision trees. International Journal of Computer Applications **36**(11), 8–12 (2011)
16. Pandey, M., Sharma, V.K.: A decision tree algorithm pertaining to the student performance analysis and prediction. International Journal of Computer Applications **61**(13), 1–5 (2013)
17. Sachin, R.B., Vijay, M.S.: A survey and future vision of data mining in educational field. In: Proceedings of International Conference on Advanced Computing & Communication Technologies, pp. 96–100 (2012)
18. Ahmed, A.E.D., Elaraby, I.S.: Data mining: A prediction for student's performance using classification method. World Journal of Computer Application and Technology **2**, 43–47 (2014)
19. Abaya, S.A., Gerardo, B.D.: An education data mining tool for marketing based on c4.5 classification technique. In: Proceedings of International Conference on e-Learning and e-Technologies in Education, pp. 289–293 (2013)

20. Kumar, V., Singh, S.: Classification of students data using data miningtechniques for training & placement department intechnical education. In: Proceedings of International Conference on Computer Science and Networks, pp. 121–126 (2012)
21. Pumpuang, P., Srivihok, A., Praneetpolgrang, P.: Comparisons of classifier algorithms: bayesian network, C4.5, decision forest and NBTree for course registration planning model of undergraduate students. In: Proceedings of IEEE International Conference on Systems, Man and Cybernetics, pp. 3647–3651 (2008)
22. Bunkar, K., Singh, U.K., Pandya, B., Bunkar, R.: Data mining: prediction for performance improvement of graduate students using classification. In: Proceedings of Ninth International Conference on Wireless and Optical Communications Networks, pp. 1–5 (2012)
23. Siddiqui, M.A., Gemalel-Din, S.: Evaluation of academic plans of study using data mining techniques. In: Proceedings of IEEE International Conference on Advanced Learning Technologies, pp. 224–228 (2013)
24. Marquez-Vera, C., Morales, C.R., Soto, S.V.: Predicting school failure and dropout by using data mining techniques. IEEE Revista Iberoamericana de Tecnologias del Aprendizaje **8**(1), 7–14 (2013)
25. Chawla, N.V., Bowyer, K.W., Hall, L.O., Kegelmeyer, W.P.: SMOTE: Synthetic Minority Over-sampling Technique. Journal of Artificial Intelligence Research **16**, 321–357 (2002)
26. Erdogan, S.Z., Timor, M.: A data mining application in a student database. Journal of Aeronautic and Space Technologies **2**(2), 53–57 (2005)

The Optimization of Parallel DBN Based on Spark

Juan Yang and Shuqing He

Abstract Deep Belief Network (DBN) is widely used for modelling and analysis of all kinds of actual problems. However, it's easy to have a computational bottleneck problem when training DBN in a single computational node. And traditional parallel full-batch gradient descent exists the problem that the speed of convergence is slow when we use it to train DBN. To solve this problem, the article proposes a parallel mini-batch gradient descent algorithm based on Spark and uses it to train DBN. The experiment shows the method is faster than parallel full-batch gradient and the convergence result is better when batch size is relatively small. We use the method to train the DBN, and apply it to text classification. We also discuss how the size of batch impacts on the weights of network. The experiments show that it can improve the precision and recall of text classification compared with SVM when batch size is small.

Keywords Deep learning · DBN · Spark · Parallel algorithm · Text classification

1 Introduction

Deep Learning consists of a series of algorithm which can effectively abstract date [1]. It has been widely paid much attention to as a hot subject of academic nowadays [2]. Deep Belief Network (DBN), as one of the earliest Deep Learning frameworks, is a deep neural network consist of a visual layer and multiple hidden layers [3]. DBN has a complex network structure. Meanwhile, the increasing of the number of network layer leads to the expansion of the number of cells in the network and that the number of the parameters between cells and cells increase rapidly. As the number of parameters expands sharply, it is necessary for computer to have enormous computer processing power.

J. Yang · S. He(✉)
Key Lab of Intelligent Telecommunication Software and Multimedia,
Beijing University of Posts and Telecommunications, Beijing 100876, China
e-mail: yangjuan@bupt.edu.cn, shvqinghe@gmail.com

K. Lavangnananda et al. (eds.), *Intelligent and Evolutionary Systems*,
Proceedings in Adaptation, Learning and Optimization 5,
DOI: 10.1007/978-3-319-27000-5_13

Nowadays what is generally used to train deep neural network is Graphics Processing Unit (GPU), which can shorten the training time effectively by its excellent computer processing ability. However, when confronted with massive data, a single computer is still faced with the problem of insufficient computing capacity. Some experts try to solve this problem by making use of the parallel network framework formed of multiple GPU structures or thousands of computational modes [4] [5]. However, the methods above are based on proprietary software, not general purpose computing framework.

In respect of general framework, the methods in [6] [7] focus on describing the process of aggregate the weights generated by computational nodes. They do not optimize the method used to train the model in the computational node. Therefore we propose a novel parallel gradient descent algorithm based on a general computing framework named Spark and uses it train DBN. The algorithm optimize the training process in each computational node by adding a control factor which can control the batch size.

We realize the parallel algorithm based on Spark. We show that the algorithm can faster the convergence of the network by taking a small batch size, and we also give a reasonable explanation for the huge difference in the convergence of later sub-network of parallel DBN. In addition, we apply the DBN trained by the algorithm proposed in the article to text classification and compare the result with SVM.

The rest of this article is organized as follow. We will introduce the DBN and Spark in section 2. In section 3, we present the parallel algorithm of training DBN based on Spark. In section 4, we summarize and analyze the experiments performed of the DBN training by proposed algorithm in the reconstruction error and text classification respectively. Finally, we conclude the article and discuss future work in section 5.

2 Related Work

2.1 Deep Belief Network

DBN is a deep neural network contains multiple layers of hidden cells, it also can be regarded as a graphical model which can be viewed as stacked by multiple Restricted Boltzmann Machines (RBMs). For every hidden layer of RBM, it's always be regarded as the visible layer of later RBM. The structure chart of DBN is shown as Fig. 1 [8], where the gray arrows don't belong to DBN when DBN is regarded as a generative model, W_{n-1} means the weight matrix between layer n-2 and layer n-1 and W^T_{n-1} is the transposed matrix of W_{n-1}. In Fig. 1, the hidden layer n and the hidden layer n-1 build up a RBM; the hidden layer n-1, the hidden layer below and the visible layer build up a directed belief network.

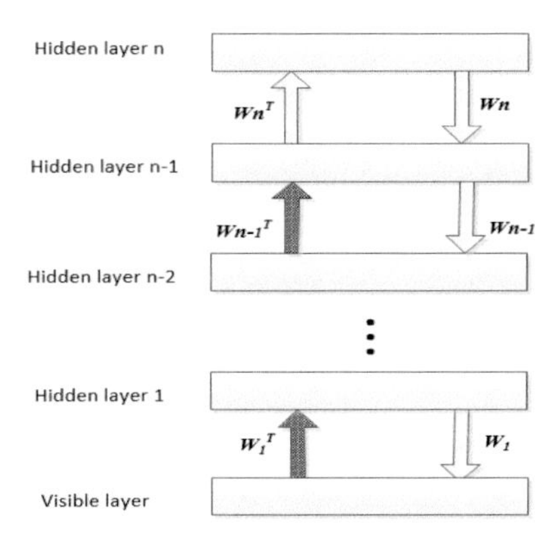

Fig. 1 Structure chart of DBN

A greedy and layer-wise unsupervised training procedure is always used to train the DBN [9], then the neural network is fine-tuned with the data labelled. During the training procedure, contrastive divergence (CD) algorithm is applied to each RBM which is regarded as sub-network of DBN [3].

RBM is a kind of energy model, also can be looked as a kind of probability graph model. As a bipartite graph, RBM has a visible layer which represent the observations, and a hidden layer. Fig. 2 is a diagram of RBM [10], consisting of m hidden cells and n visible cells, where we represent hidden layer as $\boldsymbol{h} = (h_1, h_2, h_3, \ldots, h_m)$ and visible layer as $\boldsymbol{v} = (v_1, v_2, v_3, \ldots, v_n)$. Besides, $(b_1, b_2, b_3, \ldots, b_m)$ represents the bias of the hidden layer and$(c_1, c_2, c_3, \ldots, c_n)$ represents the bias of the visible layer. In Fig. 2, $\boldsymbol{w}$ means weight matrix between the hidden layer and the visible layer; h_i means the i th cell in the hidden layer; v_j means the j th cell in the visible layer; w_{ij} means the weights between h_i and v_j. In this article, we make h_i, v_j to take values {0, 1}.

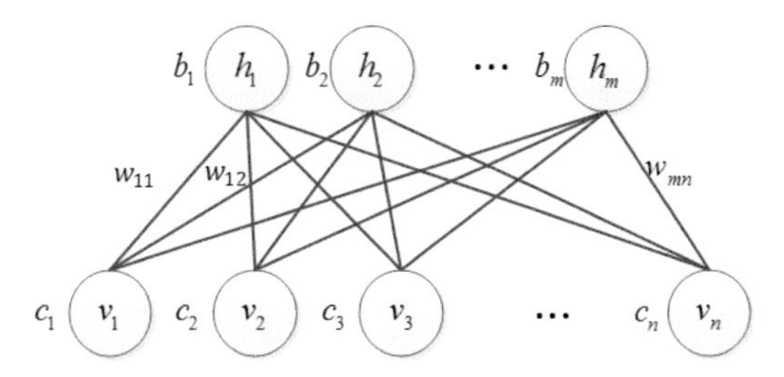

Fig. 2 An undirected graph of RBM

As mentioned above, RBM is a kind of energy model, and its energy function can be defined as follow:

$$E(\boldsymbol{v}, \boldsymbol{h}) = -\sum_{i=1}^{m}\sum_{j=1}^{n} w_{ij} v_j h_i - \sum_{i=1}^{m} b_i h_i - \sum_{j=1}^{n} c_j v_j \tag{1}$$

According to (1), when the state is given with the state vector of visible layer $\boldsymbol{v}$ and the state vector of hidden layer $\boldsymbol{h}$, the joint probability distribution is as (2), where the denominator $\sum_{\boldsymbol{v},\boldsymbol{h}} e^{-E(\boldsymbol{v},\boldsymbol{h})}$ means all possible configurations. In other words, formula 2 means the probability which current state appeared in the state space.

$$P(\boldsymbol{v},\boldsymbol{h}) = e^{-E(\boldsymbol{v},\boldsymbol{h})} \Big/ \sum_{\boldsymbol{v},\boldsymbol{h}} e^{-E(\boldsymbol{v},\boldsymbol{h})} \tag{2}$$

It is straightforward to infer (3) and (4), where sigm is the sigmod tion $\mathrm{sigm}(x) = 1 / (1 + e^{-x})$. Formula 3 means the conditional probability of a configuration of the hidden cell h_i, given a configuration of the visible vector $\boldsymbol{v}$. The formula 4 means the conditional probability of a configuration of the visible cell v_i, given a configuration of the visible vector $\boldsymbol{h}$.

$$P(h_i|\boldsymbol{v}) = sigm(b_i + \textstyle\sum_j w_{ji}v_j) \tag{3}$$

$$P(v_i|\boldsymbol{h}) = sigm(c_i + \textstyle\sum_j w_{ji}h_j) \tag{4}$$

After obtaining the likelihood function of training set S, we can get the partial derivative calculation formula of likehood L_S. The formula is shown in (5), where v^m means the m th sample vector in S, θ means the parameter in RBM, n_s means the number of the samples in S.

$$\frac{\partial \ln L_s}{\partial \theta} = \sum_{m=1}^{n_s} \frac{\partial \ln P(v^m)}{\partial \theta} = \sum_{m=1}^{n_s} [\ln \sum_h e^{-E(v_m,h)} - \ln \sum_{v,h} e^{-E(v_m,h)}] \tag{5}$$

During the weights be updated iteratively, Gibbs sampling [11] is used to estimates the probability distribution of the target of the RBM. As mentioned above, the training of DBN is unsupervised and the process is from the bottom to the top one by one greedily. By use the greedy algorithm recursively, we can learn the weights of DBN one layer at a time.

After all the RBMs have been trained, the all weights of DBN have been learned which mean we can get a DBN that have been trained.

2.2 *Spark*

Spark is a distributed general purpose computing framework based on map-reduce thought [12]. Different from MapReduce [13], the calculation model of Spark is based on memory, it make Spark good at dealing with machine learning problem.

Similar with MapReduce, Spark composed by master and workers (slavers). The management framework of Spark is show as Fig. 3. In Fig. 3, Driver is the program that runs the user's main function and executes various parallel operations on the cluster. The executor responsible for the execution of tasks and store data in the worker. Cluster manager grants executors to a Spark application.

The core concept of Spark is Resilient Distributed Dataset (RDD) [14]. RDD provides a model of highly restricted and shared memory. RDD supports two types of operating driving program: conversion and action.

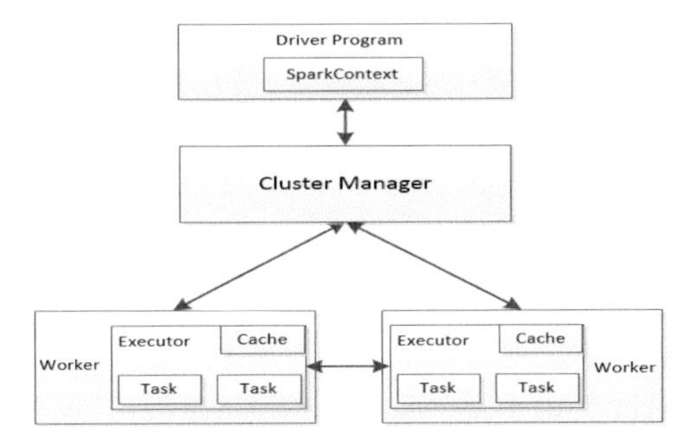

Fig. 3 The Management Framework of Spark

Data running in Spark are represented as RDD object, hence, the operating on the dataset is expressed as the operating on the RDD object. And the RDD operations defined in driver will be distributed to workers. New data fragment will be generated according to the operation received by worker. After that, the data fragment will be returned to the driver or write in the local storage system.

3 Parallelization of DBN Based on Spark

When we use the greedy layer-wise algorithm to train the DBN, we sequentially train the RBMs which make up the DBN, but for each RBM we train it parallelly. We will describe the parallel training process of RBM in section 3.1, and in section 3.2 we give the full process of parallel DBN training.

3.1 The Parallel Training Process of RBM

In the parallel training process of RBM_k which is the k th RBM in the DBN, the initial weights of RBM_k is provided to each computational node at first. After that, one epoch of training is completed on each slave node with corresponding training subset, all weights are gathered in driver. After average all weights, we use the mean to update the weight in the driver.

Given the number of partition p and training set S which contain ns instances, we divide S into p disjoint subsets $S = s_1 \cup s_2 \cup \cdots \cup s_p$. For i th slave computation node c_i, denote the training subset in c_i as s_i, there are $\lfloor n_s/p \rfloor$ instances in s_i. In the weight initialization process, the same initial weight is provided to each slaver. Denote the weight in slaver c_i as w_i^m, where m is the epoch number. The weight w_i^m is learned using corresponding subset s_i in c_i.

During the training process in slaver, we use "mini-batch" thought to update the weight. That is, after obtain the initial weight w_i^m, c_i update the weight by formula 6 where w_i^{m+1} means the $m + 1$ th epoch of training and Δw_i^m is the gradient.

$$w_i^{m+1} = w_i^m + \Delta w_i^m \tag{6}$$

Formula 6 is the gradient descent formula essentially, and the gradient Δw_i^m is calculated by formula 7 which is deduced from formula 5. Note that we ignore the bias for visible and hidden layer for convenience. In formula 7, $w_{i,j}^m$ means the weight between h_i and v_j which mention in Section 2, l_0 means the original data vector of the l th training set instance, l_k means the result of l_0 having been sampled k times by Gibbs sampling, n_b means the mini-batch size. For convenience, we use batch short for mini-batch below.

$$\frac{\partial \ln L_s}{\partial w_{i,j}^m} = \sum_{l=1}^{n_b} [P(h_i = 1|V^{l_0}) v_j^{l_0} - P(h_i = 1|V^{l_k}) v_j^{l_k}] \tag{7}$$

Unlike getting multiple results at a time by using MATLAB or GPU programming, we just can get one result at a time. So we implement mini-batch thought by "delay" the time that we update the weights of DBN. In other words, instead of update the weights immediately, that we update the weights until the number of instance having been trained is equal to the batch size.

In the driver, the common weights should be updated by all computational nodes as formula 8, where p is the parallel size.

$$w^{m+1} = \sum_{i=1}^{p} w_i^{m+1} / p \tag{8}$$

We describe the algorithm in pseudo code as Algorithm 1. In the Algorithm 1, the key is that we add the controlling factor to adjust the batch size, and the weights will update only until the number of instance having been trained is equal to the mini-batch size. In Algorithm 1, epochs is iterative times, *lr* is the learning rate.

Algorithm 1 Mini-BatchGD

1: Distribute the initial weight w^0
2: **for** m = 0 to *epochs* **do**
3:     **for** $s_i \in S$ **do**
4:         Δw_i^{m+1} = calculateGradient(s_i, w_i^m)
5:         **if** count = n_b **do**
6:             $w_i^{m+1} = w_i^m + lr * \Delta w_i^{m+1}$
7:             count = 1
8:         **else** count = count + 1
9:     **end for**
10:   Aggregate the weights: $w^{m+1} = \sum_{i=1}^{p} w_i^{m+1} / p$
11: **end for**

By controlling batch size, we make the method find the balance between full-batch gradient descent and stochastic gradient descent. Also, appropriate batch size can make the convergence speed faster and can not be caught in the turbulence in the local optimum.

3.2 The Implement of DBN Training

The training of DBN is sequential, the later RBMs could be trained after the previous RBMs have been trained. The layer by layer process start with a random initialization executed by driver, and corresponding data set is cached is the each slaver. After initialization process, the training of DBN start with the first RBM. The second RBM is trained thereafter, and the training process is go on until the last RBM is trained.

As mentioned above, each of slaver train the corresponding RBM independently. After we run Algorithm 1, the driver aggregate the result from the slaver. The whole flow of training DBN is outlined in Algorithm 2, where setParameterToZero function sets the parameters of RBM to 0, *dbn* is a DBN which had random initialized, S is the training set, p is the number of parallelization, ephs is the number of epochs, θ is the parameter of RBM.

In Algorithm 2, we describe the whole process of parallel training. In general, the driver collect the result of parameters trained, and then update the global network which is RBM here. After the sub-networks have been trained one by one, the DBN will been trained.

Algorithm 2 Parallel DBN

```
1:  for RBM^i in dbn do
2:    do epoch until ephs times
3:          Parallel s_i in S
4:              θ^l = RBM^i. θ
5:              θ^l =Mini-BatchGD(θ^l, l_r, n_b, s_i)
6:          end parrallel
7:          (W,B,C) = aggregate(w^l,b^l,c^l)
8:          RBM^i.setParameterToZero()
9:          for (w^k,b^k,c^k) in (W,B,C) do
10:                 RBM^i. θ = RBM^i. θ + θ^k/p
11:         end for
12: end for
```

4 The Experiment and Analysis

The test platform of this experiment is Spark on Yarn. In detail, the Spark version is 1.3.0 and the Hadoop version is 2.6.0. The cluster contains 5 nodes. The hardware configuration of every node is Dual CPU, 24-core, 64G of RAM and the configuration of every core is Intel(R) Xeon(R) CPU E5-2620 v2 @ 2.10GHz.

All the experimental datasets are from the text categorization corpus of Sogou. And the text categorization corpus comes from the news corpus and equivalent category message saved by Sohu news website.

The experiment will be divided into two phases. In the first phase, we train a DBN which contains three layers by using proposed algorithm. In second phase, we add an output layer containing 9 output units to the top of DBN. At the same time, we apply the network to predict the text classification results regarding DBN as a Discriminative Model and eventually analyze the results.

4.1 *The Pre-process of Data*

The dataset obtained in the experiment is the express edition, containing 9 categories and more than 10000 news records. 4100 news records selected randomly are regarding as training set, and the other 1288 news records are regarded as testing set. Specific as shown in Table 1, where C1~C9 represent Finance and Economics, IT, Health, Sport, Tourism, Education, Recruitment, Culture and Military successively.

Table 1 The number of training set and testing set in each categories

	C1	C2	C3	C4	C5	C6	C7	C8	C9
Training set	400	400	500	300	400	500	500	600	500
Testing set	100	66	200	22	100	200	200	200	200

We segment the news data based on NLPIR Chinese segmentation system. After that, the term frequency-inverse document frequency (TF-IDF) [15] is used as the weights factor. For word w_i in the text d_j, computational formula is $tfidf_{i,j} = tf_{i,j} \times idf_i$, where the $tf_{i,j}$ and idf_i is calculated as (9) (10).

$$tf_{i,j} = \frac{n_{i,j}}{\sum_k n_{k,j}} \tag{9}$$

$$idf_i = \log \frac{D}{\alpha + \{j : w_i \in d_j\}} \tag{10}$$

In (9) (10), $n_{i,j}$ means the number of times that w_i appears in the d_j, $\sum_k n_{k,j}$means the number of all the words appearing in the d_j, D means the number of texts in the dataset, $\{j: w_i \in d_j\}$means the number of texts containing w_i, α means smoothing parameter.

4.2 *The Results and Analysis of Experiment*

4.2.1 The Results and Analysis of Parallelization of DBN

In order to show the influence of the batch size to the learning time and performance of parallel DBN, we select the size of 1, 2, 5, 10, 50, 100 and 205 to illustrate the results. When the size is 205, it is full batch gradient ascent actually. In the experiment, the process of fine-tune of DBN is not added to the parallel method.

When we train a network containing the DBN of three layers, we are actually training two RBMs which make up DBN by stacking. As a performance evaluation standard of DBN, this article respectively analyzes two RBMs contained in DBN. In this DBN, RBM_1 consists of two layers, whose units respectively are 2000 and 500, and RBM_2 consists of the layer with 500 units and the layer with 200 units. We take the reconstruction error as the standard of performance, the formula is as (11).

$$R = \frac{\sum_{i=1}^{N} \sum_{j=1}^{n_v} (v_{ij}^0 - v_{ij}^1)^2}{N} \tag{11}$$

Where R means the reconstruction error, N means the number of input data, n_v means the number of cells contained in the visible layer, v_{ij}^0 means the value of j th cell in i th input data vector v_i^0, v_{ij}^1 means the value of j th cell in the vector which Gibbs sample from v_i^0 , Fig. 4 (a) and Fig. 4 (b) respectively mean the tendency that the reconstruction error of RBM_1 and RBM_2 changes with the increasing of iterations when batch size is 1, 2, 3, 10, 50, 100, 205(full-batch).

Fig. 4 (a) clearly expresses that the tendency of RBM_1's reconstruction error along with the increasing of the number of iterations when the size of batch is different. As we expected that with the decrease of the batch size, the convergence speed of network weights speeds up. When the size is 205, the speed of convergence of the network is obviously slow and even cannot reach local optimum after the network has been trained 200 iterations.

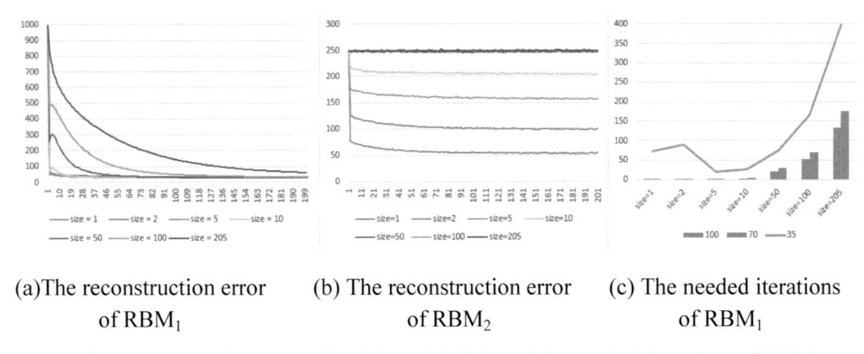

(a)The reconstruction error of RBM_1 (b) The reconstruction error of RBM_2 (c) The needed iterations of RBM_1

Fig. 4 The reconstruction error of RBM_1 and RBM_2 and the needed iterations of RBM_1

Fig. 4 (c) shows us the needed iterations of different size when the convergence criterion for stopping of the RBM_1 is its reconstruction error less than: 100, 70 and 35. When the requirement of the reconstruction error is relatively loose, satisfying convergence criterion for stopping is faster with the size of batch is smaller. However, with the reconstruction error is more and more strict, the performance of training when the size is 5 or 10 is better than the performance when the size is 1 or 2. Our experiment shows that as the network close to the local optimal, choosing appropriate size in this method can help avoid the shock of the network and speed up the speed of reaching the local optimal value.

Fig. 4 (b) expresses the changing tendency of the reconstruction error of RBM_2 with the increasing of the iterations. The error rates of the last two layers vary widely when the error rates of the first two layers are the nearly the same. When the size of batch is greater than 50, RBM_2 nearly cannot be trained. But the reconstruction error of RBM_2 appears obviously decreasing when the size of batch reduces. In order to explain this phenomenon, we choose 1, 10 and 205 as the batch size, the weights of DBN produced by training with 200 times iterative is shown as Fig. 5. Please note that in order to observe more conveniently we adjust the scope of the y-axis of each figure.

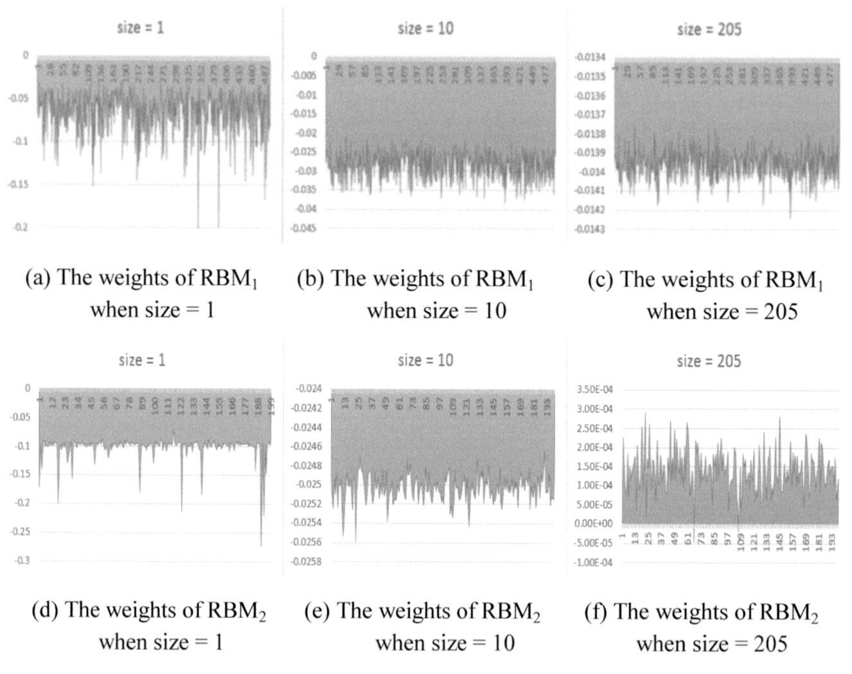

(a) The weights of RBM_1 when size = 1

(b) The weights of RBM_1 when size = 10

(c) The weights of RBM_1 when size = 205

(d) The weights of RBM_2 when size = 1

(e) The weights of RBM_2 when size = 10

(f) The weights of RBM_2 when size = 205

Fig. 5 The Weights of RBM when the size is 1, 10, 205

By observing the weights of RBM_1 in Fig. 5 (a) (b) (c), we can know that the weights of each RBM becomes smaller as the batch size increases. And the tendency does not change with the reconstruction error. When the convergence criterion for stopping isn't satisfied, the weights will become larger with the times of the adjustment of weights increasing. And from Formula (6) (7) (8), we know that it can make the training of the later RBM become more difficult. Just as Fig. 5 (d) (e) (f) shows, the network trains sufficiently when the size is 1, on the contrary, the training effect of the network is poor when size is 205. And the closer the value to the initial value, the poorer the training effect. To prove the point, we make iterations of training RBM_1 increase from 200 to 1000 and iterations of training RBM_2 not change when the size is 10. The reconstruction error of RBM_2 drops from 210 to 140 sharply while the error of RBM_1 drops from 29 to 26 slightly. It's strongly proved our hypothesis that the size of weights and the times of the updating of weights are closely linked.

In conclusion, when the batch size is small, the network of the later layers can have a good enough convergent effect though the convergent process vibrates near the local optimal value, and at the same time the speed that it get local optical is fast. But in theory, as long as the iterations are large enough, relatively large batch size can get better local optimal value.

4.2.2 The Results and Analysis of Text Classification

After training the DBN by using the proposed method, 9 output units are added to the top layer. That is, we use the weights of DBN to initialize a neural network except the last layer. The weights initialization of the last layer we use the normalized initialization [16], which make the connection weights between the last layer and the previous layer follow Uniform distribution. We adopt the traditional BP algorithm [17] in the process of fine-tuning. We use two indicators to evaluate the performance of text categorization: precision (P) and recall(R) in this article.

After fine-tuning, there are 1124 records classified correctly in the total 1288 testing records when batch size = 1, 1088 correct records when size = 10, 1073 correct records when size = 205 which is parallel full-batch gradient descent. Each category of the prediction results are as table 2.

From table 2 and table 3, we can know that the expected result that the precision and recall is increasing as the size decreasing. When the size is 1 or 10, the effect of classifying is better than that when size is 205(full-batch). Relatively, the results have little difference when the size is 1 or 10. The result when the size is 10 is even better than the result when the size is 1 in some categories.

Table 2 The precision of DBN when batch size is 1, 10, 205(full-batch)

	C1	C2	C3	C4	C5	C6	C7	C8	C9
Size = 1	0.91	0.74	0.88	0.88	0.91	0.89	0.88	0.81	0.93
Size = 10	0.87	0.61	0.87	0.96	0.84	0.88	0.88	0.78	0.88
Size=205	0.75	0.55	0.92	0.72	0.82	0.87	0.92	0.81	0.85

Table 3 The recall of DBN when batch size is 1, 10, 205(full-batch)

	C1	C2	C3	C4	C5	C6	C7	C8	C9
Size = 1	0.85	0.80	0.94	0.95	0.85	0.81	0.87	0.81	0.97
Size = 10	0.77	0.74	0.90	1.00	0.91	0.81	0.87	0.72	0.95
Size=205	0.79	0.77	0.87	0.95	0.87	0.82	0.83	0.70	0.97

To illustrate the effect of classification, this experiment is compared with LibSVM [18]. Fig. 6 show the results of precision and recall respectively. In terms of resource consumption, SVM take less than 5 minutes to complete training the model in single computing node. At the same time, training the DBN model takes about 60 minutes.

From Fig. 6, we can see that the neural network produced by DBN when size is 1 is slightly lifted compared with which by SVM when considering the precision and recall. When the size is 10, the effect of classifying is similar. However, the effect of classifying is worse than SVM when the size is 205.

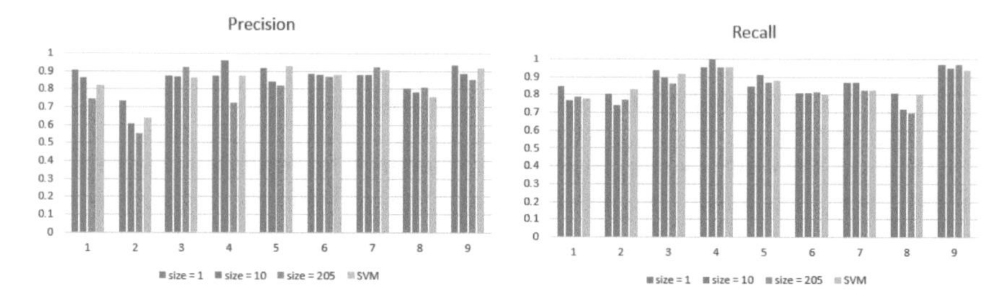

Fig. 6 Precision and Recall of each categories

5 Conclusion

The article designs and realizes a new parallel algorithm of DBN based on distributed general purpose computing framework named Spark. The algorithm control the convergence of the network in each computing node by adjusting the batch size.

The results of our algorithm show the algorithm proposed in the article is better than the traditional parallel full-batch gradient descent in convergence rate and convergence precision when the batch size is relatively smaller. We also describe the phenomenon that the latter RBMs in parallel DBN network nearly cannot be trained when the batch size is large relatively and give the explanation. In addition, compared with the SVM, the performance of DBN is better in text classification when batch size is small.

For future work, we would like to research how to enhance the sparsity of DBN and focus on improving the precision and recall of the method.

Acknowledgements This work is supported by the National High-tech R&D (863) Program of China (No. 2015AA050204).

References

1. Ng, A., Ngiam, J., Foo, C.Y.: Deep learning (2014)
2. Bengio, Y.: Learning deep architectures for AI. Foundations and trends® in Machine Learning **2**(1), 1–127 (2009)

3. Hinton, G., Osindero, S., The, Y.W.: A fast learning algorithm for deep belief nets. Neural Computation **18**(7), 1527–1554 (2006)
4. Dean, J., Corrado, G., Monga, R.: Large scale distributed deep networks. In: Advances in Neural Information Processing Systems, pp. 1223–1231 (2012)
5. Seide, F., Fu, H., Droppo, J.: 1-Bit stochastic gradient descent and its application to data-parallel distributed training of speech DNNs. In: Fifteenth Annual Conference of the International Speech Communication Association (2014)
6. De Grazia, M.D.F., Stoianov, I., Zorzi, M.: Parallelization of deep networks. In: Proceedings of 2012 European Symposium on Artificial NN, Computational Intelligence and Machine Learning, pp. 621–626 (2012)
7. Sainath, T.N., Kingsbury, B., Ramabhadran, B., et al.: Making deep belief networks effective for large vocabulary continuous speech recognition. In: 2011 IEEE Workshop on Automatic Speech Recognition and Understanding (ASRU), pp. 30–35. IEEE (2011)
8. Haykin, S.S.: Neural networks and learning machines. Pearson Education, Upper Saddle River (2009)
9. Bengio, Y., Lamblin, P., Popovici, D.: Greedy layer-wise training of deep networks. Advances in Neural Information Processing Systems **19**, 153 (2007)
10. Fischer, A., Igel, C.: An introduction to restricted Boltzmann machines. In: Progress in Pattern Recognition, Image Analysis, Computer Vision, and Applications, pp. 14–36. Springer, Heidelberg (2012)
11. Liu, J.S.: The collapsed Gibbs sampler in Bayesian computations with applications to a gene regulation problem. Journal of the American Statistical Association **89**(427), 958–966 (1994)
12. Zaharia, M., Chowdhury, M., Franklin, M.J.: Spark: cluster computing with working sets. In: Proceedings of the 2nd USENIX Conference on Hot Topics in Cloud Computing, pp. 10–10 (2010)
13. Dean, J., Ghemawat, S.: MapReduce: simplified data processing on large clusters. Communications of the ACM **51**(1), 107–113 (2008)
14. Zaharia, M., Chowdhury, M., Das, T.: Resilient distributed datasets: a fault-tolerant abstraction for in-memory cluster computing. In: Proceedings of the 9th USENIX Conference on Networked Systems Design and Implementation. USENIX Association, p. 2 (2012)
15. Salton, G., McGill, M.J.: Introduction to modern information retrieval (1983)
16. Glorot, X., Bengio, Y.: Understanding the difficulty of training deep feedforward neural networks. In: International Conference on Artificial Intelligence and Statistics, pp. 249–256 (2010)
17. Rumelhart, D.E., Hinton, G.E., Williams, R.J.: Learning representations by back-propagating errors. Cognitive Modeling **5** (1988)
18. Chang, C.C., Lin, C.J.: LIBSVM: a library for support vector machines. ACM Transactions on Intelligent Systems and Technology **2**(3), 27 (2011)

Predicting Potential Retweeters for a Microblog on Twitter

Soniya Rangnani and V. Susheela Devi

Abstract Recently, retweeting is found to be an important action to understand diffusion in microblogging sites. There have been studies on how tweets propagate in networks. Previous studies have shown that history of users interaction and properties of the message are good attributes to understand the retweet behavior of users. Factors like content of message and time are less investigated. We propose a model for predicting users who are more likely to retweet a particular tweet using tweet properties, time and estimates of pairwise influence among users. We have analyzed retweet cascades and validated that structural, social, behavioral and history of nodes are equally important for influence estimation among users. We develop a model which ranks the users based on the likelihood of the users to be potential retweeters. We have performed experiments on real world Twitter sub-graphs and our results validate our proposed work satisfactorily. We have also compared our results with existing works and our results outperform them.

Keywords Retweet behavior · Information diffusion · Predicting potential retweeters

1 Introduction

Information diffusion is a generic concept that refers to the propagation process of an object regardless of its type or nature. One of the reasons to study networks is diffusion of information over a network. In context of social networks, the models assume that people are influenced by their surrounding in the network. In other words, they model the process as an "Information cascade". Spread of particular objects like

S. Rangnani(✉) · V. Susheela Devi
Computer Science and Automation Department, Indian Institute of Science, Bangalore 560012, Karnataka, India
e-mail: {soniya.rangnani,susheela}@csa.iisc.ernet.in

K. Lavangnananda et al. (eds.), *Intelligent and Evolutionary Systems*, Proceedings in Adaptation, Learning and Optimization 5,
DOI: 10.1007/978-3-319-27000-5_14

hashtags, URLs or even broader concepts like topics are problems to be explored in case of microblogging sites.

Micro-blogging services like Twitter have given people a platform to obtain, share and spread the ideas as and when they wish. Users can post their tweets through web service or third-party applications available 24/7.

Retweeting is the key mechanism of information diffusion in Twitter. Often users give credit for a message to another user who posted the same message before. This phenomenon is known as retweeting. A user 'retweets a tweet' means he shares the tweet with his followers with acknowledgement to its source (i.e. original poster of tweet). Direction of information flow can be strongly indicated by retweeting mechanism. People are actually selective about whom they want to listen to. This makes retweet mechanism even more interesting.

Diffusion process is a time dependent process where probability is associated with each edge of a social graph. This probability represents how likely that edge is to participate in a diffusion. Interaction among users helps to quantify and estimate mutual influence appropriately. In case of social networks, user profile information contribute to understand how much and how a user tends to adopt an information. Also for the same network, different messages propagate through different parts as per the interest of users in the path.

We have developed a ranking approach where the adoption of information is explained by topology of the network and the interactions that occur between pair of users. Our proposal has been validated on real Twitter datasets. Experiments show that our approach is able to predict the retweeters of a particular tweet posted by a particular user at a particular time depending on the network of users and the content.

The paper is organized as follows. Problem is stated in Section 2. Related work is introduced in Section 3. The detailed description of our work is presented in Section 4. Experimental results are presented in Section 5. Finally the conclusions and future work are given in Section 6.

2 Problem Statement

To model a microblogging social network (e.g., Twitter), we adopt the following terminology. Let $G = (U, E)$ be the social graph with $U = \{u_1, u_2, \ldots, u_N\}$ being the set of N users (nodes) and $E \subseteq U \times U$, the set of directed edges, where the direction is in accordance with the information flow. Thus, $(u, v) \in E$ denotes that user u is followed by v, indicating that information flows from user u to user v.

We introduce functions *follower* ($F_r : U \rightarrow 2^U$) and *followee* ($F_e : U \rightarrow 2^U$), defined as follows:

$$F_r(u) = \{v \in U \mid (u, v) \in E\}, \text{ and}$$
$$F_e(v) = \{u \in U \mid (u, v) \in E\}.$$

Thus, for a given user u, $F_r(u)$ denotes the set of users who *follow* u, which means a message posted by u will reach all members in $F_r(u)$. Similarly, $F_e(u)$ denotes those whom u follows, that is, a message posted by any member of $F_e(u)$ will reach u.

Let M be a universal pool (a set) of messages that were tweeted. And let T denote a universal timeline. We define functions $tt : M \to T$ and $tu : M \to U$, which give the timestamp $tt(m)$ and tweeter $tu(m)$, respectively, of message $m \in M$ when it was tweeted (*i.e.*, first posted).

Let us now define the retweet decision function, δ as follows:

$$\delta : U \times M \to \{0, 1\}$$
$$(u, m) \mapsto \begin{cases} 1, \text{ if } u \text{ retweets } m, \\ 0, \text{ otherwise.} \end{cases}$$

For a user u, let $M_u = \{m_1, m_2, \ldots, m_{|M_u|}\}$ denote the messages that have reached u. We assume here that a user u retweets only messages in M_u, *i.e.*, $\delta(u, m) = 0$ for all $m \notin M_u$.

For experimental purposes, let T be divided into two parts, *viz.*, training and testing periods, denoted by T_{tr} and T_{te} respectively. This also partitions the set of messages M into train and test sets $M^{(tr)}$ and $M^{(te)}$ respectively, defined by

$$M^{(tr)} := \bigcup_{t \in T_{tr}} tt^{-1}(t) \quad \text{and} \quad M^{(te)} := \bigcup_{t \in T_{te}} tt^{-1}(t).$$

Finally, we define two decision functions $\delta_{tr} := \delta|_{U \times M^{(tr)}}$ and $\delta_{te} := \delta|_{U \times M^{(te)}}$. We predict future retweet decisions δ_{te}, given past retweet decisions δ_{tr}.

3 Related Work

Various studies in the context of social network have been conducted with the aim of predicting properties of the information spread. Yang and Counts [10] finds *mention* as an important action for predicting speed, scale and range of diffusion. Suh *et al.* [9] present exploratory analysis of retweetabililty with respect to properties of the user who has posted the tweet and the content of the tweet (number of URLs, hashtags in tweet). This study shows content properties like presence of URLs, hashtags and number of followers and followees are correlated to retweetability but number of past tweets of users is a poor predictor of retweetability.

Many existing works estimate number of nodes adopting the information [5, 8], lifespan of the information [4] and its popularity [1]. However, there are very few prediction models in literature which aim to predict which node will participate in diffusion of a particular information depending on time at which the tweet occured. Gulaba *et al* [3] have proposed a Linear Threshold approach to model spread of URLs in Twitter network. The parameters of the model include virality of information, influence between pair of users and receptiveness of users for information. Content of URLs has not been analyzed. In a similar study, Petrovic *et al.* [8] performed experimental work to predict whether a tweet will be retweeted or not. They developed an online learning-based algorithm [14] to make the prediction as quickly as possible. They trained a set of local models merely on different subsets of data which are generated based on the time of the day to be able to better exploit the time information of tweets. Luo *et al.* [7] consider the contents of tweets and predict the activated nodes (retweeters). However, they have not discussed explicitly about influence among the users. They showed that retweeting behavior mostly depends on diffusion history and tweet properties. Similar observations were seen in the models by Uysal, *et al.* [11] and Chen, *et al.* [12]. These works were not able to make any remarkable observations on the temporal and topical factors affecting retweetability, and were left for further investigation. This differs from the works where temporal properties are primarily studied. We compare our results with the work of Luo *et al.* [7] as it is the closest to our work.

Given the state-of-art, we propose a model which exploits all the dimensions of information diffusion process- influence, content and temporal properties to predict the potential retweeters for a tweet.

4 Methodology

User v follows user u, and u posts a tweet m through a microblogging service. We want to model the retweeting behavior of user v *i.e.*, whether v will retweet the tweet m or not.

We model the task of finding retweeters as a ranking task. We formulate a function which ranks the followers of the initiator of the test tweet according to how receptive they are to the test tweet. The function generates a confidence score for the edges joining the initiator and the followers. As two different pieces of information propagate through different users in the network, we divide the task of calculating confidence score. First part is independent of the test tweet. It models transmission rate of the tie between the initiator and the follower. We call this as 'Pairwise Influence Estimation'. The second part incorporates the tweet properties and user activeness as per time in the ranking function. We call it 'Score Calculation'.

4.1 Features

Retweeting is a time dependent process. We model how likely a user is to participate in a diffusion. Interaction among users helps to quantify and estimate mutual influence appropriately. In case of social networks, user profile information contributes to understand how much and how a user tends to adopt an information. Also for the same network, different messages propagate through different parts as per the interest of users in the path. Retweet actions of a user can be explained by following three factors- pairwise influence, content and time.

Pairwise Influence. Pairwise influence is independent of the tweet under consideration. Influence can be seen as transmission rate of information on that edge. In practice, we only observe log of actions of users or cascade data originating from users. Influence among them is not available. Various works explain factors causing users' participation in the spread of information. We use three sets of features: *structural*, *user-profile based* and *user-pair based*, which characterize static properties of pair of users with respect to diffusion. These are explained in detail below.

1. **Structural features**
   Many topological properties of nodes of a social graph explain the information spread. These include *eigenvector centrality*, *in-degree*, *out-degree*, and *clustering coefficient* of user nodes.
2. **User-profile based features**
   These features are collected from a user's profile information and observing the tweet/retweet activity of the user in the training period. *Number of tweets*, *retweet ratio*, *social pressure* and *number of tweets which either mention or retweet the user* comprise this set of features.
   Let us define $R_{tr}(u) := \{m \in M_u \mid \delta_{tr}(u, m) = 1\}$ as the set of retweets by user u during the training period, and $rt(u, m)$ represents the timestamp of a retweet $m \in R_{tr}(u)$.
   Retweet ratio $\rho(u)$ of user u is the ratio of number of retweets by the user and total number of his original tweets in the training period.
   $$\rho(u) := \frac{|R_{tr}(u)|}{|tu^{-1}(u) \cap M_{tr}|} \tag{1}$$
   Social pressure [6] $\psi(u)$ for user u can be estimated as the expected number of active neighbors of u right before u himself gets active. It can be formulated as follows:
   $$\psi(u) = \frac{\sum_{m \in R_{tr}(u)} |\{v \in F_e(u) : rt(v, m) < rt(u, m)\}|}{|R_{tr}(u)|} \tag{2}$$
3. **User-pair based features**
   Homophily and *social influence* are two aspects being captured in this set of features. The phenomenon of people tending to communicate with those similar to them in socially significant ways is called homophily. It is found to have an important implication in information flow in social networks [2]. Many studies have proved that both aspects are equally important and are very different as concepts. Homophily is captured by the number of common URLs and common hashtags found in tweets of pairs of users, and by the number of common followees. For an edge from user u to user v, *number of tweets of user u retweeted by user v* and *number of times user v has mentioned user u* are the features characterizing social influence.

An edge (u, v) denotes the spread of message from user u to user v. We classify edges of the social graph in two classes $(1, -1)$. Positive class corresponds to the edges which have participated in at least one diffusion sequence in the training data (*aka retweet* class). An edge is classified as a negative instance (*aka no retweet* class), if it has not participated in any of the diffusion sequences. We have extracted twenty one features (first two set of features for both the users u and v and the third set for the pair (u,v) Table 1), and Logistic Regression was applied to the data. Influence can be seen as the confidence score of an edge to estimate participation in diffusion. We define the supervised classification task as follows.

Let $PLR_{v,u}$ be the estimate of Logistic regression for a pair of users (u, v) whose feature vector is X.

$$PLR_{v,u} = P(Y = 1 \mid X). \tag{3}$$

The parametric equation for P(Y=1 | X) is formulated as follows:

$$P(Y = 1 \mid X) = \frac{1}{1 + \exp(\beta_0 + \sum_i(\beta_i X_i)} \tag{4}$$

$$P(Y = -1 \mid X) = 1 - P(Y = 1 \mid X) \tag{5}$$

Maximum likelihood estimation is used to learn the parameters β_i of the model. For vector X, $P(Y = 1 \mid X)$ is calculated by substituting estimated β_i in equation 4 above.

Content. Content of tweets shared by a user highly signifies the topics he follows. It is intuitive to assume that a user tends to share those tweets which are similar in content to the past ones. A content-based similarity score $K_\sigma(u, m)$ between the words in the history of user u and words of tweet m can be formulated to capture u's willingness to retweet (diffuse the content). Let $\kappa(m)$ denote the content (set of distinct words) of tweet m and $vocab_u$ denote distinct words in the posts of user u.

$$K_\sigma(u, m) = \frac{|\kappa(m) \cap vocab_u|}{|\kappa(m)|} \tag{6}$$

To model thematic interest among a connected pair of users (u, v) and tweet m, we formulate a combined content factor $K_f(u, v, m)$ as

$$K_f(u, v, m) = K_\sigma(u, m) \cdot K_\sigma(v, m) \tag{7}$$

Time. We assume a hourly time resolution for modeling a user. It is important to consider how receptive (to retweet) a user generally is at any given hour. We assume that an individual's activity level (total number of posts) in a particular hour follows a Poisson distribution. Let $x_{dh}^{(u)} \in \mathbb{N}$ be the number of tweets (including retweets) posted by user u on day d and during hour $h \in \{1, 2, \ldots, 24\}$. For an hourly analysis over multiple days, we consider, for each hour h, the sequence $x_{1h}^{(u)}, x_{2h}^{(u)}, \ldots, x_{nh}^{(u)}$ of number of tweets posted by user u for n consecutive days. The maximum likelihood estimate of the activity level of the user u at hour h may be represented as

$$\lambda_{uh} = \frac{1}{n} \sum_{d=0}^{n} x_{dh}^{(u)} \tag{8}$$

We define the receptivity of a user for a piece of information as *average activation time* ($\overline{\tau}$) for user u. It is the average number of hours a user takes to retweet a tweet. Let h_m be the hour of the day at which tweet m is posted. The temporal aspect is captured for the tweet m and the user u using $\lambda_{u(h_m+\overline{\tau})}$.

We analyse activeness of users per hour as tweets posted in that hour. The hours of a day at which user u is more active to adopt information are called as Active Hours (H_u) of user u. It can be defined as -

$$H_u = \{\ h \in [1, 24] \cap \mathbb{Z} \mid \lambda_{uh} \neq 0\ \} \tag{9}$$

Table 1 Features for influence estimation

	user u	**user v**
Structural features	eigenvector centrality	eigenvector centrality
	in-degree	in-degree
	out-degree	out-degree
	clustering coefficient	clustering coefficient
User-profile based	Number of tweets	Number of tweets
	retweet ratio	retweet ratio
	social pressure	social pressure
	number of tweets which either mention or retweet the user	number of tweets which either mention or retweet the user
User-pair based	number of common URLs	
	number of common common hashtags	
	the number of common followees	
	number of tweets of user u retweeted by user v	
	number of tweets of user v retweeted by user u	

4.2 *Score Function*

Given a test tweet, its time of post and its initiator, we calculate confidence score of the edges of the social graph. Following are the three main steps in calculating score for an edge:

Data: m, v, $\overline{\tau}$, α_1, α_2 PLR, H_v, K_f
Result: score(u,v)
$u = tu(m)$
$t_0 = tt(m)$
score(u,v) = $PLR_{vu} \times 0.5$;
$t_v = t_0 + \overline{\tau}$;
$\mathbb{I} = [t_v \, , t_v + 1]$;
if ($\mathbb{I} \bigcap H_v \neq \phi$) **then**
  score(u,v) = score(u,v)+(1-score(u,v)).α_1;
else
  score(u,v) = score(u,v)+(1-score(u,v)).α_2;
end
score(u,v) = score(u,v)+(1-score(u,v)).$K_f(u,v,m)$;

Algorithm 1. Score Function

Step 1: We initialise score with static probability PLR_{uv}. The likelihood of activation of edge with respect to history and properties of user nodes is considered to be half of score.

$$score(u, v) = 0.5 \times PLR_{vu} \tag{10}$$

Step 2: Temporal aspect is characterised by active hours of the follower, his average time of activation and time of post of test tweet. This step penalises non active followers. For a test tweet posted at hour t_0, we check activity of user in the interval $[t_0 + avgacttime, t_0 + avgacttime + 1]$ and perturb $score(u,v)$ calculated from the previous step appropriately.

$$score(u, v) = \begin{cases} score(u, v) + (1 - score(u, v)).\alpha_1 \text{ when follower } v \text{ is active.} \\ score(u, v) + (1 - score(u, v)).\alpha_2 \text{ when follower } v \text{ is not active} \end{cases} \tag{11}$$

Here, α_1 and α_2 are two parameters which are like weights for activeness of follower v. If the follower is active, we increase probability by $(1 - score(u, v)) * \alpha_1$ and if he is not active, we increase probability by $(1 - score(u, v)) * \alpha_2$. Their values are chosen appropriately depending on data. These values address the trade-off between penalising the inactive users and crediting the active users ($\alpha_1 \geq \alpha_2$).

Step 3: We increase probability as per content similarity $K_f(u, v, m)$ (see equation 7) between test tweet m and users forming an edge as shown below.

$$score(u, v) = score(u, v) + (1 - score(u, v)) \cdot K_f(u, v, m) \tag{12}$$

The three steps are applied to all the edges and scores are calculated. Algorithm 1 briefly explains score function. It takes a test tweet m and user v as input and outputs score between user v and the user u who posted message m. The other inputs $\overline{\tau}$, α_1, α_2, PLR_{uv}, H_v, K_f are explained in Table 2. These inputs are calculated while training the model. The next section describes the prediction phase of the model.

Table 2 Notations

Notations	Description
$G = (U, E)$	social graph with users U and edges E
m	tweet under consideration
v	user under consideration
$F_r(v)$	Followers of user v
$F_e(v)$	Followees of user v
$tt(m)$	timestamp at which the tweet m was posted
$tu(m)$	user who posted the tweet m
$\delta(v, m)$	Estimated retweet decision
$M^{(tr)}$	Messages by users during training period
$Y^{(tr)}$	Retweet decisions for $M^{(tr)}$
PLR_{vu}	Influence of u on v (ref. Section 4.1)
k	Parameter k
H_v	hours at which user v is observed to be active (ref. Section 4.1)
$\overline{\tau}$	Average activation time (ref. Section 4.1)
λ_{vh}	Temporal feature for user v and hour h (ref. Section 4.1)
α_i	Parameter α_i's
S	list of confidence scores of users
$rankedS$	sorted list of scores of users

4.3 Prediction

We model the task of finding retweeters as a ranking task. The scores generated for the edges reflect the receptiveness of users for the particular tweet under consideration. Given a test tweet, the score values of edges are calculated and ordered in a non increasing fashion with respect to scores. The model reports top k scoring users as retweeters, k being the size of prediction. There could be different ways of fixing size of prediction.

One way to fix k is to use a validation set of tweets. The properties of diffusion sequences in the validation set are found to be similar to that of diffusion sequences in the training set. For a user, it has been seen that the average of retweeters size or audience size of his tweets in the training method is approximately the same as the audience size of the tweets in the validation set. A possible value of users who retweet (k) is twice the average of retweeters count.

Another way to fix prediction size is to fix a threshold for the scores. We analyse the scores of the tweets in the validation set. The edges with scores greater than threshold is reported as potential paths of information flow and the users to which the edges are directed to, are the potential retweeters. The threshold is selected by analysing the scores of edges for the tweets in the validation set and an optimal threshold is selected which gives better prediction results.

Data: v, $M^{(tr)}$, $Y^{(tr)}$

Result: (PLR_{vu}), $(\lambda_{vh})_{h=1}^{24}$

Step 1: Calculate Influence component for edges using $M^{(tr)}$ and $Y^{(tr)}$ as $(PLR_{vu})_{u \in Fe(v)}$

Step 2: Estimate hourly parameters for v as $(\lambda_{vh})_{h=1}^{24}$

Algorithm 2. Training for user v

Data: m, v, k
Result: $\delta(v, m)$
$u = tu(m)$
$t_0 = tt(m)$
$F = F_r(u)$
for $v' \in F$ **do**
  Calculate $S_{v'}$ using Algorithm 1
end
$rankedS$ = sort scores S in non increasing order
if $v \in$ *top k scoring users in* $rankedS$ **then**
  $\delta(v, m) = 1$
else
  $\delta(v, m) = 0$
end

Algorithm 3. Prediction

4.4 Algorithms for Learning and Testing the Model

Notations used in the algorithms are shown in Table 2. For training the model, tweets posted in the training period $M^{(tr)}$ and their labels $Y^{(tr)}$ are taken to estimate PLR_{uv} for all pairs of users (u, v) constituted as edges in the network. We estimate hourly parameters λ_{vh} during training (See Algorithm 2).

To predict the retweeters of a test tweet m, the major step is to calculate scores of all the followers of the user who posted m. Followers are ranked according to their scores. If user v is in the top k ranked followers, we predict that v has retweeted m. Otherwise, we predict that he hasnot retweeted. The prediction phase is explained briefly in Algorithm 3.

5 Model Evaluation

5.1 Data Collection and Preprocessing

As per our best knowledge, there is no annotated dataset for retweeters available with the full set of features used by our approach. We collected data of Twitter graph using web service calls of Twitter REST API[1]. To be assured about availability of diffusion sequences in data, we selected one seed node. We collected hundred recent tweets of the seed node and their retweeters. These retweeters and the seed node constitute the social graph. The links between user nodes are requested using Twitter API service routines to complete the graph. We collected recent 3200 tweets of users.

A tweet is a JSON (Javascript Object Notation) object which contains abundant meta-data about its status. It include *created_at* (time of post), *user* (details of initiator) and *text* (the message) as major fields. A retweet is also a JSON object with additional field *retweeted_status* which contains the object of original tweet which is retweeted. Fig. 1 shows the structure of JSON object which retweets the tweet shown in the shaded region as the value of field 'retweeted_status'. The major part of preprocessing is extracting diffusion sequence from tweets. The tweet and all retweets with same embedded tweet are collected and ordered chronologically. These ordered set of users form a diffusion sequence.

```
{
"created_at": "Thu Oct 23 16:02:46 +0000 2010",
"id": 65214042119,
"text": " what we've are to at @holay happy customers",
"user": {
        "name": "Jack Brown",
        "id": 18756958,
        }
"retweeted_status":
 {
 "created_at": "Thu Oct 21 16:02:46 +0000 2010",
 "id": 28039652340,
 "text": "what we've are to at @holay happy customers",
 "user": {
         "name": "Holay, Inc.",
         "id": 16958872,
         }
 ...}
...
}
```

Fig. 1 Retweet structure example

We have done experiments on two Twitter subgraphs. Dataset 1 contains 341 users with 1997 edges. We collected 1.9 lakhs tweets spanning from 1st June, 2014 to 4th August, 2014. We divide the tweets into training and test data with respect to time. We exacted 481 training diffusion sequences and 169 test diffusion sequences. Dataset 2 is bigger then dataset 1. It has 3327 nodes and approximately 38.9 lakhs tweets spanning from 1st July, 2014 to 30th November, 2014. Diffusion sequence

[1] https://dev.twitter.com/rest/public

Table 3 Top Fscore values on features of edge (u,v)

Features	F score
#Retweet from u to v	0.1904
Outdegree of u	0.1838
eigenvector centrality	0.1813
Retweet ratio of u	0.1619
#CommonFollowees	0.0848
#Retweet from u to v	0.0599
Indegree of v	0.0531

of 2046 tweets from training data were considered and 248 tweets from test data. Number of non-retweeters (i.e. number of followers of initiators) are 60 to 261 for dataset 1 and 700 to 3300 in dataset 2. The seed node for dataset 1 is 'PMOffice'. It is the account managed by Prime Minister Office of India. The seed node of dataset 2 is 'CNNBreakingNews'. It is a news account managed by CNN news.

5.2 Features Analysis

We analyze the significance of the features used for calculating Influence component PLR explained in Section 4.1 in two ways:

1. We calculated the Fisher score (Fscore) of the features to understand the contribution of the features used by us. Fisher Score [13] gives a higher score to features that assign similar values to the samples from the same class and different values to samples from different classes. The evaluation criterion used in Fisher Score can be formulated as

$$FScore(f_i) = \frac{\sum_{j=1}^{c}(n_j(\mu_{i,j} - \mu_i)^2)}{\sum_{j=1}^{c}(n_j(\sigma_{i,j}^2))} \tag{13}$$

where μ_i is the mean of the feature f_i, n_j is the number of samples in the jth class, and $\mu_{i,j}$ and $\sigma_{i,j}^2$ are the mean and the variance of f_i on class j, respectively.

Table 3 shows top Fscore features in decreasing order.

2. After applying Logistic Regression, we analyse the parameter β for every feature. Features corresponding to top absolute normalized parameters in descending order are

1. *Outdegree of u*
2. *#CommonHashtags*
3. *#Retweet from u to v*
4. *Clustering Co-efficient of u*
5. *#RetweetsMentions of u*
6. *Retweet Ratio of u*
7. *Social pressure of v*

Both the above methods found many common features which helped us in understanding the significance and contribution of features.

By analysing both the methods, we conclude that the categories of features mentioned below contribute to probability estimation of pairwise influence between users

1. Structural features:
   Outdegree of u, eigenvector centrality of u , Clustering co-efficient of u
2. User-profile based features:
   #RetweetsMentions of u, Social Pressure of v, retweet ratio of u
3. Homophily:
   #Commonhashtags, #CommonFollowees
4. Influence:
   #Retweet from u to v

5.3 Baseline Methods

We compare our model with two models- the work of Luo *et al* [7] 'RankSVM' and Random model. RankSVM generates a ranking function which ranks followers according to how likely they are to retweet the tweet. They develop feature families based on retweet history, follower status, follower active time and follower interests. They report that follower interest, retweet history and follower status are good predictors of retweetability. We use these features to predict retweeting of users. Random model generates random ranking for the followers.

5.4 Performance Measures

We show results with Mean precision average (Map@n) performance measure. Map@n is evaluation measure for ranked retrieval results. For a test tweet, appropriate sets of retrieved users are naturally given by top k retrieved users. This means that the k users with higher scores are chosen as the predicted retweeters. Suppose top k scoring users are predicted as retweeters. We define $P(k)$ as precision of set of predicted retweeters. For a test tweet m, average precision $(AP@k)$ is the average of precision values obtained for the set of top k predicted retweeters for some k. We calculate average precision as

$$AP@k(m) = \frac{\sum_{i=1}^{i=k} 1_{\{u_i \in Retweeters_m\}}(u_i) P(i)}{|Retweeters_m|}$$

where $Retweeters_m$ gives the set of retweeters. For the entire set of test tweets $M = \{m_1, m_2, \ldots, m_{|M|}\}$ and prediction count k_i for all $i \in [1, |M|]$, we calculate $Map(M)$ as

$$Map(M) = \frac{\sum_{j=1}^{j=|M|} AP@k_j(m_j)}{|M|}$$

5.5 Experimental Results

We performed experiments on both the datasets. On validation data, we record the performance of our model, for different values of parameters α_1 and α_2. The parameters corresponding to which best performance are recorded is used for prediction of test tweets. For evaluating the ranks of the users, we show results with respect to number of original retweeters $|Retweeters_m|$ for message m. Table 4 presents the performance of ranking methods for finding retweeters based on the approaches. We report Mean Precision Average at twice and thrice of original number of retweeters of a tweet i.e $k_i = 2 \times |Retweeters_{m_i}|$ and $k_i = 3 \times |Retweeters_{m_i}|$. Tables 5 and 6 show average precision with $k = 2 \times |Retweeters_m|$ for twenty test tweets from

Table 4 Performance of Methods

Methods	Dataset no.	Map@n (n=k×#retweeters)	
		k=2	k=3
	Our Method	0.1881	0.2280
1	RankSVM	0.0459	0.0603
	Random	0.0237	0.0342
	Our Method	0.0846	0.2113
2	RankSVM	0.0680	0.0743
	Random	0.0307	0.0419

Table 5 Average Precision values for dataset 1

Sr. no.	Our Method	RankSVM	Random
1	0.0535	0.0781	0.0395
2	0.1927	0.1059	0.0738
3	0.6561	0.0959	0.0000
4	0.7202	0.1290	0.0039
5	0.6163	0.2222	0.0000
6	0.0902	0.0369	0.0102
7	0.2014	0.0277	0.0059
8	0.3763	0.0586	0.0235
9	0.0574	0.0526	0.0000
10	0.3238	0.1279	0.0078
11	0.3026	0.0526	0.0146
12	0.7462	0.1378	0.0000
13	0.3029	0.0259	0.0000
14	0.5446	0.0676	0.0083
15	0.4333	0.0537	0.0000
16	0.2314	0.0661	0.0258
17	0.0678	0.0458	0.0249
18	0.1046	0.0994	0.0310
19	0.1530	0.0752	0.0143
20	0.1572	0.0246	0.0000

Table 6 Average Precision values for dataset 2

Sr. no.	Our Method	RankSVM	Random
1	0.0976	0.0338	0.0227
2	0.1534	0.0564	0.0206
3	0.1535	0.0560	0.1182
4	0.2222	0.0447	0.0000
5	0.0831	0.0566	0.0476
6	0.0508	0.0178	0.0280
7	0.0626	0.1099	0.0745
8	0.0899	0.0372	0.0417
9	0.0752	0.0039	0.0392
10	0.1369	0.0311	0.0175
11	0.1106	0.0113	0.0338
12	0.0565	0.0018	0.0556
13	0.1052	0.0124	0.0708
14	0.0933	0.0064	0.0443
15	0.1032	0.0157	0.0000
16	0.2009	0.0000	0.0000
17	0.2238	0.0419	0.0318
18	0.0778	0.0079	0.4319
19	0.0509	0.0337	0.3131
20	0.1314	0.0174	0.0000

dataset 1 and 2. The values of average precision on the test tweets calculated by our method is much higher compared to the values of other two methods in both the datasets. Our method shows a significant improvement over RankSVM and Random model. It can be seen from Tables 4 that mean average precision of our method is significantly more than that obtained by RankSVM and Random. From Table 5 and 6, it can be seen that in almost all the cases, the average precision is higher for our method.

6 Conclusion

Finding retweeters in Twitter can help deliver information to other people more efficiently and effectively. Many recent diffusion models are based on the dynamics of interactions between neighbor nodes in the network, and largely ignores important dimensions as content of a piece of information. We propose a model that predicts how a message will be diffused by making use of additional dimensions such as the content of the piece of information diffused, user's profile and attentiveness of user towards the upcoming messages. Our results show promising predictions on potential retweeters among followers. We capture local aspects of diffusion. Future work can include studying and incorporating global aspects of the graph such as virality of the content. More preprocessing and different formulations for content can be used for calculation of scores and modeling of the task.

References

1. Hong, L.G., Dan, O., Davison, B.D.: Predicting popular messages in twitter. In: Proceedings of the 20th International Conference Companion on World Wide Web, pp. 57–58. ACM (2011)
2. Jadbabaie, A., Molavi, P., Sandroni, A., Tahbaz-Salehi, A.: Non-bayesian social learning. Games and Economic Behavior **76**(1), 210–225 (2012)
3. Galuba, W., Aberer, K., Chakraborty, D., Despotovic, Z., Kellerer, W.: Outtweeting the twitterers - predicting information cascades in microblogs. In: WOSN 2010 Proceedings of the 3rd Wonference on Online Social Networks, p. 33. USENIX Association Berkeley, CA (2010)
4. Kong, S.B., Feng, L., Sun, G.Z., Luo, K.: Predicting lifespans of popular tweets in microblog. In: Proceedings of the 35th International ACM SIGIR Conference on Research and Development in Information Retrieval, pp. 1129–1130. ACM (2012)
5. Kupavskii, A., Ostroumova, L., Umnov, A., Usachev, S., Serdyukov, P., Gusev, G., Kustarev, A.: Prediction of retweet cascade size over time. In: Proceedings of the 21st ACM International Conference on Information and Knowledge Management, pp. 2335–2338. ACM (2012)
6. Lagnier, C., Denoyer, L., Gaussier, E., Gallinari, P.: Predicting information diffusion in social networks using content and users profiles. In: Advances in Information Retrieval, pp. 74–85. Springer (2013)
7. Luo, Z.C., Osborne, M., Tang, J.T., Wang, T.: Who will retweet me?: finding retweeters in twitter. In: Proceedings of the 36th International ACM SIGIR Conference on Research and Development in Information Retrieval, pp. 869–872. ACM (2013)
8. Petrovic, S., Osborne, M., Lavrenko, V.: Rt to win! predicting message propagation in twitter. In: Fifth International AAAI Conference on Weblogs and Social Media (ICWSM) (2011)
9. Suh, B.W., Hong, L.C., Pirolli, P., Chi, E.H.: Want to be retweeted? large scale analytics on factors impacting retweet in twitter network. In: 2010 IEEE Second International Conference on Social Computing (Socialcom), pp. 177–184. IEEE (2010)
10. Yang, J., Counts, S.: Predicting the speed, scale, and range of information diffusion in twitter. In: Fourth International AAAI Conference on Weblogs and Social Media (ICWSM), pp. 355–358 (2010)
11. Uysal, I., Croft, W.B.: User oriented tweet ranking: a filtering approach to microblogs. In: Proceedings of the 20th ACM International Conference on Information and Knowledge Management, pp. 2261–2264. ACM (2011)
12. Chen, K., Chen, T., Zheng, G., Jin, O., Yao, E., Yu, Y.: Collaborative personalized tweet recommendation. In: Proceedings of the 35th International ACM SIGIR Conference on Research and Development in Information Retrieval, pp. 661–670. ACM (2012)
13. Duda, R.O., Hart, P.E., Stork, D.G.: Pattern Classification, 2nd edn. Wiley, New York (2001)
14. Crammer, K., Dekel, O., Keshet, J., Shai, S.S., Singer, Y.: Online passive-aggressive algorithms. The Journal of Machine Learning Research **7**, 551–585 (2006)

Part III
Evolutionary Algorithms and Optimization

An Implementation of Tree-Seed Algorithm (TSA) for Constrained Optimization

Mustafa Servet Kıran

Abstract One of the recent proposed population-based heuristic search algorithms is tree-seed optimization algorithm, TSA for short. TSA simulates the growing over on a land of trees and seeds and it has been proposed for solving unconstrained continuous optimization problems. The trees and their seeds on the D-dimensional solution space correspond to the possible solution for the optimization problem. At the beginning of the search, the trees are sowed to the land, and a number of seeds for each tree are produced during the iterations. The tree is removed from the stand and its best seed is added to the stand if the fitness of the best seed is better than the fitness of this tree. In the present study, a constraint optimization problem, the well-known pressure vessel design-PVD problem, is solved by using TSA. To overcome the constraints of the problem, a penalty function is used and the problem is considered as a single objective optimization problem. The experimental results obtained by the TSA are compared with the results of state-of-art methods such as artificial bee colony (ABC) and particle swarm optimization (PSO). Based on the solution quality and robustness, the promising and comparable results are obtained by the proposed approach.

Keywords Heuristic search · Population-based search · Tree-seed · Constrained optimization

1 Introduction

In last decades, many nature-inspired optimization algorithms have been proposed to solve discrete or continuous optimization problems. Ant colony optimization

M.S. Kıran(✉)
Computer Engineering Department, Faculty of Engineering,
Selcuk University, 42075 Konya, Turkey
e-mail: mskiran@selcuk.edu.tr

K. Lavangnananda et al. (eds.), *Intelligent and Evolutionary Systems*,
Proceedings in Adaptation, Learning and Optimization 5,
DOI: 10.1007/978-3-319-27000-5_15

(ACO) has been developed by imitating behaviors of real ants between nest and food source [1]. Particle swarm optimization (PSO) has been proposed by simulating social behaviors of birds or fishes [2]. Artificial bee colony algorithm (ABC) has been investigated by considering the intelligent behaviors of real honey bee colonies, such as waggle dance and foraging nectar sources [3]. By being inspired Newtonian Gravity Rules, the gravitational search (GSA) algorithm has been proposed by Rashedi et al. [4] . The some disadvantages of these algorithms have been improved by the researchers since their inventions [5-14] . A new population-based search algorithm named as Tree-Seed Algorithm-TSA has been developed by using some components of these algorithms by Kıran [15]. Kıran [15] investigated the performance of TSA on optimizing unconstrained numeric benchmark functions and a multi-level thresholding problem. In this approach, there are two type of searcher called as trees and seeds. In the initialization of the algorithm, the trees are sowed to the land. The tree locations correspond to the possible solution for the optimization problem and the land represents the search space of the optimization problems. A pre-defined number of seeds are produced for each tree at the each iteration of the algorithm. If the fitness of the best seed is better than the current tree, the tree removed from the stand and the new seed is added to the stand as a new tree. This search that is performed with tree and seeds is run until a predetermined number of iteration is met.

In this study, TSA algorithm is applied to solve a constrained optimization problem, and obtained results are compared with the results PSO and ABC. The paper is organized as follows. Section 1 introduces the study, TSA is detailed in Section 2 and Section 3 explains the problem dealt with the study. The application of the algorithm to the problem is presented in Section 4 and the experimental study and comparisons are given in Section 5. The obtained results are presented and discussed in Section 6 and the finally the study is concluded and a future direction is presented in Section 7.

2 The Tree-Seed Algorithm

TSA is one of the population-based algorithms, and the population will be named as stand hereafter. The stand is obtained by using Eq. 1.

$$T_{i,j} = L_j + r_{i,j} \times (U_j - L_j)\ ,\quad i = 1,2,...,N \quad and \quad j = 1,2,...,D \tag{1}$$

where, $T_{i,j}$ is the value of jth dimension of ith tree, U_j and L_j is the upper and lower bounds of jth dimension, respectively. N is the number of trees in the stand and D is the dimensionality of the optimization problem. $r_{i,j}$ is a random number produced in range of [0,1].

After the trees in the stand are produced, the fitness values of the trees are calculated by using an objective function which is specific for the optimization problem.

While a predetermined termination condition is met, a number of seeds are produced for each tree by using Eq. 2 or 3.

$$S_{k,j} = T_{i,j} + \alpha_{i,j} \times (B_j - T_{r,j}) \tag{2}$$

$$S_{k,j} = T_{i,j} + \alpha_{i,j} \times (T_{i,j} - T_{r,j}) \tag{3}$$

where, $S_{k,j}$ is jth dimension of the kth seed produced for $T_{i,j}$ that is jth dimension ith tree. B_j is the jth dimension of best tree obtained so far and $T_{r,j}$ is the jth dimension of rth tree which is randomly selected. $\alpha_{i,j}$ is the scaling factor which is randomly produced in range of [-1, 1]. The selection of Eq. 2 or Eq. 3 is depended on a control parameter for TSA. This control parameter is named as ST (search tendency) in range of [0,1]. If the intensification of stand around the best tree location is demanded, ST parameter should be chosen a high value. If the exploration capability of the stand is strengthened, the ST parameter should be selected a low value. For selection Eq. 2 or Eq. 3, a random number is generated in range of [0,1]. If this value is lower than ST, the Eq. 2 is selected; otherwise Eq. 3 is selected. Briefly, ST parameter is used for controlling the exploration and exploitation capability of the stand. The local search capability around the tree locations can be also improved by using the number of seeds. In the TSA algorithm, minimum number of seeds is 10% of the number of trees and maximum number of seeds is 25% of the number trees (stand). The number of seeds is therefore obtained by depending on the stand.

After the seeds are generated for a tree, the best seeds are selected by using Eq. 4 and the best tree is selected by using Eq. 5.

$$S_b = \arg Min\{f(S_k^i)\} \tag{4}$$

where, S_b is the best seed of the k number of seeds produced for ith tree. If the fitness of S_b is better than the fitness of ith tree. This tree is removed from the stand and S_b is added to the stand.

$$B = \arg Min\{f(T)\} \tag{5}$$

where, B is the best tree location in the T stand.

After the initialization of the algorithm, while a predetermined termination condition is met, the search procedure and equations given above are iterated consecutively.

3 The Pressure Vessel Design-PVD Problem

There are four design variables in PVD problem and the aim for PVD problem is to optimize these design variables for obtaining most appropriate pressure vessel design. PVD problem has widely used to compare the performance of metaheuristic search algorithms [16-22]. In the test problem used in the experiments are directly taken from [23] and a pressure vessel with the volume of 750 ft^3 and a working pressure of 3000 psi is tried to design in this study. The schematic diagram of the pressure vessel is given in Fig. 1.

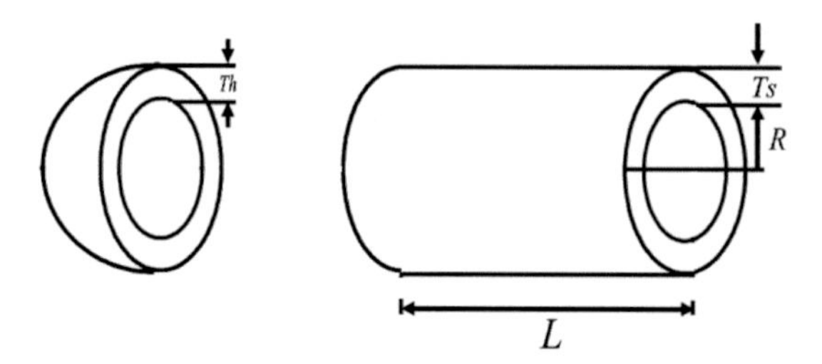

Fig. 1 The schematic diagram of pressure vessel

As seen from the Fig. 1, the design variables are the thickness of the shell (Ts), the thickness of the head (Th), the radius (R) and length of the shell (L). It is expected to be that Ts and Th are in multiplies of 0.0625 inches. R and L are continuous values for PVD problem.

According to explanations above, the mathematical formulation of the PVD problem is given as follows:

The decision variables

$$\vec{X} = (x_1, x_2, x_3, x_4) = (T_s, T_h, R, L) \tag{6}$$

The objective function

$$f(\vec{X}) = 0.6224x_1x_3x_4 + 1.7781x_2x_3^2 + 3.1611x_1^2x_4 + 19.84x_1^2x_3 \tag{7}$$

Subject to

$$g_1(x_1, x_3) = 0.0193x_3 - x_1 \le 0 \tag{8}$$

$$g_2(x_2, x_3) = 0.0095x_3 - x_2 \le 0 \tag{9}$$

$$g_3(x_3, x_4) = 1296000 - \pi x_3^2 x_4 - (4/3)\pi x_3^3 \le 0 \tag{10}$$

$$g_4(x_4) = x_4 - 240 \le 0 \tag{11}$$

$$g_5 = 1.1 - x_1 \le 0 \tag{12}$$

$$g_6(x_2) = 0.6 - x_2 \le 0 \tag{13}$$

The objective of PVD is to find $\vec{X}$ values those minimizes cost of $f(\vec{X})$ under g_1, g_2, g_3, g_4, g_5 and g_6 constraints. In constraints, g_3 provides to be the pressure vessel having volume of 750 ft^3. The other constraints are in design code of PVD problem and detailed explanations can be seen in [23].

4 Application of TSA to PVD Problem

To cope with constraints of PVD problem, a penalty function is used. If a constraint is violated, a penalty value is added to the objective function. For g_6 constraint, the penalty value for is calculated as follows:

Let be $x_2 = 0.2$, g_6 constraint functions is calculated as 0.4, and the constraint is therefore violated. New $f(\vec{X})$ is obtained by using Eq. 14

$$nF(\vec{X}) = f(\vec{X}) + \sum_{i=1}^{6} v_i \tag{14}$$

where, v_i is the violation of ith constraint and is calculated as follows:

$$v_i(\vec{X}) = \begin{cases} k \times (g_i(\vec{X}))^2, \; if(g_i(\vec{X}) > 0) \\ 0 \;\; otherwise \end{cases} \tag{15}$$

where, k is a high positive constant number.

Briefly, the violation value (Eq. 14 and 15) of each constraint is added to objective function value. While the method minimizes the objective function, it is tried to cope with the constraints by using the penalty function (Eq. 15)

By considering TSA, PVD problem and explanations given above, the pseudo code of the application of TSA to PVD is presented in Fig. 2.

```
Initialization of the algorithm
UNTIL a termination condition is met
   FOR each tree
       Produce seeds from the tree by using Eq. 2 or 3 by depending on ST
       Calculate fitness of the seeds by using Eq. 14
       IF fitness of the best seeds is better than the fitness of the tree THEN
       Begin
                     Remove the tree from the stand.
                     Add the seed to the stand
       End(IF)
   End(FOR)
   Determine the best tree location
End(Until)
       Report the best solution
```

Fig. 2 The pseudo code of application of TSA to PVD problem

5 Experiments

In order to analysis the performance of TSA on PVD problem, the method is run 30 times with random initialization for each test case and mean of obtained results and standard deviations are reported. The size of stand is taken as 10, 20, 30, 40 and 50 and ST parameter is taken as 0.1, 0.2, 0.3, 0.4 and 0.5. The number of seeds that will be produced for each seed is controlled by the size of stand. Minimum and maximum numbers of seeds for each tree are 10% and 25% of the size of the stand, respectively. The maximum number of function evaluations is used for termination of the algorithm and it is used as 4.00E+4 in the analysis of control parameters of TSA and comparisons for all algorithms. Based on these conditions, the effect of the control parameters (size of stand and ST) to the performance of TSA on the PVD problem is analyzed, and the obtained results are given in Table 1, and the best results for each row are written in boldface font type.

Table 1 The effects of control parameters to the performance of TSA

ST	Pop_Size=10		Pop_Size=20		Pop_Size=30		Pop_Size=40		Pop_Size=50	
	Mean	Std.	Mean	Std.	Mean	Std.	Mean	Std.	Mean	Std.
0.1	7213.86	32.23	7201.30	9.31	7199.59	3.39	7201.28	6.52	7203.10	5.15
0.2	7201.77	7.58	7199.47	1.28	7200.90	4.89	7199.83	1.75	7204.27	5.27
0.3	7199.69	2.57	7201.33	5.43	7199.12	2.02	7201.53	3.30	7202.63	3.50
0.4	7201.92	7.03	7199.35	1.72	7200.25	1.89	7202.70	4.10	7205.42	4.28
0.5	7201.51	8.70	7198.75	0.77	7201.90	4.29	7203.19	3.11	7207.33	5.42

As seen form Table 1, when the size of stand and ST are taken as 20 and 0.5, respectively, the best results are obtained. When 20 or 30 trees are in the stand and ST parameter is between 0.2 and 0.5, the promising results can be obtained by the TSA for PVD problem.

To compare the TSA with the PSO and ABC, these algorithms are implemented and run 30 times with the same conditions. For PSO and ABC algorithms, the population size is taken as 40 and the maximum number of function evaluations used for termination criterion for the algorithms is set to 4.00E+4. The mean and standard deviation of the results obtained by PSO and ABC are compared with results of TSA. Because the best performance is obtained by TSA while there are 20 trees in the stand and 0.5 is for ST, the results are directly taken from Table 1. The comparison is given in Table 2 and the algorithm with the best results is written as boldface font type.

Table 2 Comparison of ABC, PSO and TSA on solving PVD problem

Metrics	ABC	PSO	TSA
Best Cost	7201.68	7197.98	7197.99
Worst Cost	7287.61	7205.49	7203.84
Mean Cost	7229.42	7198.27	7198.75
Std.Dev	21.72	1.37	0.77
x1 of Best Cost	1.125	1.125	1.125
x2 of Best Cost	0.625	0.625	0.625
x3 of Best Cost	58.2379413565	58.2909762512	58.2907205752
x4 of best Cost	43.9905728518	43.6881425550	43.6897067538

In the comparisons, it is shown that the methods produce very promising and comparable results. Based on the mean and best solutions, the PSO algorithm and TSA are better than the ABC and PSO is slightly better than TSA. According to the standard deviations, TSA and PSO algorithm are better than ABC and TSA is slightly better than the PSO algorithm.

6 Results and Discussion

Experimental studies show that all the methods are successful in solving PVD problem. It is shown that the performances of TSA and PSO algorithm are better than the performance of ABC because all the design parameters of a solution in TSA and PSO at the each iteration are updated. But in ABC algorithm, only one dimension of a solution is updated at the each iteration. Therefore, iteration time (4.00E+4 maximum number of function evaluation) is not seen enough for ABC on solving PVD problem. Due to the fact that TSA and ABC algorithm are designed to overcome some characteristics (such as multimodality) of the optimization problems, there is randomness in search strategies of these algorithms. All the agents in these algorithms are not directly affected by the best solution in the population. But PSO algorithm uses the best solution in the population to effectively search the solution space. Therefore, different algorithms can be successful in solving the optimization problems by depending on the characteristics of the optimization problem. For this problem, PSO algorithm is better than ABC algorithm but slightly better than TSA algorithm because both all the design parameters of a solution are updated in TSA algorithm at the each iteration and TSA algorithm uses the best solution in the stand by using a control parameter (ST) to search solution space.

7 Conclusion and Future Works

A well-known constrained benchmark problem (Pressure Vessel Design Problem) is studied in this paper. For solving this problem, a newly developed optimizer (TSA) is proposed and the effects of its parameters to the performance are investigated. The performance of TSA is also compared with the performance of PSO and ABC algorithms. Based on the experimental study, it is shown that TSA algorithm is an alternative approach to solve constrained optimization. In near future, the TSA will be applied to solve a large set of constrained benchmark problems and its performance will be compared with the swarm intelligence algorithms and evolutionary computation methods.

Acknowledgements The author wishes to thank the Coordinatorship of Scientific Research Projects at Selcuk University for institutional supports.

References

1. Dorigo, M., Maniezzo, V., Colorni, A.: Ant system: Optimization by a colony of cooperating agents. IEEE Transactions on Systems Man and Cybernetics Part B-Cybernetics **26**(1), 29–41 (1996)
2. Kennedy, J., Eberhart, R.: Particle swarm optimization. In: IEEE International Conference on Neural Networks Proceedings, vol. 1-6, pp. 1942–1948 (1995)
3. Karaboga, D., Basturk, B.: A powerful and efficient algorithm for numerical function optimization: artificial bee colony (ABC) algorithm. Journal of Global Optimization **39**(3), 459–471 (2007)
4. Rashedi, E., Nezamabadi-Pour, H., Saryazdi, S.: GSA: A Gravitational Search Algorithm. Information Sciences **179**(13), 2232–2248 (2009)
5. Kiran, M.S., Hakli, H., Gunduz, M., Uguz, H.: Artificial bee colony algorithm with variable search strategy for continuous optimization. Information Sciences **300**, 140–157 (2015)
6. Gao, W.F., Liu, S.Y., Huang, L.L.: A Novel Artificial Bee Colony Algorithm Based on Modified Search Equation and Orthogonal Learning. IEEE Transactions on Cybernetics **43**(3), 1011–1024 (2013)
7. Zhu, G.P., Kwong, S.: Gbest-guided artificial bee colony algorithm for numerical function optimization. Applied Mathematics and Computation **217**(7), 3166–3173 (2010)
8. Karaboga, D., Gorkemli, B., Ozturk, C., Karaboga, N.: A comprehensive survey: artificial bee colony (ABC) algorithm and applications. Artificial Intelligence Review **42**(1), 21–57 (2014)
9. Alam, S., Dobbie, G., Koh, Y.S., Riddle, P., Rehman, S.U.: Research on particle swarm optimization based clustering: A systematic review of literature and techniques. Swarm and Evolutionary Computation **17**, 1–13 (2014)
10. Jordehi, A.R., Jasni, J.: Parameter selection in particle swarm optimisation: a survey. Journal of Experimental & Theoretical Artificial Intelligence **25**(4), 527–542 (2013)
11. Kameyama, K.: Particle Swarm Optimization - A Survey. IEICE Transactions on Information and Systems **E92d**(7), 1354–1361 (2009)

12. Janacik, P., Orfanus, D., Wilke, A.: A survey of ant colony optimization-based approaches to routing in computer networks. In: Fourth International Conference on Intelligent Systems, Modelling and Simulation (ISMS 2013), pp. 427–432 (2013)
13. Mohan, B.C., Baskaran, R.: Survey on Recent Research and Implementation of Ant Colony Optimization in Various Engineering Applications. International Journal of Computational Intelligence Systems **4**(4), 566–582 (2011)
14. Mohan, B.C., Baskaran, R.: A survey: Ant Colony Optimization based recent research and implementation on several engineering domain. Expert Systems with Applications **39**(4), 4618–4627 (2012)
15. Kiran, M.S.: TSA: Tree-Seed Algorithm for Continuous Optimization. Expert Systems with Applications **42**(19), 13 (2015)
16. Cai, J.B., Thierauf, G.: Evolution strategies in engineering optimization. Engineering Optimization **29**(1–4), 177–199 (1997)
17. Chickermane, H., Gea, H.C.: Structural optimization using a new local approximation method. International Journal for Numerical Methods in Engineering **39**(5), 829–846 (1996)
18. Coelho, L.D.: Gaussian quantum-behaved particle swarm optimization approaches for constrained engineering design problems. Expert Systems with Applications **37**(2), 1676–1683 (2010)
19. Coello, C.A.C.: Use of a self-adaptive penalty approach for engineering optimization problems. Computers in Industry **41**(2), 113–127 (2000)
20. Gandomi, A.H., Yang, X.S., Alavi, A.H.: Cuckoo search algorithm: a metaheuristic approach to solve structural optimization problems. Engineering with Computers **29**(1), 17–35 (2013)
21. He, S., Prempain, E., Wu, Q.H.: An improved particle swarm optimizer for mechanical design optimization problems. Engineering Optimization **36**(5), 585–605 (2004)
22. Garg, H.: Solving structural engineering design optimization problems using an artificial bee colony algorithm. Journal of Industrial and Management Optimization **10**(3), 18 (2014)
23. Onwubolu, G.C., Babu, B.V.: New Optimization Techniques in Engineering. Springer, Berlin (2004)

Utilization of Bat Algorithm for Solving Uncapacitated Facility Location Problem

İsmail Babaoğlu

Abstract The uncapacitated facility location problem (UFLP) is a location-based binary optimization problem investigated by using various methods in the literature. This study demonstrates a solution methodology for UFLP by a binary version of a novel swarm intelligence method namely bat algorithm (BA). BA is an optimization method employed for solving continuous optimization problems in the literature, suggested by inspiring the echolocation of microbats in nature. As implemented within some studies, sigmoid function is used in BA in order to obtain binary version of the algorithm (BBA) in this study, and then BBA is used for solving UFLP. According to the experimental results, BBA acquires successful results for solving UFLP in terms of solution quality.

Keywords Binary bat algorithm · Uncapacitated facility location problem · Binary optimization · Swarm intelligence

1 Introduction

With growing phenomena, the population-based metaheuristic iterative methods have been used in solving many real world optimization problems by simulating biological or physical systems as in nature. This inspiration has huge positive effects on solving various kinds of problems as seen from literature. Most of the nonlinear optimization problems have been investigated by using these population-based methods like particle swarm optimization (PSO), ant colony optimization (ACO), artificial bee colony algorithm (ABC), gravitational search algorithm (GSA) and bat algorithm (BA). PSO simulates the social behaviors of birds or fishes [1], and ACO algorithm has been developed by inspiring the behavior of real ants between the nest and the hive [2]. The ABC algorithm has

İ. Babaoğlu(✉)
Department of Computer Engineering, Faculty of Engineering, Selcuk University, Konya, Turkey
e-mail: ibabaoglu@selcuk.edu.tr

K. Lavangnananda et al. (eds.), *Intelligent and Evolutionary Systems*,
Proceedings in Adaptation, Learning and Optimization 5,
DOI: 10.1007/978-3-319-27000-5_16

been proposed by simulating intelligent behavior of the honey bees [3]. Besides, the GSA has been suggested for optimizing nonlinear optimization problems by modeling the Newtonian gravitation rules [4]. One of the recent swarm intelligence algorithms is BA which is proposed by echolocation behavior of microbats [5]. BA has been utilized in various applications and on solving various problems. Bora et al. used BA to solve the brushless DC wheel motor problems [6], and also Premkumar et al. used BA optimized adaptive neuro-fuzzy inference system for speed control of brushless DC motor [7]. In a different field, Ye et al. suggested fuzzy entropy-based thresholding for image segmentation using BA [8], and Zhang et al. utilized mutation based BA for image matching [9]. In another study, BA was used for solving constrained optimization problems in [10].

Although BA was early suggested for solving continuous optimization problems, it is also modified and enhanced for solving binary or discrete problems by the researchers [11]. This paper presents an implementation of binary version of BA (BBA) for solving the uncapacitated facility location problem (UFLP), and its performance is compared with the performances of some swarm intelligence optimization algorithms.

The rest of paper is organized as follows: Section 1 introduces the study, and the continuous and binary BAs are detailed in Section 2. The mathematical model of UFLP is given in Section 3, and Section 4 presents the experimental study and results. Finally, the study is concluded in Section 5.

2 BAT Algorithm

In order to give an analysis of the suggested approach, BA, its formulation and pseudo code are presented in this section. After the description of BA, binary version of BA namely BBA is also presented.

2.1 The Basic Version of Bat Algorithm

BA is a novel optimization algorithm in swarm intelligence concept. The algorithm was suggested by Yang [5] and developed by inspiring echolocation capability of microbats. The prime behavior of the microbats during movement to their prey is using echolocation even in complete darkness. Echolocation, also called as bio sonar, is defined as emitting very loud sound pulses and listen the echo bounces of these sounds in order to find prey, avoid obstacles and locate their roosting crevices.

The loudness varies from the loudest while searching to quieter as the bats move towards to prey. According to the studies [12], a microbat detects the 3D environment via the time delay, which occurs in emission of echoes, and time difference, which occurs in loudness variations of the echoes between self-ears.

With respect to these behaviors of microbats BA was formulated in order to find optimal solutions in solution space. For generalization of the BA algorithm, three rules must be considered [5];

1. Bats utilize echolocation in order to detect distance, and also know the difference between prey and obstacles.
2. Bats fly randomly during their way to prey having velocity v_i at position x_i with a fixed frequency f_{min}, varying wavelength λ and loudness A_0. They can adjust the wavelength or frequency of the emitted pulses, the rate of pulse emission r depending on the distance to the prey.
3. The loudness is accepted to be decreasing from a large positive value A_0 to a minimum constant value A_{min}.

Another espousal is to use frequency f within range $[f_{min}\ f_{max}]$ which corresponds to a range of wavelengths $[\lambda_{min}\ \lambda_{max}]$. Because λf is constant (Eq. 1), f can be used varying by fixing the wavelength λ.

$$\lambda = \frac{v}{f} \tag{1}$$

where v is velocity of sound which is typically 340m/s in air.

According to these definitions, the position update is formulated [5]. The positions x_i are accepted as solutions, and velocities v_i are used for updating the positions. The new positions x_i^t and velocities v_i^t at time t can be given as

$$f_i = f_{max} + (f_{max} - f_{min})\beta \tag{2}$$

$$v_i^t = v_i^{t-1} + (x_i^t - x_b)f_i \tag{3}$$

$$x_i^t = x_i^{t-1} + v_i^t \tag{4}$$

where f_i is the frequency of ith bat, f_{min} and f_{max} are minimum and maximum frequency values, $\beta \in [0,1]$ is a random vector generated by uniform distribution and x_b is the global best location among all bats at time t. Because the product $\lambda_i f_i$ is the velocity increment, f_i can be used by fixing λ_i. Initially, frequency of each bat is assigned randomly within a range $[f_{min}\ f_{max}]$.

Candidate solutions related to each bat are obtained by random walk after the selection of global best solution at the local search process. The candidate solution generation is implemented by following equation;

$$x_c = x_i^t + \epsilon A^t \tag{5}$$

where ϵ is a random number within a range $[-1,1]$, A^t is the average loudness of all bats at time t.

During the iteration process, loudness A_i is usually decreased and pulse emission rate r_i is increased while a bat gets closer to its prey. The update formulation of loudness and pulse emission rate are implemented as;

$$A_i^{t+1} = \alpha A_i^t\ , r_i^{t+1} = r_i^0[1 - exp(-\lambda t)] \tag{6}$$

where α and λ are constants. Each bat has different values of loudness and pulse emission rate given randomized initially. As bats get closer to its prey these values are updated. In other words, loudness and pulse emission rate of each bat are updated only if the solutions are improved and solutions get closer to the optimal solution. According to the definitions, pseudo code of the BA is given in Figure 1.

```
Define objective function f(x), x=(x^1,..x^d)
Initialize population x_i and v_i (i=1..n)
Calculate objective values of the population
Define pulse frequency f_i at x_i
Initialize loudness r_i and pulse emission rates A_i
While t<maximum number of iterations do
   For each bat (solution) x_i do
   Generate candidate solution x_c using Eq.2, Eq.3 and Eq.4
   (Generate candidate solution using frequency, current solution
   and best solution)
   If rand >r_i do
   Generate candidate solution x_c using Eq.4
   (Generate candidate solution around best solution)
   End if

   If rand <A_i and f(x_i)<f(x_c) do
   Update solution x_i, (x_i= x_c)
   Update A_i and r_i using Eq.6 (increase A_i and decrease r_i)
   End if

   End for
   Calculate objective values of all solutions
   Find the current best solution
   End while
```

Fig. 1 Pseudo code of BA

2.2 *Binary Version of Bat Algorithm*

With some modifications or adoptions for development binary versions of concerned algorithms, many studies have been carried out [13-16]. In order to obtain binary version of BA, sigmoid transfer function was used as in binary PSO [17] and also suggested by Nakamura et al. [11] for BA.

The binary version of PSO (BPSO) updates the positions of the particles according to Eq. 7 and Eq. 8 [17].

$$x_{ij}(t+1) = \begin{cases} 0, & rand \geq S(v_{ij}(t)) \\ 1, & rand < S(v_{ij}(t)) \end{cases} \tag{7}$$

$$S(v_{ij}(t)) = \frac{1}{1+e^{-v_{ij}(t)}} \tag{8}$$

where x_{ij} is j^{th} position of ith particle, *rand* is a random value sampled from a uniform distribution within a range [0,1], *S* is sigmoid function, v_{ij} is the velocity of the jth position of ith particle.

The binary version of BA is obtained by using sigmoid function given in Eq. 8. According to this assumption, Eq. 4 in BA can be replaced by

$$x_{ij}^{t+1} = \begin{cases} 0, & rand \geq S(v_{ij}^{t}) \\ 1, & rand < S(v_{ij}^{t}) \end{cases} \tag{9}$$

where *rand* is a random value sampled from a uniform distribution within a range [0,1].

3 The Uncapacitated Facility Location Problem

Being a discrete facility location problem, the UFLP involves the location of an undetermined number of facilities for minimizing the fixed setup costs and variable costs of serving the market demand from these facilities. This discrete economical decision problem is one of the most fundamental problems in location theory, and it is also known as simple plant or simple warehouse location problem [18-20].

The UFLP in a simple form can be represented as follows [21, 16];

$$\text{Min} \sum_{i=1}^{m} \sum_{j=1}^{n} c_{ij} x_{ij} + \sum_{j=1}^{n} f_j y_j \tag{10}$$

$$\text{s.t.} \quad \sum_{j=1}^{m} x_{ij} = 1 \ , \quad i = 1, \dots, m \tag{11}$$

$$x_{ij} \leq y_j \ , \ i = 1, \dots, m \ \text{and} \ j = 1, \dots, n \tag{12}$$

$$x_{ij} \geq 0 \ , \ i = 1, \dots, m \ \text{and} \ j = 1, \dots, n \tag{13}$$

$$x_{ij} \in \{0,1\} \ , \ i = 1, \dots, m \ \text{and} \ j = 1, \dots, n \tag{14}$$

$$y_j \in \{0,1\} \ , \ j = 1, \dots, n \tag{15}$$

where $i=1,\dots,m$ stands for customer demand points, $j=1,\dots,n$ stands for possible facility locations, c_{ij} is the cost for supplying customer at point *i* from the facility location *j*, x_{ij} is 1 if customer *i* is assigned to facility location *j*, and otherwise 0, f_j stands for fixed cost for opening facility *j*, y_j is 1 if facility *j* is opened, and otherwise 0. Eq.10 ensures to satisfy all customer demands, and Eq.11 provides to supply customer *i* only if the facility *j* is opened.

The main aim of UFLP is to obtain the optimum solution which corresponds to all customer demand and minimizes the total cost. Many studies were presented in literature for solving UFLP [20]. The earlier studies such as development of branch-and-bound procedures [22], branch-and-bound procedures using linear programming relaxation [23] and Lagrangian relaxation [24] are presented for conventional solutions for UFLP. Besides, some other studies such as using genetic algorithms [25], tabu search [26,27], continuous and discrete particle swarm optimization algorithm [28] and continuous artificial bee colony algorithm [16] for solving UFLP were presented in last decades.

This study demonstrates a utilization of BBA for solving UFLP. Each bat represents a possible solution in UFLP in the proposed approach. Objective function which calculates gap is generated using Eq. 10 – Eq. 15, and the gap value which is obtained by Eq. 16 is used for a clearly demonstration of the results of the methods.

$$\text{gap} = \frac{f_{sol} - f_{opt}}{f_{opt}} \times 100 \quad (16)$$

4 Experimental Results and Discussion

The performance of the proposed approach is examined on 15 test problems taken from OR-library [29], and descriptions of these UFLPs are given in Table 1. *UFLP name* stands for the common name of the problem, *Problem size* stands for number of facility locations and the number of the customer demand points and *Cost of optimal solution* stands for the cost of the optimal solution of the problem in Table 1. In the proposed approach, the aim is to find the number of opened facilities with minimum cost by corresponding demands of the customers.

According to the given UFLPs, BBA is implemented in order to minimize the cost by using Eq. 10. The number of the population size is used as 40 and the number of the generations is employed as 2000. Therefore, total function evaluation is equal to 80000. For each UFLP, the number of positions of the bats is used equal to the value given in *Problem size* column of Table 1. The loudness and pulse rate limits are utilized as 0.9 and 0.9, and constant values α and λ which are used for updating loudness and pulse rate limits are chosen as 0.9 and 0.9, respectively. The maximum and minimum values of the frequency are chosen as 1 and 0, respectively. In order to achieve more robust results, the suggested approach was run 30 times for each UFLP with random seeds. According to Eq. 16, the obtained mean results of 30 runs are reported in Table 2.

For presenting a clear analysis of the results of the proposed approach, results of the suggested approach are compared with results of some binary structured optimization algorithms which were used for solving UFLP having the same number of function evaluations and stopping criteria, and the results of the compared algorithms are directly taken from [16]. The comparative results are given in Table 2 where the better results are written in bold font type. BPSO column stands for results of the binary particle swarm optimization algorithm, IBPSO column stands for results of the improved version of binary particle swarm optimization algorithm, DisABC column stands for results of the discrete artificial bee colony algorithm, ABC_{bin} column stands for results of the binary artificial bee colony algorithm and BBA column stands for results of the binary bat algorithm in Table 2.

Table 1 Description of the UFLP dataset

UFLP name	Problem size	Cost of optimal solution
Cap71	16×50	932,615.75
Cap72	16×50	977,799.40
Cap73	16×50	1,010,641.98
Cap74	16×50	1,034,976.98
Cap101	25×50	796,648.44
Cap102	25×50	854,704.20
Cap103	25×50	893,782.11
Cap104	25×50	928,941.75
Cap131	50×50	793,439.56
Cap132	50×50	851,495.33
Cap133	50×50	893,076.71
Cap134	50×50	928,941.75
CapA	100×1000	17,156,454.48
CapB	100×1000	12,979,071.58
CapC	100×1000	11,505,594.33

Table 2 Comparison of BBA with some binary structured optimization algorithms

UFLP name	BPSO	IBPSO	DisABC	ABC_{bin}	BBA
Cap71	**0.0000**	0.0374	**0.0000**	**0.0000**	**0.0000**
Cap72	**0.0000**	0.2749	**0.0000**	**0.0000**	**0.0000**
Cap73	0.0242	0.198	**0.0000**	**0.0000**	**0.0000**
Cap74	0.0088	0.4031	**0.0000**	**0.0000**	**0.0000**
Cap101	0.0462	0.5968	**0.0000**	**0.0000**	0.0359
Cap102	0.0148	0.7317	**0.0000**	**0.0000**	**0.0000**
Cap103	0.0422	0.6410	**0.0000**	0.0051	0.0582
Cap104	0.0810	0.9964	**0.0000**	**0.0000**	**0.0000**
Cap131	**0.1317**	2.4236	0.6196	0.1967	0.2132
Cap132	0.0914	3.6014	0.0945	**0.0199**	0.0914
Cap133	**0.1115**	5.2626	0.0309	0.0747	0.1349
Cap134	0.1346	7.6338	**0.0000**	**0.0000**	0.0501
CapA	2.1785	37.8862	**0.1522**	3.1723	8.4934
CapB	**1.9490**	55.2701	3.3027	2.8154	4.3459
CapC	**1.4870**	45.5561	4.6968	2.0374	4.2456

According to the comparison table, it can be obviously seen that BBA acquires more successful results than both BPSO and IBPSO in most cases. Besides, BBA has successful results which are almost equal to the results of both DisABC and ABC_{bin} algorithms.

For high dimensional UFLPs, the performance of the BPSO is better than the other methods because the particles in BPSO algorithm is affected global best and personal best solution at the each iteration and all the dimension of the problem is updated. Therefore, this behavior is useful for solving the high dimensional problems in this study. On the other hand, BPSO is ineffective for UFLPs when the dimension of the problem is low because the perturbation in BPSO algorithm is high for low dimensional problems.

Even though BBA has comparable results with both DisABC and ABC_{bin}, future work includes parameter optimization of BBA and utilization of some hybridization techniques which are implemented for PSO in order to obtain better results on binary optimization problems.

5 Conclusion

This study presents a utilization of BBA for solving UFLP. The performance of the proposed approach is investigated on 15 pure binary test problems taken from OR-library. Obtained results are compared with the results of some known swarm intelligence optimization algorithms. It can be seen from the comparisons that BBA can be effectively and successfully usable in solving UFLPs. The results also indicate that the proposed approach is an alternative optimizer for binary optimization. As seen from the experimental results, the performance of the BBA depends on the dimensionality of the problem. To overcome this issue, future work includes parameter optimization of BA and utilization of some hybridization techniques which are implemented for PSO for obtaining better results.

Acknowledgements The author wishes to thank the Coordinatorship of Scientific Research Projects at Selcuk University for institutional supports.

References

1. Kennedy, J., Eberhart, R.C.: Particle swarm optimization. In: IEEE International Conference on Neural Networks, vol. 4, pp. 1942–1948 (1995)
2. Dorigo, M., Gambardella, L.M.: Ant Colony System: A Cooperative Learning Approach to the Traveling Salesman Problem. IEEE Transactions on Evolutionary Computation **1**, 53–66 (1997)
3. Karaboga, D.: An idea based on honey bee swarm for numerical optimization. Technical Report, TR06 Erciyes University, Engineering Faculty, Department of Computer Engineering (2005)

4. Rashedi, E., Nezamabadi-Pour, H., Saryazdi, S.: GSA: A Gravitational Search Algorithm. Information Sciences **179**(13), 2232–2248 (2009)
5. Yang, X.-S.: A new metaheuristic bat-inspired algorithm. In: Gonzalez, J.R., et al. (eds.) Nature Inspired Cooperative Strategies for Optimization (NISCO 2010), vol. 284, pp. 65–74. Studies in Computational Intelligence. Springer (2010)
6. Bora, T., Coelho, L., Lebensztajn, L.: Bat-inspired optimization approach for the brushless DC wheel motor problem. IEEE Trans. Magnet. **48**, 947–950 (2012)
7. Premkumar, K., Manikandan, B.V.: Speed control of Brushless DC motor using bat algorithm optimized Adaptive Neuro-Fuzzy Inference System. Applied Soft Computing **32**, 403–419 (2015)
8. Ye, Z.-W., Wang, M.-W., Liu, W., Chen, S.-B.: Fuzzy entropy based optimal thresholding using bat algorithm. Applied Soft Computing **31**, 381–395 (2015)
9. Zang, J., Wang, G.: Image matching using a bat algorithm with mutation. Applied Mechanics and Materials **203**, 88–93 (2012)
10. Gandomi, A.H., Yang, X.-S., Alavi, A.H., Talatahari, S.: Bat Algorithm for Constrained optimization tasks. Neural Computing and Applications **22**, 1239–1255 (2013)
11. Nakamura, R.Y.M., Pereira, L.A.M., Costa, K.A., Rodrigues, D., Papa, J.P.: BBA: a binary bat algorithm for feature selection. In: XXV SIBGRAPI Conference on Graphics, Patterns and Images, pp. 291–297 (2012)
12. Richardson, P.: Bats. Natural History Museum, London (2008)
13. Rashedi, E., Nezamabadi-Pour, H., Saryazdi, S.: BGSA: binary gravitational search algorithm. Natural Computing **9**, 727–745 (2010)
14. Banati, H., Bajaj, M.: Fire Fly Based Feature Selection Approach. International Journal of Computer Science Issues **8**(4), 473–480 (2011)
15. Kashan, M.H., Nahavandi, N., Kashan, A.H.: DisABC: a new artificial bee colony algorithm for binary optimization. Applied Soft Computing **12**(1), 342–352 (2012)
16. Kiran, M.S.: The continuous artificial bee colony algorithm for binary optimization. Applied Soft Computing **33**, 15–23 (2015)
17. Kennedy, J., Eberhart, R.C.: A discrete binary version of the particle swarm algorithm. In: IEEE International Conference on Systems, Man, and Cybernetics, vol. 5, pp. 4104–4108 (1997)
18. Verter, V.: Foundations of location analysis, uncapacitated and capacitated facility location problems. In: Eiselt, H.A., Marianov, V. (eds.) International Series in Operations Research & Management Science, pp. 25–37. Springer Science (2011)
19. Beltran-Royo, C., Vial, J.P., Alonso-Ayuso, A.: Semi-Lagrangian relaxation applied to the uncapacitated facility location problem. Computational Optimization and Applications **51**(1), 387–409 (2012)
20. Galvâo, R.D., Raggi, L.A.: A method for solving to optimality uncapacitated location problems. Annual Operations Research **18**(1), 225–244 (1989)
21. Ardjmand, E., Park, N., Weckman, G., Amin-Naseri, M.R.: The discrete Unconscious search and its application to uncapacitated facility location problem. Computers & Industrial Engineering **73**, 32–40 (2014)
22. Efroymson, M.A., Ray, T.L.: A branch-bound algorithm for plant location. Operational Research **14**, 361 (1966)
23. Holmberg, K.: Exact solution methods for uncapacitated location problems with convex transportation costs. European Journal of Operational Research **114**(1), 127–140 (1999)

24. Barcelo, J., Hallefjord, A., Fernandez, E., Jörnsten, K.: Lagrangian relaxation and constraint generation procedures for capacitated plant location problems with single sourcing. Operations Research Spektrum **12**(2), 78–79 (1990)
25. Jaramillo, J.H., Bhadury, J., Batta, R.: On the use of genetic algorithms to solve location problems. Computers & Operations Research **29**(6), 761–779 (2002)
26. Al-Sultan, K.S., Al-Fawzan, M.A.: A tabu search approach to the uncapacitated facility location problem. Annual Operations Research **86**, 91–103 (1999)
27. Sun, M.H.: Solving the uncapacitated facility location problem using tabu search. Computers & Operations Research **33**(9), 2563–2589 (2006)
28. Güner, A.R., Şevkli, M.: A discrete particle swarm optimization algorithm for uncapacitated facility location problem. J. Artif. Evol. Appl., 1–9 (2008)
29. Beasley, J.E.: OR-library – distributing test problems by electronic mail. J. Oper. Res. Soc. **41**(11), 1069–1072 (1990)

Implementation of Bat Algorithm on 2D Strip Packing Problem

Ahmet Babalik

Abstract This paper suggests utilization of a novel metaheuristic method namely bat algorithm (BA) in order to solve 2D rectangular strip packing problem. Although BA is proposed for solving continuous optimization problems, a discrete version of BA is developed by being used neighborhood operators to solve the problem dealt with this study. Firstly, bottom left approach is used as the placement algorithm in the problem, then, discrete BA is used for obtaining the proper sequence of the rectangular object list. The performance of the proposed approach is investigated on 9 different problems on well-known 2D rectangular problem literature. Experimental results show that discrete BA is effective and alternatively usable in solving 2D rectangular strip packing problems.

Keywords Bat algorithm · 2D strip packing problem · Discrete optimization

1 Introduction

A packing problem is a combinatorial optimization problem encountered in various forms such as wood, glass, paper and textile packing and container loading. Being a NP-hard problem, packing problem can be classified into different forms as one, two and three dimension considering dimension and rotatable and non-rotatable considering orientation. 2D strip packing problem (2DSPP) considers placing N objects into a space which has a fixed width and infinite length without overlapping and with minimum height [1]. There are many different solution techniques proposed for solving 2DSPP such as bottom left (BL), bottom left fill (BLF) and best fit (BF) in literature. Usually, achievement of these methods depends on a proper sequence of rectangles.

There are various placement heuristics for solving 2DSPP. One of the well-known is BL algorithm proposed by Baker et al [2]. BLF heuristic, the improved version of BL, which suggests placement of the new objects to the gaps occurred during the placement process is developed by Chazelle [3]. Burke et al. suggested

A. Babalik(✉)
Department of Computer Engineering, Faculty of Engineering,
Selcuk University, Konya, Turkey
e-mail: ababalik@selcuk.edu.tr

K. Lavangnananda et al. (eds.), *Intelligent and Evolutionary Systems*,
Proceedings in Adaptation, Learning and Optimization 5,
DOI: 10.1007/978-3-319-27000-5_17

a new placement heuristic called as best fit (BF) in which the candidate objects could be dynamically selectable from the object list [4].

As well as researchers develop many heuristic placement algorithms as mentioned above, metaheuristic algorithms were combined with these heuristics or new metaheuristic methods were developed by researchers. Many metaheuristic methods like genetic algorithm and particle swarm optimization are suggested for obtaining the proper sequence [4,5]. Improved BL algorithm was developed by Liu and Teng in order to use with genetic algorithm for orthogonal packing of rectangles [6]. Hopper and Turton suggested a new placement approach and described two variants of genetic algorithm for 2D packing problems [7], and hybridized two heuristics with three metaheuristic methods in a different study [8]. One another metaheuristic algorithm which was based on heuristic recursive strategy and simulated annealing algorithm was presented for solving 2D strip rectangular packing problem by Zhang at al. [9]. Leung et al. applied simulated annealing and genetic algorithm metaheuristics by hybridizing them for solving 2D packing problem [10], and Leo et al. suggested combination of a novel allocation method and genetic algorithm for solving 2D packing problem [11].

Because metaheuristic methods have better search capability in solution space, many other metaheuristic methods can be used in 2DSPP. One of the novel metaheuristic algorithms in literature is bat algorithm (BA) which was developed by inspiring from the echolocation behavior of the microbats during flight to their prey. Although BA is a novel algorithm, it is used in many optimization problems in literature. Hasancebi et al. used BA for structural optimization [12]. Prekumar and Manikandan implemented speed control of brushless DC motor using BA optimized adaptive neuro-fuzzy inference system [13]. Yilmaz and Kucuksille suggested improved version of BA in order to enhance its local and global search capabilities [14]. Gandomi and Yang developed chaotic BA for improving performance of BA for solving continuous optimization problems [15].

In this study, being a novel population based metaheuristic approach; BA is utilized for solving 2DSPP. This is the first study that implements BA for solving 2DSPP considering the literature search.

The rest of the paper is organized as follows; 2D strip packing problem is briefly explained in section 2, BA is explained in section 3, proposed method is presented in section 4, experimental results are given and discussed in section 5, and conclusion is given in section 6.

2 2D Strip Packing Problem

In the literature, authors usually deal with two objectives in 2DSPP's. One of the objectives is to minimize the packing height and the other is to obtain minimum gap size between objects within the package. Minimizing the packing height is the prime aim of this study, and mathematical model of 2DSPP is given as follows [1];

Minimize H

Subject to

$$x_i + w_i \leq W, \quad \forall i \in N \tag{1}$$

$$y_i + h_i \leq H, \quad \forall i \in N \tag{2}$$

$$x_i + w_i \leq x_j \ or \ x_j + w_j \leq x_i \ or \tag{3}$$

$$y_i + h_i \leq y_j \ or \ y_j + h_j \leq y_i, \quad \forall (i,j) \in N, \ i \neq j \tag{4}$$

$$x_i + y_i \geq 0, \ \forall i \in N \tag{5}$$

where N is the rectangular problem set which includes n rectangles inside, H is the height of the packaged rectangles, W is the width of the package and it's a fixed value, i is the rectangles indices and $\forall i = 1,..,n$, h_i is the height of the ith rectangle, w_i is the width of the ith rectangle, x_i, y_i are the coordinates of the bottom left corner of rectangles.

There are some heuristic placement strategies implemented by the authors which can be seen from literature. Most common of these are BL, BLF and BF. This study is focused on BL placement strategy, and the definition of BL is given below.

Bottom left (BL) algorithm is heuristic placement strategy developed Baker et al [2]. In BL algorithm, each rectangle is placed upper right corner on the board and shift down as low as possible then shift left as far as possible to the left position on the board [2]. A brief illustration of BL algorithm is given in Figure 1.

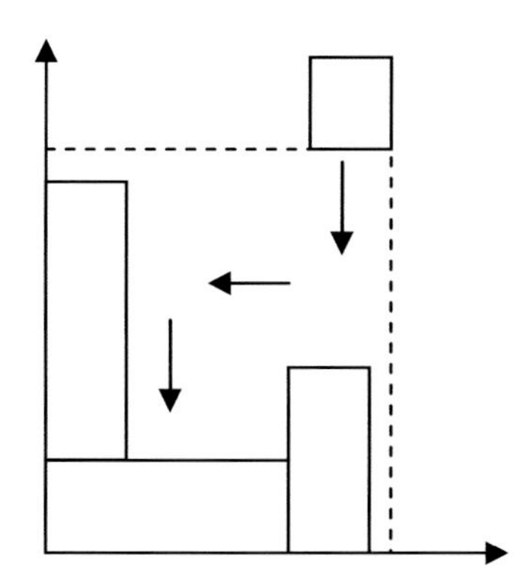

Fig. 1 BL heuristic

3 BAT Algorithm

BA is a novel nature-inspired metaheuristic algorithm based on echolocation behavior of bats, which is developed by Yang [16]. Bats have advanced capability of echolocation which is a type of sonar. Microbats usually use echolocation to detect objects around them, and they also produce three dimensional images of the environment. Microbats use this information for prey and navigation. Microbats emit loud sound pulse and listen to the echo which returns from obstacles. In other words, microbats obtain three dimensional map of the surrounding by means of the time delay for this signal between their two ears. These pulses properties are varying depend on species and hunting strategies [16].

Microbats emit short pulses throughout flying. If a probable prey is nearby, their pulse emitting rate increase, and so the frequency of sound emitted from the bats increases, and the wavelength of emitted sounds is shortened. Actually, bat algorithm use this idea of frequency tuning .

The bat algorithm tries to simulate the ability of microbats during their way on finding prey. Each bat represents a possible solution in BA which is formulated as positions, and each bat also has a frequency and velocity for generating new solutions and updating the positions. BA assumes following three rules [16]:

- All bats use echolocation to sense disance, and they also "know" the difference between food/prey and background barriers in some magical way;
- Bats fly randomly with velocity v_i at position x_i with a fixed frequency f_{min}, varying wawelength λ and loundness A_0 to search their prey. They can automatically adjust the wavelength (or frequency) of their emitted pulses and adjust the rate of pulse emission $r \in [0, 1]$, depending on the proximity of their target;
- It is assumed that, the loudness varies from a large positive A_0 to a minimum constant value $\tilde{A}_{min}$

Pseudo code of the bat algorithm is shown in Figure 2 [16].

```
Bat Algorithm
Initialize the bat population x_i and v_i
Define pulse frequency f_i at x_i
Initialize pulse rates r_i and the loudness A_i
while(termination condition is not met)
Generate new solutions by adjusting frequency [Eq.(6)]
Update velocities [Eq.(7)]
Update locations/solutions [Eq.(8)]
if(rand>r_i)
Select a solution from among the best solutions
Generate a local solution around the selected best
solution
endif
Generate a new solution by flying randomly
if(rand < A_i & f(x_i) < f(x_*))
Accept the new solutions
Increase r_i and reduce A_i
endif
Rank the bats and find the current best x_*
endwhile
```

Fig. 2 Pseudo code of the BA

According to the assumptions, the formulization of BA can be given as follows;

$$f_i = f_{min} + \beta(f_{max} - f_{min}) \tag{6}$$

$$v_i^t = v_i^{t-1} + (x_i^t - x_*)f_i \tag{7}$$

$$x_i^t = x_i^{t-1} + v_i^t \tag{8}$$

where f_i is the frequency value of ith bat, f_{min} and f_{max} are minimum and maximum frequency values, $\beta \in [0\ 1]$ is a random vector, v_i^t is velocity of ith bat, v_i^{t-1} is the velocity of ith bat in previous iteration, x_i^t is the positions/solution of each bat and x_* is the positions of the current global best bat. A new solution for each bat are calculated as given in Eq. (9) by using random walk.

$$x_{new} = x_{old} + \epsilon \overline{A}^t \tag{9}$$

where x_{new} is the new candidate solution, x_{old} is the old position of the bat, ϵ is a random number within a range [-1,1], $\overline{A}^t$ is the average of the loudness value of whole population at time t.

In each iteration, with some conditions aforementioned, pulse emission rate r_i and loundness A_i are updated. Generally, while pulse emission rates increases, loudness values decreases. Pulse emission rates and loudness values are updated as given below;

$$A_i^{t+1} = \alpha A_i^t \tag{10}$$

$$r_i^{t+1} = r_i^0[1 - exp(-\gamma t)] \tag{11}$$

where α and γ are constants, $0 < \alpha < 1$ and $\gamma > 0$.

4 Proposed Method

BA is a continuous optimization algorithm. According to the algorithm, the update rules (equations Eq. (8) and Eq. (9)) generate continuous values. Thus, basic form of BA could not be applicable to solve discrete optimization problems. To come with this issue, the neighborhood operators are used instead of update rules in order to obtain discrete form of BA for creating a feasible solution for the problem. Three different neighborhood operators are experimented in this study [17];

Random Swap (RS): This operator changes two different position of the sequence which is randomly selected. Only two position of the sequence is changed, and the demonstration is given in Figure 3.

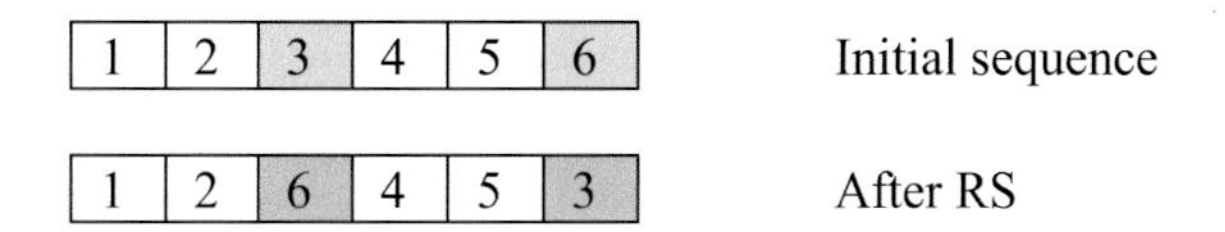

Fig. 3 Random swap function

Random Insertion (RI): According to this operator, one element is selected and inserted into a random position (repositioned) in the sequence. The rest of the elements in sequence are shifted. Demonstration of RI is given in Figure 4.

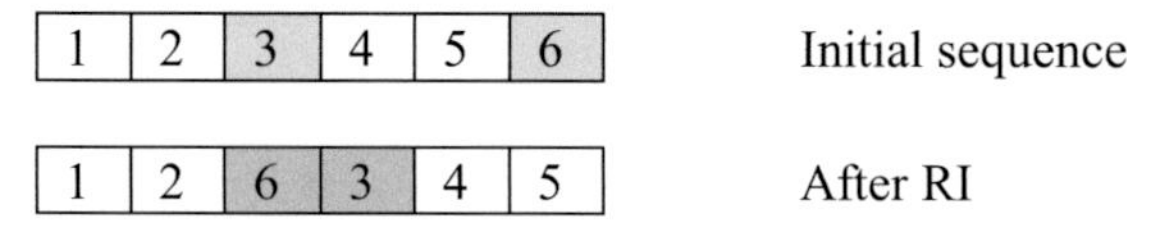

Fig. 4 Random insertion function

Random Reversing Subsequence (RRS): This operator changes the sequence by reversing the subsequence which is between randomly selected two points. Demonstration of RRS is given in Figure 5.

1	2	3	4	5	6	Initial sequence
1	5	4	3	2	6	After RRS

Fig. 5 Random reversing subsequence function

By modifying the update rules using neighborhood operators, discrete BA is obtained. In the proposed discrete BA, positions of each bat represent the sequence of rectangles. At the beginning of the discrete BA, the positions of bats are obtained by random permutation of the sequences of rectangles. And the variations in the whole iteration cycle are carried out by using mentioned neighborhood operators.

The prime aim of the proposed discrete BA is to obtain proper sequences which are used in the placement strategy for 2DRSSP. The objective value which is the fitness function of the proposed system of each bat is obtained by calculating the height of the package after packing the given sequence (bat) with bottom left approach. The height is defined as the distance of the top rectangle to the bottom of the package.

According to the definitions mentioned, the pseudo code of the proposed system can be given in Figure 6.

```
Proposed Algorithm
Initialize the bat population x_i by random permutation
Initialize pulse rates r_i and the loudness A_i
while(termination condition is not met)
Generate new solutions by using neighborhood operator (RI,
RS or RRS)
if(rand>r_i)
Select a solution from among the best solutions
Generate a local solution around the selected best
solution by using neighborhood operator (RI, RS or RRS)
endif
if(rand < A_i & f(x_i) < f(x_*))
Accept the new solutions
Increase r_i and reduce A_i
endif
Rank the bats and find the current best x_*
endwhile
```

Fig. 6 Pseudo code of the proposed algorithm

5 Results and Discussion

In this study, discrete BA is obtained by using neighborhood operators. Three different neighborhood operators are tested separately at equal circumstances. These neighborhood operators are RI, RS and RRS. In order to compare the results of the study with the results in literature [4], the population size and the iteration number is used as 50 and 1000 respectively. The loudness value and pulse emission rate are used as both equal to 0.9. The constants α and γ which are used to update the loudness value and pulse emission rate are also used as both equal to 0.9. BA is implemented to 2DSPP's with given parameters and this evaluation is run 5 times for comparison with the literature also [4].

The proposed algorithm is evaluated on nine well-known benchmark problems [8]. The description of the problems is given in Table 1. The number of the rectangles in each problem is used as the dimension of the bat in BA. According to the proposed algorithm for the given experimental problems, BL placement heuristic is used in the fitness function. The objective value of each bat is the height of the package in which the objects are packaged with the given sequence (bat) using BL. The obtained results for three different neighborhood operators are given in Table 2. Table 3 presents the comparison of the results of proposed algorithm with the results of heuristics given in [4]. Table 4 presents the comparison of the results of proposed algorithm with the results of metaheuristic algorithms given in [4], also.

Table 1 Specifications of benchmark problems [8]

Problem	Number of Rectangles	Sheet Width	Opt. Height
C1-P1	16	20	20
C1-P2	17	20	20
C1-P3	16	20	20
C2-P1	25	40	15
C2-P2	25	40	15
C2-P3	25	40	15
C3-P1	28	60	30
C3-P2	28	60	30
C3-P3	28	60	30

According to the mean results given in Table 2, it can be seen that BA-RS is slightly better mean results than mean results of the other neighborhood operator variants. Nevertheless, BA-RS achieved at least equal of better results than results of the other operator variants. According to the results given in Table 3, all of the three neighborhood operator variants of BA obtain better results than the results of the placement heuristics.

Table 2 Comparison of the performance of neighborhood operators used with BA

	BA-RRS			BA-RI			BA-RS		
Problem	Best	Worst	Mean	Best	Worst	Mean	Best	Worst	Mean
C1-P1	20	21	20,8	21	21	21	20	20	20
C1-P2	21	22	21,2	22	22	22	21	22	21,4
C1-P3	20	21	20,4	21	22	21,4	20	21	20,6
C2-P1	16	16	16	16	16	16	16	16	16
C2-P2	16	16	16	16	16	16	16	16	16
C2-P3	16	16	16	16	16	16	16	16	16
C3-P1	32	32	32	32	33	32,2	31	32	31,8
C3-P2	32	33	32,8	33	34	33,6	32	33	32,4
C3-P3	32	33	32,8	32	33	32,8	31	33	32,4

Table 3 Comparison of BA-RS with BL and BLF (% over optimal)

	C1-P1	C1-P2	C1-P3	C2-P1	C2-P2	C2-P3	C3-P1	C3-P2	C3-P3
BL	45	40	35	53	80	67	40	43	40
BL-DW	30	20	20	13	27	27	10	20	17
BL-DH	15	10	5	13	73	13	10	10	13
BLF	30	35	25	47	73	47	37	50	33
BLF-DW	10	15	15	13	20	20	10	13	13
BLF-DH	10	10	5	13	73	13	10	6,7	13
BF	5	10	20	6,7	6,7	6,7	6,7	13	10
BA-RRS	4	6	2	6,7	6,7	6,7	6,7	9,3	9,3
BA-RI	5	10	7	6,7	6,7	6,7	7,3	12	9,3
BA-RS	0	7	3	6,7	6,7	6,7	6	8	8

BL stands for Bottom Left; BL-DW stands for Bottom Left Decreasing Width; BL-DH stands for Bottom Left Decreasing Height; BLF stands for Bottom Left Fill; BLF-DW stands for Bottom Left Fill Decreasing Width; BLF-DH stands for Bottom Left Fill Decreasing Height; BF stands for Best Fit.

According to the comparison given in Table 4, it can be seen that the proposed algorithm obtains better or at least equal results than the results of GA+BLF and SA+BLF.

Table 4 Comparison of obtained results with GA+BLF and SA+BLF

	BA-RRS		BA-RI		BA-RS		GA+BLF		SA+BLF	
Problem	Best	Worst	Best	Worst	Best	Worst	Best	Worst	Best	Worst
C1-P1	20	21	21	21	20	20	20	21	20	21
C1-P2	21	22	22	22	21	22	21	21	21	21
C1-P3	20	21	21	22	20	21	20	21	20	21
C2-P1	16	16	16	16	16	16	16	16	16	16
C2-P2	16	16	16	16	16	16	16	16	16	16
C2-P3	16	16	16	16	16	16	16	16	16	16
C3-P1	32	32	32	33	31	32	32	32	32	33
C3-P2	32	33	33	34	32	33	32	32	32	32
C3-P3	32	33	32	33	31	33	32	32	32	33

6 Conclusion and Future Works

This study presents an implementation of BA with three different neighborhood operators on 2DSPP. Because 2DSPP is a combinatorial problem, discrete version of BA is obtained by utilizing neighborhood operators on the update mechanism of BA. The proposed algorithm is evaluated on 9 well-known 2DSPP. According to the results, the proposed algorithm is capable of solving 2DSPP's and obtains better or at least equal results than both results of the placement heuristics or metaheuristic algorithms when compared to [4].

BL heuristic is utilized in the suggested algorithm. It is expected that the suggested algorithm can give better results when BLF strategy is used instead of BL. This will be assessed in the future works. Bigger 2DSPP's can be investigated with the proposed algorithm in the future works. Parameters of the BA can be evaluated also.

In the study, a new discrete optimization algorithm by using BA is introduced, and a comprehensive research and comparisons based on BA and other metaheuristics for solving discrete problems will be investigated in near future.

Acknowledgments The author wishes to thank the Coordinatorship of Scientific Research Projects at Selcuk University for institutional supports.

References

1. Riff, M.C., Bonnaire, X., Neveu, B.: A revision of recent approaches for two-dimensional strip-packing problems. Engineering Applications of Artificial Intelligence **22**, 823–827 (2009)
2. Baker, B.S., Coffman, E.G., Rivest, R.L.: Orthogonal packings in two dimensions. Society for Industrial and Applied Mathematics **9**(4), 846–855 (1980)
3. Chazelle, B.: The bottom-left bin-packing heuristic: An efficient implementation. IEE Transactions on Computers **32**(8), 697–707 (1983)
4. Burke, E.K., Kendall, G., Whitwell, G.: A new placement heuristic for the Orthogonal stock-cutting problem. Operations Research **52**(4), 655–671 (2004)
5. Shalaby, M.A., Kashkoush, M.: A particle swarm optimization algorithm for a 2D irregular strip packing problem. American Journal of Operations Research **3**, 268–278 (2013)
6. Liu, D., Teng, H.: An improved BL-algorithm for genetic algorithm of the orthogonal packing of rectangles. European Journal of Operational Research **112**, 413–420 (1999)
7. Hopper, E., Turton, B.: A genetic algorithm for a 2D industrial packing problem. Computers & Industrial Engineering **37**, 375–378 (1999)
8. Hopper, E.: Turton, B.C.H: An empirical investigation of meta-heuristic and heuristic algorithms for 2D packing problem. European Journal of Operational Research **128**, 34–57 (2001)
9. Zhang, D., Liu, Y., Chen, S., Xie, X.: A meta-heuristic algorithm for the strip rectangular packing problem. LNCS, vol. 3612, pp. 1235–1241 (2005)

10. Leung, T.W., Chan, C.K., Troutt, M.D.: Application of a mixed annealing-genetic algorithm heuristic for the two-dimensional orthogonal packing problem. European Journal of Operational Research **145**, 530–542 (2003)
11. Yeung, L.H.W., Tang, W.K.S.: Strip-packing using hybrid genetic approach. Engineering Applications of Artificial Intelligence **17**, 169–177 (2004)
12. Hasançebi, O., Teke, T., Pekcan, O.: A bat-inspired algorithm for structural optimization. Computers and Structures **128**, 77–90 (2013)
13. Prekumar, K., Manikandan, B.V.: Speed control of brushless DC motor using bat algorithm optimized adaptive neuro-fuzzy inference system. Applied Soft Computing **32**, 403–419 (2015)
14. Yılmaz, S., Küçüksille, E.U.: A new modification approach on bat algorithm for solving optimization problems. Applied Soft Computing **28**, 259–275 (2015)
15. Gandomi, A.H., Yang, X.S.: Chaotic bat algorithm. Journal of Computer Science **5**, 224–232 (2014)
16. Yang, X.S.: A new metaheuristic bat-inspired algorithm. In: Gonzalez, J.R., et al. (eds.) Nature Inspired Cooperative Strategies for Optimization (NISCO 2010), vol. 284, pp. 65–74. Studies in Computational Intelligence. Springer Berlin, Springer (2010)
17. Kiran, M.S., İscan, H., Gündüz, M.: The analysis of discrete artificial bee colony algorithm with neighborhood operator on travelling salesman problem. Neural Computing and Application **23**, 9–21 (2013)

Base Hybrid Approach for TSP Based on Neural Networks and Ant Colony Optimization

Carsten Mueller and Niklas Kiehne

Abstract This research article presents a hybrid approach based on an intelligent combination of artificial ants and neurons. Research on different parameter combinations are performed, in order to find the best performing parameter settings. The obtained insights are then subsumed into an intelligent architecture consisting of Ant Colony Optimization and Self Organizing Map.

Keywords Ant colony optimization · Neural network · Self organizing map · Hybrid approach · Traveling salesman problem

1 Introduction

1.1 Neural Networks and Self Organizing Maps

In biological neural networks the observed topology of neurons is often planar, whereas the input is of multiple dimensions [1, 2]. Neuron topologies do not map the exact input, but rather map the phase space of it. In result, close neurons process those stimuli which are similar. This behaviour is made use of in a special type of artificial neural networks (ANN), called Self Organizing Map (SOM) [3].

The application of the biological inspiration is to have a layer of n interconnected artificial neurons that represent the map. Each neuron is associated with a weight vector w_i of the same dimensions as the expected input and a location l_i, typically in the Euclidean plane. The location is used to model the topology of the net. The weights are initialized to either random values or close representations of the expected inputs. The goal is to train the net to respond to similar input vectors within the same

C. Mueller(✉) · N. Kiehne
Department of Information Technologies, Faculty of Informatics and Statistics, University of Economics, W. Churchill Sq. 4, 130 67 Prague 3, Czech Republic
e-mail: {carsten.mueller,niklas.kiehne}@itg-research.net

K. Lavangnananda et al. (eds.), *Intelligent and Evolutionary Systems*, Proceedings in Adaptation, Learning and Optimization 5,
DOI: 10.1007/978-3-319-27000-5_18

region of neurons on the plane. For this purpose, a set of input stimuli M is needed, which is applied successively to the neurons. During training step t, for each $m_i \in M$ the neuron n_s^t with the closest Euclidean distance between stimulus and weight is selected, and called the excitation centre.

Moreover, a set of neurons that are within a range σ^t around the centre are chosen to adjust their weights to the stimulus according to the following formula:

$$w_i^{t+1} = w_i^t + \phi * e^{\frac{-d(l_s,l_i)^2}{2*\sigma^t}} * \left(m_i - w_i^t\right) \quad (1)$$

where ϕ is interpretable as the learning rate.

The training consists of one or more epochs in which every $m_i \in M$ is applied exactly once, but in a random order. With each stimulus presentation the time dependent σ^t is updated as $\sigma^{t+1} = \sigma^t * momentum$. The momentum is an adjustable parameter to control how fast the neighbourhood radius decreases over time. After a specified amount of epochs the training is complete.

1.2 Ant Colony Optimization

Ant Colony Optimization (ACO) is a proposed metaheuristic approach for solving hard combinatorial optimization problems [4–6]. One important behaviour pattern of ants for ACO is stigmergy, the indirect communication by manipulating the environment [7–10]. Pheromone trails in ACO serve as distributed, numerical information which the ants use to probabilistically construct solutions to the problem being solved and which the ants adapt during the algorithm's execution to reflect their search experience. The behaviour is determined by the parameters α and β [4] - $\alpha > \beta$: there is bigger influence on the choice of path, which is more often explored; $\alpha < \beta$: there is bigger influence on the choice of path, which offers better solution; $\alpha = \beta$: there is balanced dependency between quality of the path and degree of its exploration; $\alpha = 0$: there is a heuristics based only on the quality of passage between consecutive points.

2 Architecture of the Hybrid Approach

The general idea of combining ACO and ANN is to let the ants construct a tour which is then improved by applying a Self Organizing Map.

As the ACO algorithm is faster in converging towards a good, but not a very good, solution, the thought is to use the ANN as a kind of local search.

The procedure is as follows:

1. Initialize ACO and SOM with the given parameters
2. Solve the given TSP with the initialized ACO
3. Extract the best found tour in ACO and insert it into the SOM
4. Solve the SOM
5. Return the solution when SOM training is finished

At first, both ACO and SOM are set up with the user specified parameters and the selected TSP case. Then, ACO is started which rapidly scans the search space and finds a useful solution. The solution provided by ACO is extracted as a list of cities that depicts the found tour.

Subsequently, the list is handed over to the already initialized ANN, that spreads the neuron's weights evenly along the solution. So in direct opposition to the circular layout used in standalone SOM, the weights are distributed across the tour found by ACO.

At this point, a critical review of the SOM's parameter σ which represents the neighbourhood radius is needed. Once the neurons are dispersed on a valid tour, the usual values, like $\sigma = 3$, would render the inserted solution useless, since the whole structure would be severely deformed. Therefore, a hybrid specific parameter is introduced as the start iteration of the SOM.

The idea is to simulate an advanced progress in the neural net, in which, due to the momentum parameter, changes are only applied to smaller groups of weights. This is done by computing σ_x with x being the start iteration:

$$\sigma_x = \sigma_0 \cdot momentum^x \tag{2}$$

Different settings of the start iteration effectively change the amount of weights that are slightly detached from the inserted tour, as shown in Figure 1b.

The SOM algorithm is then proceeding without further adjustments. As shown in Figure 1, the net first relocates the neurons off their initial positions, but only to a limited extent. One effect is, that most of the intersections and overlapping are straightened out, which is due to the simultaneous movement of multiple weights. In this spirit, the initial softening is some kind of local optimization, whose degree of locality is controllable through a parameter. Once a specific neighbourhood radius is reached, the behaviour of the algorithm is seemingly reversed.

As the amounts of neurons that are moved during one training step is decreasing over time, a behavioural turning point is observed. After the softening reached a peak, the weights are moved back to the nodes, but now without the errors the ants made (see transition from Figure 1c to 1d).

The algorithm terminates after the weights are successfully distributed on the cities, which is the known procedure as seen in standalone SOM. In addition, a tournament selection is used, so that the best of both available tours is returned.

In summary, the hybrid approach consists of the sequential processing of a TSP by ACO and SOM. The improved performance emerges from the special application of the neural net, which leads to the local straightening of crossings in the tour supplied by ACO. This behaviour is accomplished through simulating an advanced state in the overall procedure of the SOM algorithm, such that smaller amounts of neurons are moved as compared to the original approach.

The SOM is therefore used as a local optimization technique to refine the tour found by ACO.

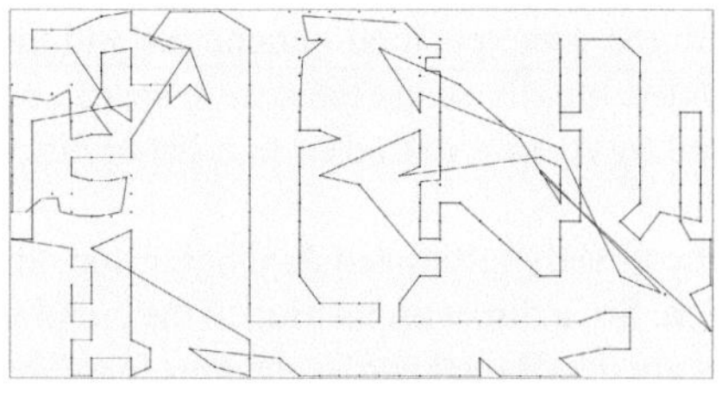

(a) Second SOM iteration. With two exceptions, the ACO tour is visible, only two regions differ from it.

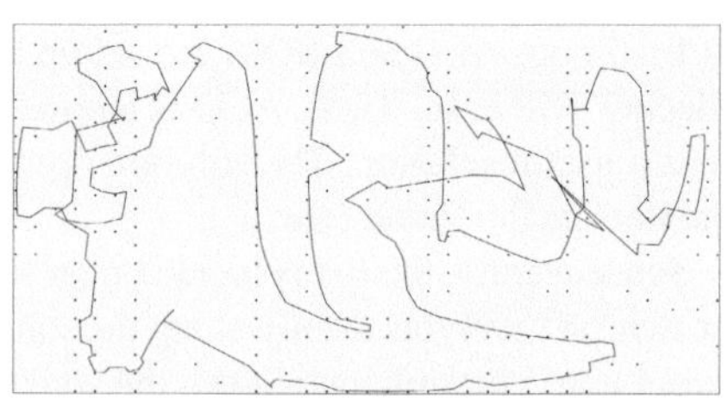

(b) After 100 iterations the general character of the ACO tour is still observable, but nearly all weights were moved from their original position.

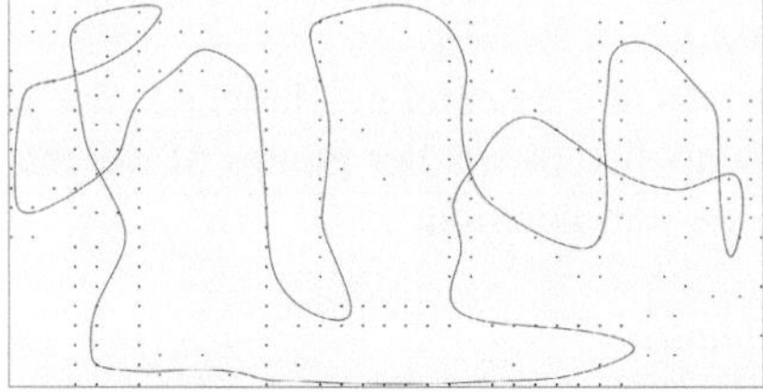

(c) At iteration 1000 the neurons form a smooth and vague representation of the handed over tour.

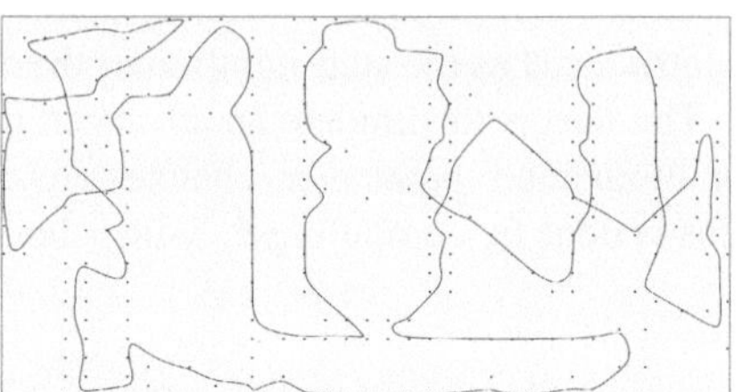

(d) With iteration 3000 the algorithm starts to move the weights back to the cities.

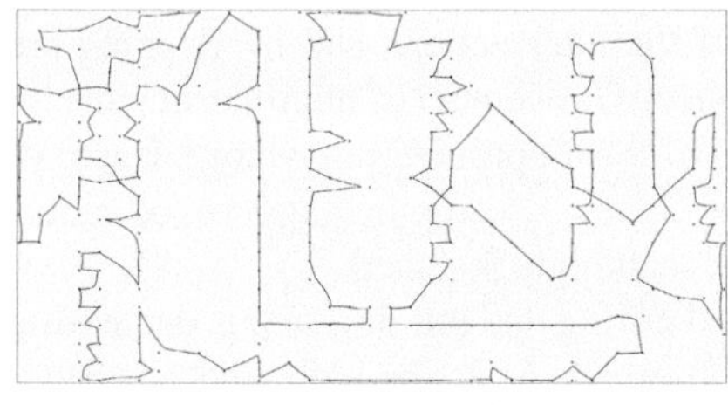

(e) The algorithm is nearly finished in the 5000th iteration, only few cities are not directly connected.

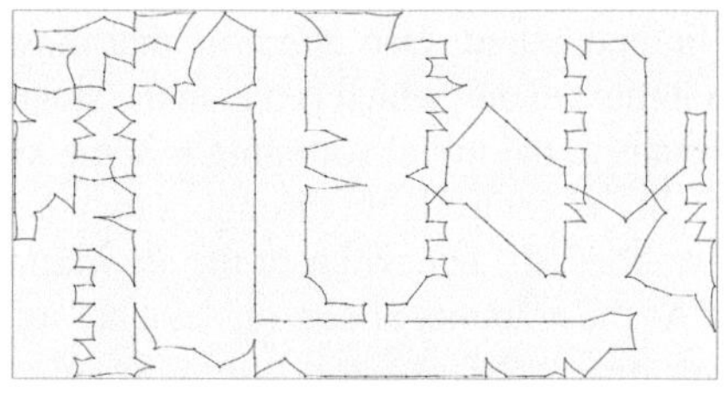

(f) After 7000 training steps the handed over tour is successfully refined by an ANN. The tour found by ACO had a relative error of 22.9%, whereas the tour after applying SOM achieves 8.0%.

Fig. 1 Six excerpts of one run of the hybrid approach.

3 Parameter Dependencies and Optimization

In the following, the parameters used for fine grained control over the algorithm's performance are explained. Since the proposed architecture uses both ACO and SOM, the parameters available are theoretically the sum of both algorithms.

But examining the influences of all these parameters, and especially their dependencies, is an extensive task. So before the actual evaluation, a logical analysis of the parameters at hand is carried out with the goal of identifying the substantial influences.

At first, parameters concerning the ACO algorithm, such as α and β, are not necessarily parts of the investigation. This is mostly due to the fact, that ACO is used to find the best possible solution in a reasonable time. But since the settings to achieve this behaviour were already figured out in related research, a repeated evaluation can be omitted.

The evaluated standard parameters for the ACO component are $\alpha = 1$, $\beta = 3$, an initial pheromone of 30, 5 ants and 1000 iterations. Furthermore, the examination of the SOM's parameters yielded, that the momentum and the number of neurons factor have good standard values that can be applied through all test cases. Therefore, the chosen values are a momentum of 0.999 and a number of neuron factor of 6. The remaining parameters are σ, ϕ and the number of iterations of the neural net, as well as the hybrid specific start iteration.

- σ: Similar to the functionality and influence of σ in standalone SOM, this parameter represents the neighbourhood radius. But since the ANN is used in a different way than before, good settings are to be discussed. As the neural net is simulated to start in a later iteration, the hybrid approach changes σ depending on the stated start iteration. Graph 2c shows, that best values for σ are between 2 and 8, with different starting iterations respectively. Considering also 2b scales the values down to $2 < \sigma \leq 5$.
- ϕ: The meaning of ϕ does not change, but it is still an influential parameter. As seen in figures 2a and 2d, valuable settings are $0.2 < \phi < 0.5$.
- Number of iterations: The used number of iterations which represents the number of times a training stimulus is presented to the neural net has probably the most obvious ranges. All related graphs in Figure 2 state, that the number of iterations should be not less than 2000, with the hint that even higher numbers could result in better solutions.
- Start iteration SOM: The start iteration parameter is controlling how far the neural net is delayed, and therefore closely connected to the behaviour of the hybrid approach. It is responsible for the degree of locality of the local search character of the SOM, since it influences how far neurons are detached from the ACO tour. Choosing a start iteration of zero and a high value of σ would move the weights to such an extent, that the information provided by the ants is lost. But a value too high would cause no improvement at all, since only single weights are adapted per iteration.
- Hence the closeness of minima and maxima as seen in Figure 2c. Interestingly, the area in the upper half shows the results of the ACO tour, caused by the mentioned single weight adjustment. So the parameter has to be chosen carefully, since even small changes of a few hundred iterations or less may decide on best or worst performance. In general, the start iteration should not be greater 2000 and not lower than 750.

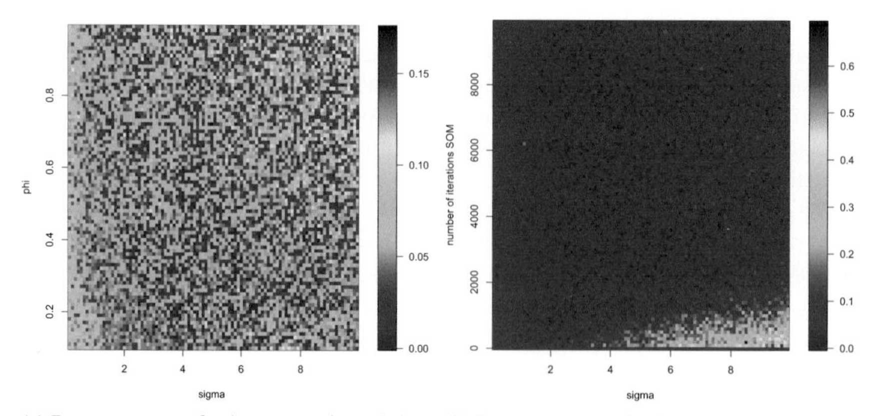

(a) Parameter test of σ in a range from 0.1 to 10 with a step size of 0.1 and ϕ ranging from 0.1 to 1 with a step size of 0.01.

(b) Parameter test of σ in a range from 0.1 to 10 with a step size of 0.1 and the number of iterations in a range from 0 to 10000 with a step size of 100.

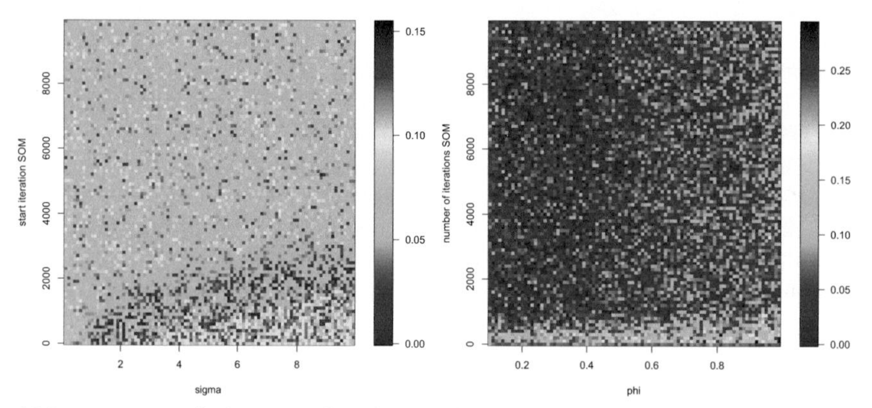

(c) Parameter test of σ in a range from 0.1 to 10 with a step size of 0.1 and the start iteration of the SOM in a range from 0 to 10000 with a step size of 100.

(d) Parameter test of ϕ in a range from 0.1 to 1 with a step size of 0.01 and the number of iterations in a range from 0 to 10000 with a step size of 100.

Fig. 2 Parameter tests of the hybrid approach of combinations of σ, ϕ, the number of iterations and the start iteration.

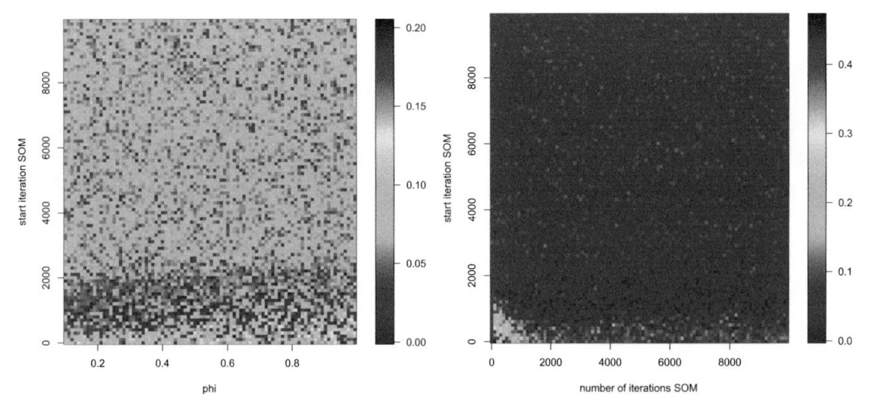

(e) Parameter test of ϕ in a range from 0.1 to 1 with a step size of 0.01 and the start iteration in a range from 0 to 10000 with a step size of 100.

(f) Parameter test of the number of iterations in a range from 0 to 10000 with a step size of 100 and the SOM's start iteration in a range from 0 to 10000 with a step size of 100.

Fig. 2 *Continued*

4 Statistical Analysis

In this section the hybrid approach is compared to its components ACO and SOM regarding their performances. For this purpose, each of the three algorithm computes 10000 results on the TSP berlin52, which then are statistically evaluated. The parameter settings applied were $\alpha = 1$, $\beta = 3$, $1 - \rho = 0.975$, an initial pheromone of 30 and 1000 iterations for the ACO as well as $\sigma = 3$, $\phi = 0.4$, a momentum of 0.999, a neuron factor of 6 and 5000 iterations for the SOM. The hybrid approach uses the same settings as specified for ACO and SOM, but ran with $\sigma = 4$ and a start iteration of 1500 instead.

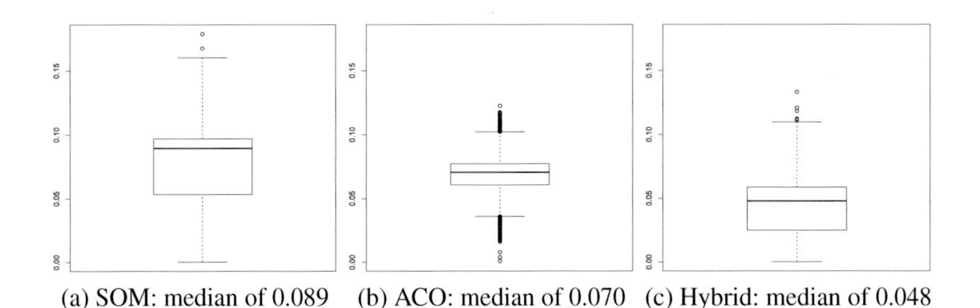

(a) SOM: median of 0.089 (b) ACO: median of 0.070 (c) Hybrid: median of 0.048

Fig. 3 Performance benchmark of ACO, SOM and the hybrid approach.

As shown in Figure 3, the hybrid approach seems to outperform its components ACO and SOM. In order to ensure the significance of the results, a welch-test based on the acquired results is computed. Since no information about the distribution of the samples is obtained, sufficiently large samples are required. With the available data of 10000 runs per algorithm, the welch-test can be safely applied. Since the hybrid approach's median is the lowest among the algorithms, the statistical test is used to validate the significance of the differences.

The p-values of the t-tests are

1. t.test(SOM, HYBRID) $\rightarrow$ p-value $< 2.2 * 10^{-16}$
2. t.test(ACO, HYBRID) $\rightarrow$ p-value $< 2.2 * 10^{-16}$

which allows the statement, that the hybrid approach outperforms both ACO and SOM on a level of significance of 5%.

5 Conclusions

In this research article a hybrid approach based on an intelligent combination of artificial ants and neurons was developed. At first, the performance of the hybrid's components was evaluated separately. For this purpose, the respective parameters, their dependencies and influences on the algorithm's behaviour where investigated and explained. Multiple tests of different parameter combinations were studied, in order to find the best performing parameter settings. The obtained insights subsumed into an intelligent architecture consisting of ACO and SOM. It was shown, that the proposed hybrid approach outperformed both its components with a high statistic significance.

References

1. Dayan, P., Abbott, L.F.: Theoretical neuroscience. MIT Press, Cambridge (2001)
2. Bear, F.M., Connors, B.W.: Neuroscience. Williams & Wilkins, Lippincott (2007)
3. Verleysen, C.M.: Special Issue on Advances in Self-Organizing Maps. Neural Networks **19**(5–6), 721–976 (2006)
4. Dorigo, M., Birattari, M., Stutzle, T.: Ant colony optimization. IEEE Comput. Intell. Mag. **1**, 28–39 (2006)
5. Toksari, M.D.: Ant colony optimization for finding the global minimum. Appl. Math. Comput. **176**(1), 308–316 (2006)
6. Belal, M., Gaber, J., El-Sayed, H., Almojel A.: Swarm intelligence. In: Handbook of Bioinspired Algorithms and Applications. Chapman & Hall, London (2006)
7. Theraulaz, G., Bonabeau, E.: A Brief History of Stigmergy. Artificial Life **5**(3), 97–116 (1999)
8. Bonabeau, E., Dorigo, M., Theraulaz, G.: Swarm intelligence - from natural to artificial systems. Oxford University Press, Oxford (1999)
9. Dorigo, M., Bonabeau, E., Theraulaz, G.: Ant algorithms and stigmergy. Future Generation Computer Systems **16**(9), 851–871 (2000)
10. Parpinelli, R.S., Lopes, H.S.: New inspirations in swarm intelligence: a survey. International Journal of Bio-Inspired Computation **3**, 1–16 (2011)

The Analysis of Migrating Birds Optimization Algorithm with Neighborhood Operator on Traveling Salesman Problem

Vahit Tongur and Erkan Ülker

Abstract Migrating birds optimization (MBO) algorithm is a new meta-heuristic algorithm inspired from behaviors of migratory birds during migration. Basic MBO algorithm is designed for quadratic assignment problems (QAP) which are known as discrete problems, and the performance of MBO algorithm for solving QAP is shown successfully. But MBO algorithm could not achieve same performance for some other benchmark problems like traveling salesman problem (TSP) and asymmetric traveling salesman problem (ATSP). In order to deal with these kinds of problems, neighborhood operators of MBO is focused in this paper. The performance of MBO algorithm is evaluated with seven varieties of neighborhood operators on symmetric and asymmetric TSP problems. Experimental results show that the performance of MBO algorithm is improved up to 36% by utilizing different neighborhood operators.

Keywords Migrating Birds Optimization · Traveling Salesman Problem · Neighborhood operators

1 Introduction

Travelling Salesman Problem (TSP), takes its name from the approach of a salesman who has to visit all his customers in all cities in a whole tour. The aim of TSP is finding the shortest path which the salesman travels some specific number of cities

V. Tongur(✉)
Department of Computer Engineering, Faculty of Engineering and Architecture,
Necmettin Erbakan University, Konya, Turkey
e-mail: vtongur@konya.edu.tr

E. Ülker
Department of Computer Engineering, Faculty of Engineering,
Selçuk University, Konya, Turkey
e-mail: eulker@selcuk.edu.tr

K. Lavangnananda et al. (eds.), *Intelligent and Evolutionary Systems*,
Proceedings in Adaptation, Learning and Optimization 5,
DOI: 10.1007/978-3-319-27000-5_19

by starting from a point and visits each city only one time and get back to that starting point. Such problems are also applied into various areas like vehicle routing, printed circuit positioning problems, tabulation charts and flexible production systems. For this reason, these kinds of problems have attracted the attention of many researchers from mathematics, biology, engineering and many other areas.

Until now, TSP problem is investigated for obtaining better solutions than solutions obtained by many meta-heuristic algorithms. TSP problem is tried to solve with discrete artificial bee colony algorithm in [1], with ant colony optimization method in [2] and by applying search space smoothing technique using an existing local search algorithm in [3]. Performance of TSP is compared with the performance of basic genetic algorithm, hopfield neural network and basic ant colony algorithm in [4]. Also, other studies in literature are given in [5]-[10]. TSP can be formulated as below; Considering a scalar graph; while $G = (N, A)$, $N = \{0, 1, ..., n\}$ shows cities, $A = NxN$ shows the roads between cities. Permutation of a matrix which is given as $D = (d_{ij})_{nxn}$ is shown as $\pi = \{\pi_0, \pi_1, ..., \pi_n\}$. In TSP, it is targeted to minimize travel costs which can be defined as;

$$minf(\pi) = \sum_{i=0}^{n-1} d_{\pi_i,\pi_{i+1}} + d_{\pi_n,\pi_0} \tag{1}$$

where, d shows matrix of couples between cities. d_{ij} shows the distance between ith and jth cities. $j = \pi_i$ shows the jth city which is visited in ith step. While permutation is generally called as tour,$(\pi_0, \pi_1, ..., \pi_i, \pi_{i+1}, ..., \pi_n, \pi_0)$ is called as borders [1]. In symmetric TSP problems, graphs are scalar. Also $d_{ij} = d_{ji}$ equality is always provided. But coming and going between cities are not same for asymmetric TSP problems. In other words $d_{ij} \neq d_{ji}$.

Next sections are organized as follows; In the second section, MBO algorithm is presented. Afterwards, solution of TSP utilizing MBO algorithm is given in the third section. In the fourth section, new neighborhood operators which are implemented in MBO algorithm are described. In the final section, experimental results are evaluated and discussed.

2 Migrating Birds Optimization Algorithm

MBO algorithm is firstly proposed by [11]. MBO algorithm is inspired from energy saving of migrating birds during flying in V formation. It is developed for discrete problems and tested on QAP problems which are based on real life problems.

Most common flying style of the migrating birds is V formation which provides flying longer distances. It is also known that basic instinct of this formation is energy saving. Leader bird is the one which spends the most energy in V formation. The other birds can ride longer by the wind energy created by frontier birds wing moves. Some of the flocks can be in different orders in V formation. The reason of this situation is that different birds have different wing lengths. In order to use wind energy created by leader bird, the other birds follow the leader in a specific distance and angle.

Fig. 1 A V formation

MBO algorithm starts with initial solutions and then develops these solutions. Initial solutions are randomly generated, and tried to converge to ideal results according to the objective function. This convergence is achieved by using real life V formation of birds. Each bird in V formation represents a solution in search space. Neighbor solutions are produced from these solutions in order to make local searches in the search space. Therefore, the neighbor search method is used by MBO algorithm. Details of MBO algorithm can be obtained from [11].

Duman et al. created neighborhood production by changing only one couple of existing solution permutation for QAP problems [11]. Each solution produce predefined number of neighbors, and keep the best neighbor for comparing with itself, and predefined sharing number (x) of the rest of the solutions are transmitted to the following bird. This neighbor solution sharing process is the most important feature of MBO algorithm which separates it from other meta-heuristic algorithms. Thus, each bird in flock is in contact with the other birds and converges to the optimal solution in a faster way. Solution producing and sharing processes of the birds are repeated as much as flapping number of the leader bird which is given as a parameter. When the flapping value is reached, the leader bird gets tired and goes to the end of left wing. The bird which follows the leader in the left side gets the lead and the birds on the left wing takes each others place in order. Next leader change occurs in

```
Generate n random solutions and place into V formation
Repeat
  Repeat
        Generate k neighbours for leader and generate k-x neighbours for other birds with swap method then
        sort these negihbours for each bird according to objective function.
        share unused best x neighbours for each bird
        if(neighbour solution better than current solution)
           current solution =  neighbour solution
  Until number of flap
  Replace leader
Until number of sum all generated neighbours > (problem size)^3
Return best result
```

Fig. 2 The pseudo code of MBO algorithm

the right wing. This process is repeated until total number of the produced neighbor birds reaches the iteration number, and iteration value should be big enough that each bird can take the lead once at least. Duman et al. determined the iteration value equals to the cube of the problem size for QAP problems [11].

3 Solution of TSP with MBO Algorithm

MBO algorithm has a permutation structure because it is designed for discrete problems. Each city for traveling salesman problem (TSP) or asymmetric traveling salesman problem (ATSP) is placed into this permutation sequence. Afterwards, this permutation sequence is evaluated according to Eq.(1). Permutation sequence and cities that placed into permutation are presented in Figure 3. Size of the permutation

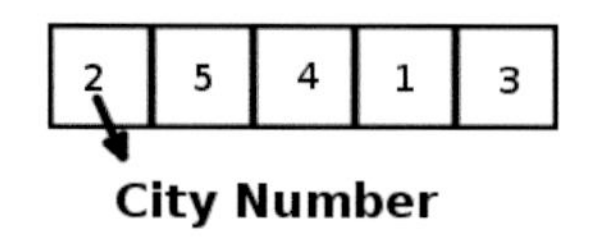

Fig. 3 Cities layout in permutation sequence

equals to the number of the cities, and each bird has a permutation. In Figure 3, five cities are randomly placed into permutation.

MBO algorithm uses neighborhood method for local search process. This method is known as swap. In this method, two randomly selected cities are replaced.

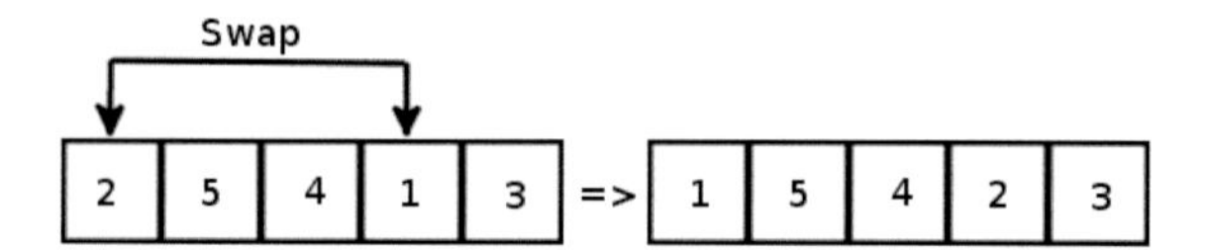

Fig. 4 Swap method for generate neighbor

Due to the algorithm structure, at least three neighbors are generated. Generated neighbors are sorted from best value to worst value according to the objective function. Then, the best neighbor solution is compared with current solution. Current solution is replaced with the neighbor solution if neighbor solution is better than current solution. These processes are conducted for all birds in the flock.

4 Neighborhood Methods

Neighborhood methods in meta-heuristics algorithms are used for local search. Eight neighborhood operators are investigated on artificial bee colony (ABC) algorithm in [12]. Seven neighborhood operators are applied to simulated annealing algorithm in [13]. In this study, seven neighborhood operators are tested for TSP and ATSP with MBO algorithm.

4.1 Swap Method

In this method, two randomly selected cities from current sequence are replaced. Other positions are not changed.

Table 1 Swap

Current sequence	Randomly selected two city	New sequence
2-5-4-1-3	2,1	1-5-4-2-3

4.2 Insertion Method

In this method, a randomly selected city is inserted to a randomly selected position.

Table 2 Insertion

Current sequence	Randomly selected city	Randomly selected position	New sequence
2-5-4-1-3	5	4	2-4-1-5-3

4.3 Swap Greedy Method

In this method, a randomly selected city is replaced with rest of the other cities. Thus, $citiesnumber - 1$ neighbors are generated. These neighbors are evaluated according to objective function, and best neighbor is selected for neighbor solution.

Table 3 Swap greedy

Current sequence	Randomly selected city	Selected city	New sequence
		2	4-5-2-1-3
		5	2-4-5-1-3
2-5-4-1-3	4	1	2-5-1-4-3
		3	2-5-3-1-4

4.4 Insertion Greedy Method

In this method, a randomly selected city is inserted to all other positions. In this way, $citiesnumber - 1$ neighbors are generated. These neighbors are evaluated according to objective function, and best neighbor is selected for neighbor solution.

Table 4 Insertion greedy

Current sequence	Randomly selected position	Selected position	New sequence
		1	1-2-5-4-3
		2	2-1-5-4-3
2-5-4-1-3	4	3	2-5-1-4-3
		5	2-5-4-3-1

4.5 Random Insertion Perturbation Method

In this method, c is assumed as number of the city. First city in the current sequence is inserted to any position at its right. In other words, first city is inserted to any position between 2 and c. On the other hand, last city in current sequence is inserted to any position at its left. In other words, last city is inserted to any position between 1 and $c - 1$. Rest of the cities is inserted into two different positions. j th city in

Table 5 Random insertion perturbation

Current sequence	Position of selected city	Selected city	Randomly selected position	New sequence
	1	2	4	5-4-1-2-3
	2	5	1	5-2-4-1-3
			3	2-4-5-1-3
	3	4	1	4-2-5-1-3
2-5-4-1-3			5	2-5-1-3-4
	4	1	2	2-1-5-4-3
			5	2-5-4-3-1
	5	3	2	2-3-5-4-1

Table 6 Starting parameters of MBO algoritm

Parameters	Values
Number of Bird (n)	51
Number of Neighbor (k)	3
Number of Neighbor Sharing (x)	1
Number of Flap (m)	20

Table 7 TSP and ATSP problems and best known solutions

Problem	Best known solutions
berlin52	7542
eil76	538
br17	39
ft53	6905

Table 8 Results of neighbourhood methods on tested problems

Problem	Method	Minimum(Avg.)	Success rate(%)	Time(Sec.)
berlin52	Swap	8955.24	81.27	0.67
	Insertion	8096.87	92.65	0.69
	Swap Greedy	8719.93	84.39	1.1
	Insertion Greedy	8056.66	93.18	1.73
	Rnd.Ins.Perturb.	8163.51	91.76	1.48
	Swap Best	8812,08	83.12	1.86
	Insertion Best	**7975.85**	94.25	3.22
eil76	Swap	653.8	78.48	2.48
	Insertion	596.55	89.12	2.53
	Swap Greedy	657.7	77.76	7.37
	Insertion Greedy	583.43	91.56	12.56
	Rnd.Ins.Perturb.	605.55	87.45	5.43
	Swap Best	656.55	77.97	6.07
	Insertion Best	**582.17**	91.79	10.03
br17	Swap	39.1	99.75	0.01
	Insertion	**39**	100	0.01
	Swap Greedy	**39**	100	0.01
	Insertion Greedy	**39**	100	0.02
	Rnd.Ins.Perturb.	**39**	100	0.02
	Swap Best	**39**	100	0.01
	Insertion Best	**39**	100	0.01
ft53	Swap	9031.7	69.21.75	0.71
	Insertion	7740	87.91	0.75
	Swap Greedy	9021.4	69.35	0.01
	Insertion Greedy	7686	88.69	2.1
	Rnd.Ins.Perturb.	7878.5	85.91	4.6
	Swap Best	8521.2	76.6	2.27
	Insertion Best	**7297.5**	94.32	2.05

current sequence is inserted to any position between 1 and $j-1$. Also this city is inserted to any position between $j+1$ and c. In this way, $2(c-1)$ neighbor solution is obtained. All generated neighbor solutions are evaluated according to objective function, and best neighbor solution is selected as neighbor solution.

4.6 Swap Best Method

In this method, *n* (number of city) neighbors are generated using swap method. All generated neighbor solutions are evaluated according to objective function, and best neighbor solution is selected as neighbor solution.

4.7 Insertion Best Method

In this method, *c* (number of city) neighbors are generated using insertion method. All generated neighbor solutions are evaluated according to objective function, and best neighbor solution is selected as neighbor solution.

5 Experiment Results

In this study, seven aforementioned methods are tested on some benchmark problems which are chosen from TSPLIB library [14]. These benchmark problems are berlin52, eil76, br17 and ft53. Berlin52 and eil76 problems have 52 and 76 cities respectively, and these problems are symmetric TSP problems. On the other hand, br17 and ft53 problems have 17 and 53 cities respectively, and these are asymmetric TSP problems. Experiments are evaluated on a computer having Intel(R) Core(TM) i5-3330 CPU @ 3.00GHz processor, 4 GB RAM and Linux Ubuntu 14.04 (64-bit) operating system. All of the methods are coded with QT Creator 3.0.1 gcc compiler and C++ language. Proposed methods are executed 10 times by keeping the starting parameters same and results are averaged. Starting parameters of MBO algorithm are given in Table 6. Selected problems and best known solutions of these problems are given in Table 7. Problems, neighborhood operators, average minimum tour length,

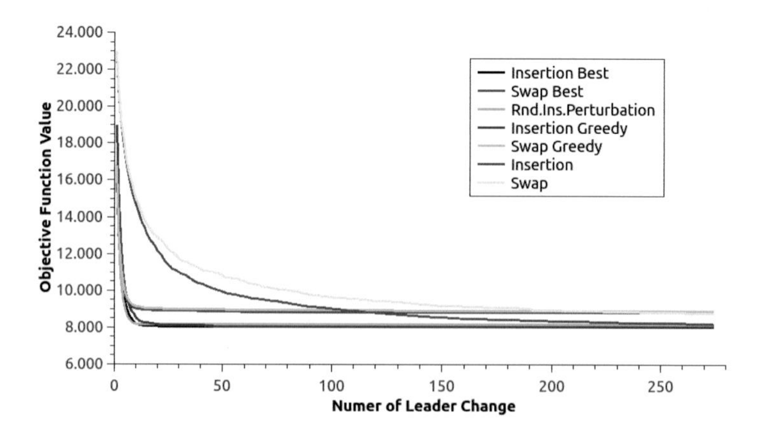

Fig. 5 Results of objective function for berlin52

success rate (%) and best result obtaining time (second) are given in Table 8. Best results are demonstrated bold in Table 8.

Since average solutions are evaluated on test results, it can not be reached to best known solutions. According to Table 8, all neighbor methods are successful in br17 problem. Furthermore, insertion best method is reached the best achievement for ft53 problem. In eil76 and berlin52 problems, success rate are 91.79% and 94.25%, respectively.

Also, the results which are obtained in leader exchange are given in Figure 5, 6, 7 and 8 for berlin 52, eil76, br17 and ft53 problems, respectively. As is seen, all neighbor methods are tendency to improvement. With regard to figures, insertion

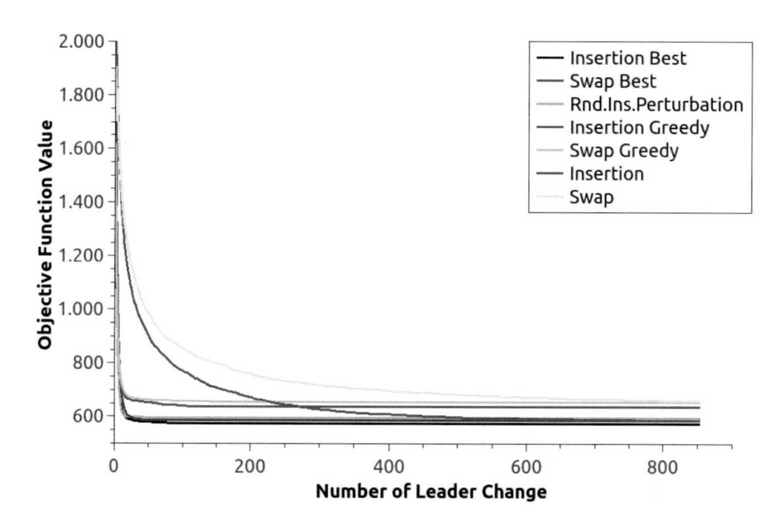

Fig. 6 Results of objective function for eil76

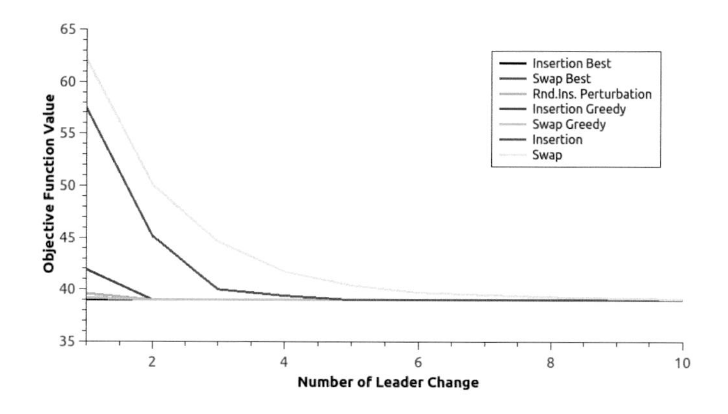

Fig. 7 Results of objective function for br17

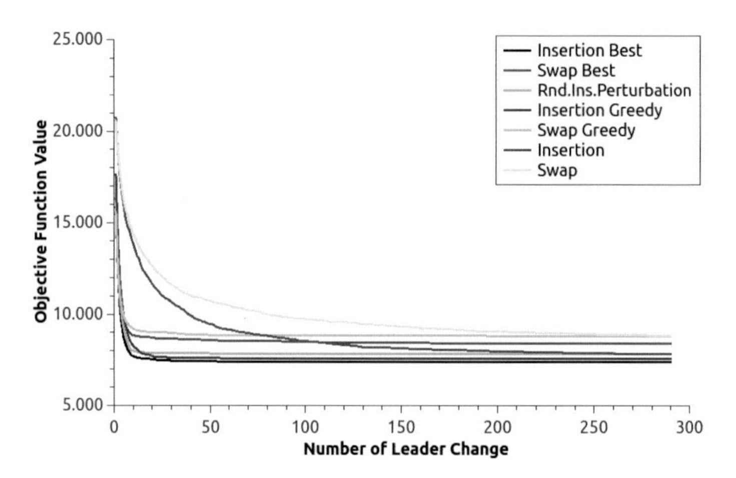

Fig. 8 Results of objective function for ft53

best method is successful in comparison to other neighbor method. Insertion best method converge faster but the step is longer.

Also observed that performance of insertion methods are better than swap methods. All neighborhood methods can be advised for small size problems (such as br17). They were given close values in terms of time and results for small size problem. But insertion methods (especially insertion best and insertion greedy) can be advised for big size problem.

5.1 Conclusion and Future Works

In this study, performance of neighborhood methods on MBO algorithm are investigated for the symmetric and asymmetric TSP problems. The obtained results showed that insertion best method were given better results than the other methods. But runtime of this method is higher than other methods. In this study, it is shown that neighborhood methods improves the performance of MBO algorithm. The selection of neighborhood method depends on the problem itself in the proposed method. However, the suggested method achieves more successful results than pure MBO. Future work include parameter optimization of MBO and enlargement of benchmark problems for presenting more general conclusion.

Acknowledgments This study has been supported by Scientific Research Project of Necmettin Erbakan University.

References

1. Karabulut, K., Tasgetiren, M.F.: A discrete artificial bee colony algorithm for the traveling salesman problem with time windows. In: 2012 IEEE Congress on Evolutionary Computation, pp. 1–7 (2012)
2. Yang, J., Shi, X., Marchese, M., Liang, Y.: An Ant Colony Optimization Method For Generalized Tsp Problem. Prog. Nat. Sci. **18**(11), 1417–1422 (2008)
3. Gu, J., Huang, X.: Efficient Local Search with Search Space Smoothing: A Case Study of the Traveling Salesman Problem (TSP). IEEE Trans. Syst. Man. Cybern. **24**(5), 728–735 (1994)
4. Hui, W.: Comparison of Several Intelligent Algorithms for Solving TSP Problem in Industrial Engineering. Syst. Eng. Procedia **4**, 226–235 (2012)
5. Yan, X., Liu, H., Yan, J., Wu, Q.: A fast evolutionary algorithm for traveling salesman problem. In: Third International Conference on Natural Computation (ICNC 2007), vol. 4, pp. 85–90 (2007)
6. Tao, Z.: TSP problem solution based on improved genetic algorithm. In: 2008 Fourth International Conference on Natural Computation, vol. 1, pp. 686–690 (2008)
7. Mrida-Casermeiro, E., Galn-Marn, G., Muoz-Prez, J.: An Efficient Multivalued Hopfield Network for the Traveling Salesman Problem. Neural Process. Lett. **14**(3), 203–216 (2001)
8. Zhu, Q., Chen, S.: A new ant evolution algorithm to resolve TSP problem. In: Sixth International Conference on Machine Learning and Applications (ICMLA 2007), pp. 62–66 (2007)
9. Yan, L., Kongyu, Y.: Immunity genetic algorithm based on elitist strategy and its application to the TSP problem. In: 2008 International Symposium on Intelligent Information Technology Application Workshops, pp. 3–6 (2008)
10. Pang, W., Wang, K.P., Zhou, C.G., Dong, L.J.: Fuzzy discrete particle swarm optimization for solving traveling salesman problem. In: The Fourth International Conference onComputer and Information Technology, CIT 2004, 796–800 (2004)
11. Duman, E., Uysal, M., Alkaya, A.F.: Migrating Birds Optimization A new Metaheuristic Approach and its Performance on Quadratic Assignment Problem. Inf. Sci. **217**, 65–77 (2012)
12. Kıran, M.S., Işcan, H., Gündüz, M.: The Analysis of Discrete Artificial Bee Colony Algorithm with Neighborhood Operator on Traveling Salesman Problem. Neural Comput. Appl. **23**(1), 9–21 (2012)
13. Güner, E., Altıparmak, F.: A Heuristic Approach for the Secondary Criterion Scheduling Problem on a Single Machine. In: Gazi Universitesi Mühendislik-Mimarlık Fakültesi Derg., Vol. 18(3), pp. 27–42 (2003)
14. TSP library for benchmark TSP problems. http://elib.zib.de/pub/mp-testdata/tsp/tsplib/tsplib.html (Accessed 22.07.2014)

Solving the IEEE CEC 2015 Dynamic Benchmark Problems Using Kalman Filter Based Dynamic Multiobjective Evolutionary Algorithm

Arrchana Muruganantham, Kay Chen Tan and Prahlad Vadakkepat

Abstract Evolutionary algorithms have been extensively used to solve static and dynamic single objective optimization problems, and static multiobjective optimization problems. However, there has only been tepid interest to solve multiobjective optimization problems in dynamic environments. It is only in the past few years that evolutionary algorithms have been used to solve dynamic multiobjective optimization problems and comprehensive benchmark suites have been proposed for testing the performance of algorithms. Prediction based algorithms may be able to provide information about the location of the changed optima and thereby assisting the evolutionary algorithm in the non-trivial task of tracking the changing Pareto Optimal Front or Set. Kalman filter is one of the widely used techniques in prediction scenarios for state estimation. A Dynamic Multi-objective Evolutionary algorithm was proposed in which the Kalman Filter was applied to the whole population to direct the search for Pareto Optimal Solutions in the decision space after a change in the problem has occurred. In this work, the Kalman Filter assisted Evolutionary Algorithm is tested on the IEEE CEC 2015 Benchmark problems set and the results are presented. It is observed that while the proposed algorithm performs well on some problems, more efficient strategies are required to supplement the algorithm in cases of high change severity, isolated and deceptive fronts.

Keywords Dynamic multiobective optimization · Kalman filtered · Evolutionary algorithm

A. Muruganantham(✉) · K.C. Tan · P. Vadakkepat
Department of Electrical and Computer Engineering, National University of Singapore, Singapore 117576, Singapore
e-mail: arrchana@u.nus.edu

K. Lavangnananda et al. (eds.), *Intelligent and Evolutionary Systems*,
Proceedings in Adaptation, Learning and Optimization 5,
DOI: 10.1007/978-3-319-27000-5_20

1 Introduction

Optimization problems are aplenty and are found in various fields such as science, engineering, economics, finance, management, scheduling, planning, design, control, etc. The list is ever growing, and scientists and industrialists alike are in the lookout for better and more efficient techniques to solve their problems. Optimization in general refers to the process of finding one or more feasible solutions which correspond to extreme values of one or more objectives. Many researchers have tend to focus on optimization problems which consider a single objective, although most real-world search and optimization problems involve more than one objective. Further, the presence of conflict in the multiple objectives makes these optimization problems (commonly termed as multiobjective optimization problems) more interesting and challenging to solve. Since no single solution can satisfy the multiple conflicting objectives simultaneously, the solution to a multiobjective optimization problem is a set of trade-off optimal solutions. Classical optimization methods such as hill climbing, simulated annealing can at best find one solution in a simulation run, thereby deeming these methods inefficient to solve multiobjective optimization problems.

Evolutionary algorithms are inspired from biological evolution and mimic nature's evolutionary principles to drive the search towards optimal solution(s). These algorithms use a population of solutions in each iteration, consequently making them ideal candidates for solving multiobjective optimization problems. Numerous Evolutionary Algorithms(EAs) have been developed in the past few decades to solve multiobjective optimization problems such as NSGA-II [1], MOEA/D [2], MOEA/D-DE [3], to name a few. The advances of Evolutionary Multiobjective Optimization(EMO) research has been drastic and has resulted in many new paradigms to be developed such as the Estimation of Distribution Algorithms(EDAs), decomposition based algorithms, and so on. Applications of EMO research have been observed in a wide variety of problems [4, 5, 6, 7, 8, 9]. However, there has only been lukewarm interest in applying Evolutionary Algorithms to solve dynamic optimization problems, where the optimum(or optima) changes with time. Furthermore, most of the EA researchers in this area have tend to focus on dynamic single-objective optimization problems, while most real-world problems are dynamic multiobjective optimization problems.

Using Evolutionary Algorithms to solve dynamic multiobjective optimization problems has started gaining attention over the past few years. Nevertheless, there is large scope for contribution and improvement in this field. In dynamic multiobjective optimization problems the fitness landscape is changing over time. Preliminary research in solving proposed benchmark problems involved applying Multiobjective Evolutionary Algorithm(MOEA) directly to solve them. However, the inherent characteristic of an MOEA is that it takes significant amount of time to converge to the Pareto Optimal Front(POF). This is an important issue in dynamic multiobjective optimization where the POF and/or the Pareto Optimal Solution(s) (POS) are continuously changing with time. In the current literature, various approaches have been proposed to solve dynamic multiobjective optimization problems. In this paper, the focus is on employing prediction techniques to solve dynamic multiobjective optimization problems. A novel Kalman Filter based dynamic multiobjective optimization algorithm was developed to solve dynamic multiobjective optimization problems [10].

Based on the IEEE-CEC 2015 Dynamic Multi-objective Optimization Benchmark problems, this paper aims to examine and discuss the performance of the Kalman Filter assisted MOEA/D-DE algorithm, MOEA/D-KF in solving the proposed benchmark set. The outline of the paper is as follows: Section 2 provides required background and outlines related work. Section 3 provides the algorithm description including a brief overview of MOEA/D-DE and Kalman Filter prediction method. Section 4 provides the experimental setup, outlines the performance metric used and the results are presented. Section 5 consists of anlaysis of the performance based on the severity and frequency of change in the problems. Section 6 outlines the discussion of the results and Section 7 concludes the work.

2 Background

This section provides the basic definitions used in the evolutionary multiobjective community together with some key concepts which are essential for understanding the work described in a more scientific manner.

2.1 Multiobjective Optimization Problem

A multiobjective problem can be expressed in its general form mathematically as

$$\begin{aligned} \text{Minimize/Maximize } & f_m(x), && m = 1, 2, \ldots, M; \\ \text{subject to } & g_j(x) \geq 0, && j = 1, 2, \ldots, J; \\ & h_k(x) = 0, && k = 1, 2, \ldots, K; \\ & x_i^L \leq x_i \leq x_i^U, && i = 1, 2, \ldots, n. \end{aligned}$$

where f_i is the i-th objective function and M is the number of objectives.

The vector, $f(x) = [f_1(x)\ f_2(x)\ \ldots\ f_m(x)]^T$ forms the objective vector, $f(x) \in \mathbb{R}^M$. A solution x is a vector of n decision variables: $x = [x_1\ x_2\ \ldots\ x_n]^T$. The above general problem is associated with J inequality and K equality constraints. The last set of constraints are called *variable bounds*, restricting each decision variable x_i to take a value within a lower $x_i^{(L)}$ and an upper $x_i^{(U)}$ bound. These variable bounds constitute the *decision variable space* $\Omega \in \mathbb{R}^n$, or simply the decision space.

In the presence of constraints g_j and h_k, the entire decision variable space Ω may not be feasible. The feasible region S is the set of all feasible solutions in the context of optimization. The feasible search space can be divided into 2 sets of solutions - pareto optimal and non pareto otpimal set. To define pareto optimality, first we need to look into the concept of domination.

Concept of Domination. There are M objective functions in a multiobjective problem. Say, we have 2 solutions, i and j. $i < j$ implies i is better than j *or* i dominates j. A solution x^1 is said to dominate another solution x^2, if both the following conditions are true.

1. The solution x^1 is no worse than x^2 in all objectives, *or* $f_m(x^1)$ is not better than $f_m(x^2)$ for all $m = 1, 2, ..., M$.
2. The solution x^1 is strictly better than x^2 in at least one objective.

Pareto Optimality. Among a set of solutions P, the non-dominated solutions, P^* are those that are not dominated by any member of the set P. When the set P comprises the entire search space, the resulting non-dominated set P^* is the *Pareto Optimal Set*(POS in the decision space). Pareto optimal solutions joined together as a curve form the *Pareto Optimal Front*(POF in the objective space). The front lies in the bottom-left corner of the search space for problems where all objectives are to be minimized.

Goals of an MOEA. The working principle for an ideal multiobjective procedure consists of finding multiple trade-off optimal solutions with a wide range of values for the objectives, and later choosing one of the obtained solutions using higher level information. In such a case it is difficult to prefer one solution over the other without any further information about the problem. If higher level information is satisfactorily available, this can be used to make a biased search. However, in the absence of any such information, all pareto optimal solutions are equally important. Therefore, there are 2 goals:

1. To find a set of solutions as close as possible to the POF, i.e. *Convergence*
2. To find a set of solutions as diverse as possible, i.e. *Diversity*

For each of the M conflicting objectives, there exists one different optimal solution. An objective vector constructed with these individual optimal objective values constitutes the ideal objective vector, z^*, which in general lies in the infeasible space. For more detailed discussion of the concepts on multiobjective optimization , please refer to [11].

2.2 *Dynamic Multiobjective Optimization Problem*

The various concepts discussed for multiobjective optimization are still essential in dynamic multiobjective optimization together with some additional issues and goal(s). In general, in a dynamic multiobjective optimization problem(DMOOP), the optimum changes with time. Mathematically, a DMOOP can be described as

$$\begin{aligned} \underset{\mathbf{x}}{\text{minimize}} \quad & \mathbf{f}(\mathbf{x}, t) = [f_1(\mathbf{x}, t) \; f_2(\mathbf{x}, t) \; \dots \; f_m(\mathbf{x}, t)]^T \\ \text{subject to} \quad & \mathbf{x} \in \Omega \end{aligned} \tag{1}$$

where t represents time index, $x \in \mathbb{R}^n$ represents the decision vector, n is the number of decision variables and $\Omega \subset \mathbb{R}^n$ represents the decision space. m is the number of objectives, $\mathbb{R}^m$ is the objective space and $f(\mathbf{x}, t)$ consists of m real-valued objective functions, each of which is continuous with respect to x over Ω. Thus, the *POF* and/or *POS* may change over time.

The goals of *convergence* and *diversity* apply to Dynamic Multiobjective Optimization Evolutionary Algorithms (DMOEAs) as well. However, it is not restricted to the above two and there is an additional goal of tracking the changing POF/POS which plays an important role in determining the overall performance.

Classification of DMOOPs. [12] have classified dynamic multiobjective optimization problems based on the possible ways a problem can demonstrate a time varying change.

Table 1 Classification of DMOOPs

Type I	POS changes, but POF does not change
Type II	Both POS and POF change
Type III	POS does not change, POF changes
Type IV	Both POS and POF do not change, although the problem can change

These four cases are summarized in the Table 1. There are other possible ways of classifying DMOOPs as well such as based on severity, predictability and visibility of change, among others [13].

3 Algorithm Description

3.1 *MOEA/D-DE*

The DMOEA used in this paper is built on the basis of Multiobjective Evolutionary Algorithm with Decomposition based on Differential Evolution (MOEA/D-DE) [3]. MOEA/D-DE decomposes a problem into several sub-problems and simultaneously optimizes them to find the pareto optimal solutions of the Multiobjective optimization problem. Each solution is assigned with a weight vector and neighbourhood relations are defined based on the weight vectors. In the context of dynamic multiobjective optimization, the usage of weight vectors enables the tracking of individual solutions in the decision space which are essential for prediction purposes. Decomposition into sub-problems is performed using the Tchebycheff approach in this paper.

3.2 *Change Detection*

Sentry particles in the population are used to observe any changes in the system, assuming there is no noise. These change detecter individuals' objective values are recomputed at the beginning of each generation to check if there has been any change since the last objective function evaluation. If there is a change in the objective function values, it is assumed that a change has occured and the Kalman filter based model is used to predict for the optimal values of solutions in the decision space. Otherwise, the optimization process proceeds as in a static MOEA.

3.3 Kalman Filter Prediction Based DMOEA

Kalman Filter is an algorithm that uses a series of measurements observed over time, containing noise and other inaccuracies, and produces statistically optimal estimates of the underlying system state [14, 15]. The algorithm works in a two-step process involving a prediction step and a measurement step. In the prediction step, the Kalman filter produces estimates of the current state variables, along with their uncertainties. Once the outcome of the next measurement is observed, these *a priori* estimates are updated to obtain the *a posteriori* estimates. The Kalman filter operates recursively in time series analysis. The fact that Kalman filter can run in real-time makes it a good candidate for the prediction model in solving DMOOP. Thus, in our study, Kalman filter is applied to the whole population to direct the search for Pareto Optimal Solutions (POS) in the decision space after a change in the problem has occured. Please refer to [10] for more details on the Kalman Filter assisted DMOEA. The 2by2 variant of the Kalman filter based algorithm from [10] is used in this paper.

4 Empirical Study

4.1 Experimental Setup

The Kalman Filter prediction based DMOEA, MOEA/D-KF is tested on the benchmark problems proposed for the IEEE CEC 2015 Competition on Dynamic Multiobjective optimization [16] in this paper. The benchmark set consists of functions from FDA [12], dMOP [17] and HE [18, 19, 20] benchmark function suites and were adapted to further test the capabilities of DMOEAs in a more comprehensible manner than currently available in the evolutionary dynamic multiobjective optimization literature. The parameter settings for the experiments are tabulated in Table 2.

4.2 Performance Metric - Modified Inverted Generational Distance

A number of performance metrics are in use for evaluation of static MOEAs which evaluate convergence and diversity quite effectively. These metrics have been modified for usage in evaluation of DMOEAs. The Inverted Generational Distance(IGD) is a unary performance indicator which provides a quantitative measurement for the proximity and diversity goal of multiobjective optimization [21]. It is mathematically given by

Table 2 Experiment Settings

Number of decision variables, n	FDA:12, dMOP and HE:10, HE2:30
Population size	100 for 2 objective problems 200 for 3 objective problems.
Neighborhood	Size: 20. Probability that parents are selected from the neighborhood is 0.9.
Decomposition method	Tchebycheff
Differential Evolution	CR = 1.0 and F = 0.5.
Polynomial Mutation	$\eta = 20$, $p_m = 1/n$. The number of solutions replaced by any child solution is at most 2.
Number of detectors	10
Percentage for RND model	20%
KF model	Process noise: Gaussian of N(0, 0.04). Observation noise: Gaussian of N(0, 0.01).
Number of changes	20

$$IGD(P^{t*}, P^t) = \frac{\sum_{v \in P^{t*}} d(v, P^t)}{|P^{t*}|} \tag{2}$$

where P^{t*} is a set of uniformly distributed Pareto optimal solutions in the POF at time $t(POF^t)$ and P^t is an approximation of the POF obtained by the algorithm in consideration. d is a distance measure between P^t and P^{t*}, given by

$$d(v, P^t) = \min_{u \in P^t} \|F(v) - F(u)\|. \tag{3}$$

A lower value of IGD implies that the algorithm has better optimization performance. To obtain a low value of IGD, it can be seen from the above 2 equations that, P^t must be very close to POF^t and cannot miss any part of POF^t, thus measuring both convergence and diversity.

To adapt the IGD metric for dynamic multiobjective optimization, the average of the IGD values in some time steps over a run is taken as the performance metric, given by

$$MIGD = \frac{1}{|T|} \sum_{t \in T} IGD(P^{t*}, P^t) \tag{4}$$

where T is a set of discrete time points(immediately before the change occurs) in a run and $|T|$ is the cardinality of T. A lower value of the MIGD metric described above would also assist in evaluating the tracking ability, as the approximated pareto front obtained from the algorithm with the changing pareto optimal front.

4.3 Results

The MOEA/D-KF algorithm is compared with a baseline of random immigrants strategy where a percentage of the population is randomly reinitialized when a change

Table 3 n_t and τ_t values for the benchmark functions

n_t	10	10	10	10	1	1	20	20
τ_t	5	10	25	50	10	50	10	50
τ_t	100	200	500	1000	200	1000	200	1000

occurs and this method is indicated by RND. The difficulty of a DMOOP is determined by the parameters n_t and τ_t which denote the severity and frequency of change respectively. The combination of parameter values used in the simulations are given in Table 3. τ_T denotes the maximum number of iterations.

Table 4 MIGD mean and standard deviation statistics for $n_t = 10$

Problem	RND	MOEA/D-KF
FDA4	0.257978 ± 0.202(+)	**0.207295 ± 0.211**
FDA5	0.538704 ± 0.222(-)	**0.383436 ± 0.193**
dMOP1	**0.173554 ± 0.461**	0.248594 ± 0.452(+)
dMOP2	0.882946 ± 0.682(+)	**0.303540 ± 0.449**
dMOP2iso	**0.009556 ± 0.020**	0.009895 ± 0.020(-)
dMOP2dec	2.739828 ± 6.218(-)	**2.671837 ± 6.109**
HE2	**0.057582 ± 0.001**	0.057710 ± 0.001(-)
HE7	**0.223214 ± 0.030**	0.333162 ± 0.073(+)
HE9	**0.403658 ± 0.035**	0.499466 ± 0.068(+)

(a) $n_t = 10, \tau_t = 5$

Problem	RND	MOEA/D-KF
FDA4	0.144859 ± 0.090(+)	**0.122009 ± 0.090**
FDA5	0.347302 ± 0.111(+)	**0.227943 ± 0.094**
dMOP1	**0.037534 ± 0.103**	0.047511 ± 0.101(+)
dMOP2	0.173277 ± 0.119(+)	**0.078387 ± 0.107**
dMOP2iso	**0.004369 ± 0.002**	0.004411 ± 0.002(-)
dMOP2dec	0.901509 ± 2.811(-)	**0.859458 ± 2.824**
HE2	0.057046 ± 0.001(-)	**0.057030 ± 0.001**
HE7	**0.168585 ± 0.018**	0.253887 ± 0.053(+)
HE9	**0.377706 ± 0.039**	0.478676 ± 0.085(+)

(b) $n_t = 10, \tau_t = 10$

Problem	RND	MOEA/D-KF
FDA4	0.085218 ± 0.016(+)	**0.077284 ± 0.014**
FDA5	0.226148 ± 0.083(+)	**0.140106 ± 0.029**
dMOP1	**0.007402 ± 0.013**	0.008307 ± 0.012(+)
dMOP2	0.022815 ± 0.013(+)	**0.013347 ± 0.012**
dMOP2iso	**0.003743 ± 0.000**	0.003747 ± 0.000(-)
dMOP2dec	**0.099968 ± 0.335**	0.106301 ± 0.337(-)
HE2	0.056916 ± 0.001(-)	**0.056910 ± 0.001**
HE7	**0.153983 ± 0.030**	0.206505 ± 0.054(+)
HE9	**0.353986 ± 0.042**	0.445529 ± 0.084(+)

(c) $n_t = 10, \tau_t = 25$

Problem	RND	MOEA/D-KF
FDA4	0.072128 ± 0.003(+)	**0.069282 ± 0.001**
FDA5	0.197262 ± 0.088(+)	**0.130583 ± 0.038**
dMOP1	**0.004237 ± 0.001**	0.004411 ± 0.001(+)
dMOP2	0.006455 ± 0.001(+)	**0.005675 ± 0.001**
dMOP2iso	**0.003730 ± 0.000**	0.003730 ± 0.000(-)
dMOP2dec	**0.029316 ± 0.099**	0.031941 ± 0.099(-)
HE2	0.056915 ± 0.001(+)	**0.056909 ± 0.001**
HE7	**0.147688 ± 0.034**	0.185881 ± 0.053(+)
HE9	**0.330218 ± 0.032**	0.420749 ± 0.083(+)

(d) $n_t = 10, \tau_t = 50$

Tables 4 and 5 provide the MIGD mean and standard deviation statistics for the different combination of parameter values for the various benchmark problems. Statistical t-test was conducted on the results at the 5% significance level and the best value is denoted in bold. (+) (and (−)) indicates that the difference between the marked entry and the best entry is statistically significant (and insignificant, respectively).

FDA4 and FDA5 are 3-objective problems, while the rest of the problems are 2-objective problems. It can be observed from Tables III-V that MOEA/D-KF performs significantly better than RND on FDA4, FDA5 and dMOP2 in all three parameter

Table 5 MIGD mean and standard deviation statistics for $n_t = 1$

Problem	RND	MOEA/D-KF
FDA4	**0.390467 ± 0.089**	0.602495 ± 0.128(+)
FDA5	**1.058697 ± 0.455**	1.193283 ± 0.450(+)
dMOP1	0.112298 ± 0.111(-)	**0.110023 ± 0.110**
dMOP2cec	**7.429464 ± 7.310**	17.893780 ± 15.640(+)
dMOP2iso	**0.084172 ± 0.073**	0.084333 ± 0.073(-)
dMOP2dec	**30.214032 ± 39.352**	31.296647 ± 39.384(-)
HE2	**0.109764 ± 0.061**	0.110784 ± 0.060(+)
HE7	**0.191397 ± 0.026**	0.237358 ± 0.046(+)
HE9	**0.317087 ± 0.110**	0.367213 ± 0.132(+)

(a) $n_t = 1, \tau_t = 10$

Problem	RND	MOEA/D-KF
FDA4	**0.068035 ± 0.001**	0.069631 ± 0.001(+)
FDA5	**0.825898 ± 0.328**	0.902508 ± 0.358(-)
dMOP1	0.083306 ± 0.073(+)	**0.082921 ± 0.072**
dMOP2cec	0.082403 ± 0.069(+)	**0.076870 ± 0.061**
dMOP2iso	**0.083391 ± 0.073**	0.083401 ± 0.073(-)
dMOP2dec	**20.317731 ± 34.716**	20.346623 ± 34.710(-)
HE2	0.107948 ± 0.060(-)	**0.107937 ± 0.060**
HE7	**0.187327 ± 0.055**	0.202417 ± 0.047(-)
HE9	**0.276737 ± 0.086**	0.349261 ± 0.123(+)

(b) $n_t = 1, \tau_t = 50$

settings. dMOP2 is a type II DMOOP and its time-varying POS is sinusoidal in nature. The Kalman filter prediction can track the changing POS better than the random immigrants strategy in this case. A similar explation for FDA5 applies as well. Though FDA4 is a type I DMOOP, wherein its POF does not change with time, its POS also follows a sinusoidal trajectory and MOEA/D-KF's better performance could be attributed to the more efficient POS tracking.

For dMOP1, the optimal values for all decision variables remain the same throughout the iteration. RND performs better than MOEA/D-KF on this problem, as a majority of RND's population is retained without any modification after a change. Once the EA converges on the POS, the RND method does not effectively disrupt the POF/POS attained.

HE7 and HE9 are type III DMOOPs and their POS is not dependent on time. This might result in the better performance of RND compared to the Kalman filter predictions. dMOP2iso and dMOP2dec consist of isolated and deceptive POF respectively. Even though MOEA/D-KF performs better than RND on dMOP2, it does not perform significantly better than RND on the isolated and deceptive POF variants of dMOP2. This maybe a result of trapping into local optima for both the algorithms.

HE2 is a type III DMOOP, similar to HE7 and HE9. However, it has discontinuous POF, with various disconnected continuous sub-regions [16]. This increases the problem complexity significantly and may lead to similar performance on RND and MOEA/D-KF as specific measures have not been taken to handle such scenarios in the Kalman Filter prediction based DMOEA. Both the algorithms seem to give similar performance on all three parameter settings for HE2.

More efficient strategies are required to enhance the performance of MOEA/D-KF on the IEEE CEC 2015 Dynamic benchmark suite, especially in the problems with isolated, deceptive and disconnected POF. Further, the state transition of the model can be modified such that it is able to better model the movement of decision variables to obtain efficient tracking performance.

5 Analysis

5.1 Effect of Severity of Change

The parameter n_t, denotes the severity of change in the DMOOP. The parameter settings for Table III-(b) and Table IV-(a) are $\tau_t = 10$ and $n_t = 10$, and 1 respectively. It can be seen from the MIGD values on the table that as n_t decreases, the severity of change in the problem increases manyfold. This gets reflected in the MIGD values obtained as high numbers as can be seen in Table IV.

Figures 1 and 2 depict the influence of severity of change on FDA4 and dMOP2 problems. The IGD box plots for RND and MOEA/D-KF are plotted for different values of severity of change. From the range of IGD in the box plots, it can be seen that in the higher n_t setting both the algorithms perform better than in the lower setting. It is also interesting to note that MOEA/D-KF performs better than RND for $n_t = 10, 20$, while it performs worser for the lowest n_t setting.

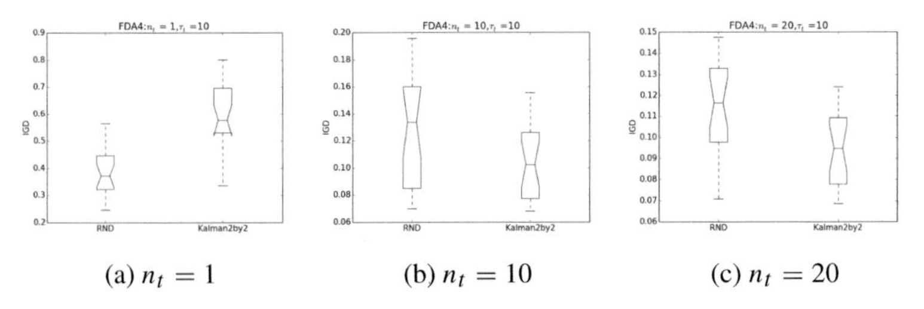

(a) $n_t = 1$ (b) $n_t = 10$ (c) $n_t = 20$

Fig. 1 Effect of severity of change: IGD box plots for FDA4 with $\tau_t = 10$ and $n_t = 1, 10$ and 20

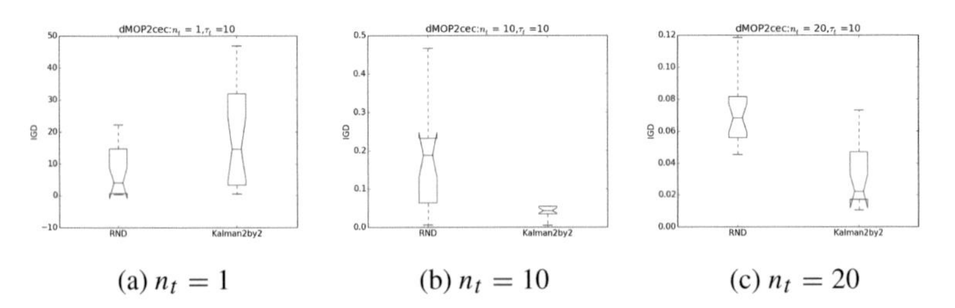

(a) $n_t = 1$ (b) $n_t = 10$ (c) $n_t = 20$

Fig. 2 Effect of severity of change: IGD box plots for dMOP2 with $\tau_t = 10$ and $n_t = 1, 10$ and 20

5.2 Effect of Frequency of Change

The parameter τ_t, denotes the frequency of change in the DMOOP. It determines the number of generations for which the problem does not change. The smaller the frequency of change the problem changes quickly. Thus, as the frequency of change increases with other parameters kept constant, the difficutly of the DMOOP would be expected to decrease.

Figures 3 and 4 depict the trend of IGD with number of changes in the DMOOP for FDA4, FDA5 and dMOP2 problems. The trends are plotted for a number of values of τ_t : 5, 10, 25 and 50. It can be observed from the figures that the MOEA/D-KF algorithm performs better than RND for the various values of τ_t. However, the problem difficulty reduces for higher values of τ_t and therefore, the distance between the trend curves of MOEA/D-KF and RND reduces as τ_t increases.

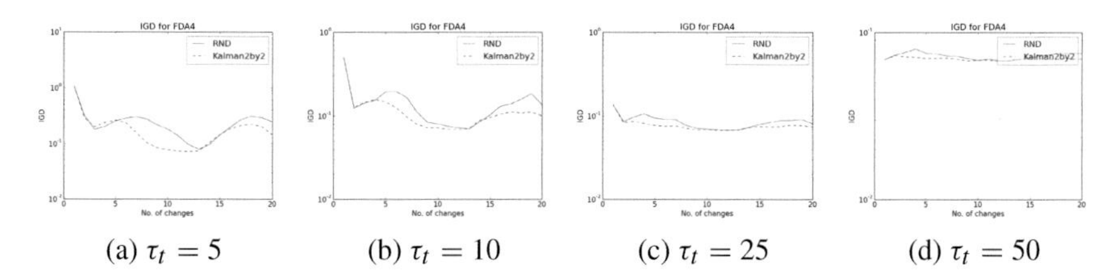

(a) $\tau_t = 5$ (b) $\tau_t = 10$ (c) $\tau_t = 25$ (d) $\tau_t = 50$

Fig. 3 Plots of IGD vs Number of Changes for FDA4 problem with $n_t = 10$ and various values of τ_t

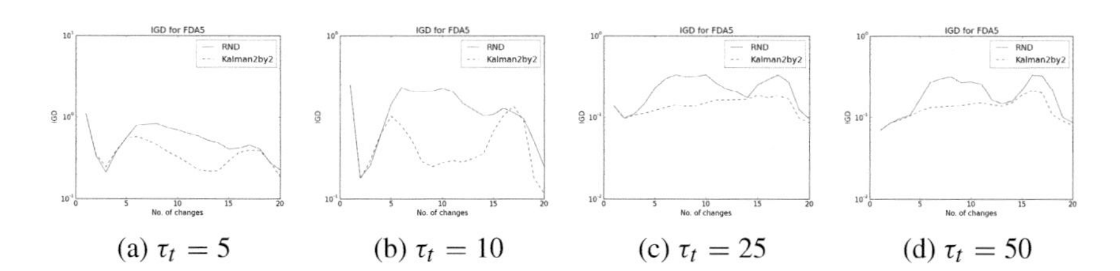

(a) $\tau_t = 5$ (b) $\tau_t = 10$ (c) $\tau_t = 25$ (d) $\tau_t = 50$

Fig. 4 Plots of IGD vs Number of Changes for FDA4 problem with $n_t = 10$ and various values of τ_t

5.3 Initial Populations Obtained Before a Change

Figure 5 shows the initial populations obtained before a change by MOEA/D-KF and RND algorithms for dMOP1, dMOP2 and dMOP2dec problems with $n_t = 10$ and $\tau_t = 10$. All three problems have similar POF. However, their Pareto Optimal Sets are very different from each other. RND is able to obtain solutions close to the POF as the problem's POS does not change with time and the algorithm does not reinitialize a major portion of the population. However, its performance is worser in the dMOP2 and dMOP2dec problems.

In dMOP2, MOEA/D-KF is able to better approximate to the POF than RND. This can be observed in Fig 5(b) as the solutions obtained by MOEA/D-KF are closer to the POF than that of RND. In the case of dMOP2dec, the problem has a deceptive POF. Further, in dMOP2dec, solutions obtained by MOEA/D-KF are within the vicinity of POF whereas only few solutions of RND are visible.

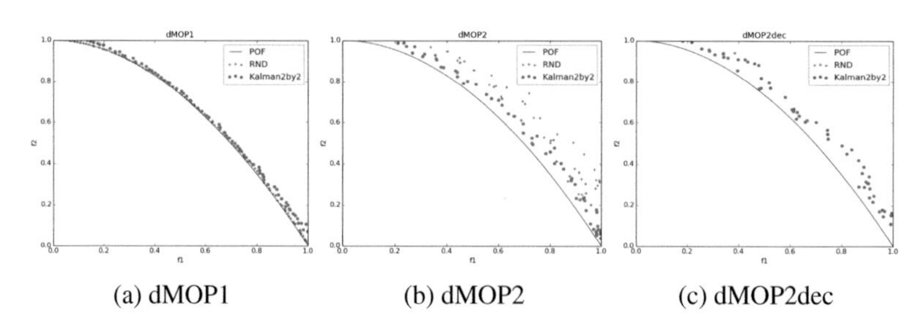

(a) dMOP1 (b) dMOP2 (c) dMOP2dec

Fig. 5 Initial populations obtained before a change by MOEA/D-KF and RND for dMOP1, dMOP2 and dMOP2dec problems with $n_t = 10$ and $\tau_t = 10$

6 Discussion

6.1 Type I DMOOPs

The POF remains static for these problems, while the POS keeps changing with time. In the benchmark problems used in this study, FDA4 falls into this category. MOEA/D-KF seems to perform quite well in these kind of problems in comparison with the RND algorithm.

6.2 Type II DMOOPs

Both the POF and POS change with time in this class of problems. FDA5, dMOP2, dMOP2iso and dMOP2dec come under this categorization. While MOEA/D-KF performs quite well in FDA5 and dMOP2, its performance is much worse in the case of dMOP2iso and dMOP2dec which have isolated POF and deceptive POF respectively. While a generalization cannot be made in this case about the performance of MOEA/D-KF or RND, it is to be noted that in problems with isolated and deceptive POF, DMOEAs may have significant difficulty in tracking the changing POF/POS.

6.3 Type III DMOOPs

The POS remains static in these problems, while the POF keeps changing with time. HE2, HE7 and HE9 problems come under this category. The discontinuous POF characteristic of HE2 leads to similar performance on RND and MOEA/D-KF while RND tends to be perform marginally better than MOEA/D-KF in the other problems as it does not disrupt the converged solutions in the decision space and thereby maintaining the solutions to be close to the POS.

7 Conclusion

In this paper, a brief overview of evolutionary algorithms in the optimization context was provided in general and the issues related to evolutionary dynamic multiobjective optimization were discussed in particular. A Kalman Filter prediction based DMOEA was also outlined. Subsequently, the algorithm was tested on the IEEE CEC 2015 Benchmark suite and its results were compared to the random immigrants strategy also based on MOEA/D-DE. While MOEA/D-KF does perform significantly better than RND in some of the problems, more effective strategies need to be observed to effectively solve them. The effect of severity of change and frequency of change was also analysed. The initial populations obtained by the two algorithms were also visualized to get a better perspective about their optimization performance.

Acknowledgments This work was supported by the Singapore Ministry of Education Academic Research Fund Tier 1 under the project R-263-000-A12-112.

References

1. Deb, K., Pratap, A., Agarwal, S., Meyarivan, T.: A fast and elitist multiobjective genetic algorithm: Nsga-ii. IEEE Transactions on Evolutionary Computation **6**(2), 182–197 (2002)
2. Zhang, Q., Li, H.: Moea/d: A multiobjective evolutionary algorithm based on decomposition. IEEE Trans. Evolutionary Computation **11**(6), 712–731 (2007)
3. Li, H., Zhang, Q.: Multiobjective optimization problems with complicated pareto sets, moea/d and nsga-ii. Trans. Evol. Comp **13**(2), 284–302 (2009)
4. Cheong, C., Tan, K., Liu, D., Lin, C.: Multi-objective and prioritized berth allocation incontainer ports. Annals of Operations Research **180**(1), 63–103 (2010)
5. Tan, W., Lu, F., Loh, A., Tan, K.: Modeling and control of a pilot ph plant using genetic algorithm. Engineering Applications of Artificial Intelligence **18**(4), 485–494 (2005)
6. Ang, J., Tan, K., Mamun, A.: An evolutionary memetic algorithm for rule extraction. Expert Systems with Applications **37**(2), 1302–1315 (2010)
7. Tan, K., Tang, H., Ge, S.: On parameter settings of hopfield networks applied to traveling salesman problems. IEEE Transactions on Circuits and Systems I: Regular Papers **52**(5), 994–1002 (2005)
8. Tan, K., Tang, H., Yi, Z.: Global exponential stability of discrete-time neural networks for constrained quadratic optimization. Neurocomputing **56**, 399–406 (2004)
9. Tan, K., Li, Y.: Grey-box model identification via evolutionary computing. Control Engineering Practice **10**(7), 673–684 (2002). Developments in High Precision Servo Systems
10. Muruganantham, A., Zhao, Y., Gee, S.B., Qiu, X., Tan, K.C.: Dynamic multiobjective optimization using evolutionary algorithm with kalman filter. Procedia Computer Science **24**, 66–75 (2013). 17th Asia Pacific Symposium on Intelligent and Evolutionary Systems, (IES2013)

11. Deb, K.: Multi-Objective Optimization Using Evolutionary Algorithms. John Wiley & Sons Inc, New York, NY, USA (2001)
12. Farina, M., Deb, K., Amato, P.: Dynamic multiobjective optimization problems: test cases, approximations, and applications. IEEE Trans. Evolutionary Computation **8**(5), 425–442 (2004)
13. Branke, J.: Evolutionary Optimization in Dynamic Environments. Kluwer Academic Publishers, Norwell (2001)
14. Kalman, R.E.: A New Approach to Linear Filtering and Prediction Problems. Transactions of the ASME Journal of Basic Engineering **82**(D), 35–45 (1960)
15. Welch, G., Bishop, G.: An introduction to the kalman filter. Technical report, Chapel Hill, NC, USA (1995)
16. Helbig, M., Engelbrecht, A.: Benchmark functions for cec 2015 special session and competition on dynamic multi-objective optimization. Technical report
17. Goh, C.K.: A competitive-cooperative coevolutionary paradigm for dynamic multiobjective optimization. IEEE Transactions on Evolutionary Computation **13**(1), 103–127 (2009)
18. Helbig, M., Engelbrecht, A.: Archive management for dynamic multi-objective optimisation problems using vector evaluated particle swarm optimisation. In: 2011 IEEE Congress on Evolutionary Computation (CEC), pp. 2047–2054 (June 2011)
19. Helbig, M., Engelbrecht, A.P.: Benchmarks for dynamic multi-objective optimisation. In: 2013 IEEE Symposium on Computational Intelligence in Dynamic and Uncertain Environments (CIDUE), pp. 84–91. IEEE (2013)
20. Helbig, M., Engelbrecht, A.P.: Benchmarks for dynamic multi-objective optimisation algorithms. ACM Comput. Surv. **46**(3), 37:1–37:39 (2014)
21. Zitzler, E., Thiele, L., Laumanns, M., Fonseca, C., da Fonseca, V.: Performance assessment of multiobjective optimizers: an analysis and review. IEEE Transactions on Evolutionary Computation **7**(2), 117–132 (2003)

Part IV
Intelligent Systems and their Applications

Ideation Support Based on Infomorphism for Designing Beneficial Inconvenience

Hiroshi Kawakami, Toshihiro Hiraoka and Setsui Riku

Abstract This paper proposes an ideation (idea generation) support process based on Channel Theory, which is a qualitative information theory. Channel Theory formulates information flow by establishing infomorphism between two classifications. By encoding a set of examples into a classification and the target of ideation into another classification, the infomorphism between two classifications guides analogical reasonings. This paper employs fuben-eki systems as the appropriate target of ideation by our proposed process. Fuben-eki denotes the benefits of inconvenience, and fuben-eki systems give users beneficial inconvenience. Our proposed process was implemented in a experimental system as a web application. The experimental result shows that the system improved not the amount but the experiment quality of ideas. Furthermore, questionnaire answers from the participants elucidated how the process helped them conceive ideas.

Keywords Idea generation · System design · Human machine system · Channel theory · Benefit of inconvenience

1 Introduction

One principle of systems design is reducing human labor. In this case, quantitative evaluation of whether a principle was achieved is easy: decreasing the amount of time, labor cost, and monetary cost. On the other hand, the harm of efficiency has also been discussed [1]. Solely pursuing efficiency causes such problems as excluding users, eliminating their ability, and depriving the pleasure of using systems.

H. Kawakami(✉)
Unit of Design, Kyoto University, Kyoto, Japan
e-mail: kawakami@design.kyoto-u.ac.jp
http://www.design.kyoto-u.ac.jp/members/kawakami/

T. Hiraoka · S. Riku
Graduate School of Informatics, Kyoto University, Kyoto, Japan

K. Lavangnananda et al. (eds.), *Intelligent and Evolutionary Systems*,
Proceedings in Adaptation, Learning and Optimization 5,
DOI: 10.1007/978-3-319-27000-5_21

Other than efficiency, several principles of systems design have been proposed, including emotional design [2] and UX design.

As one such principle, we proposed fuben-eki [3], which stands for the benefits of inconvenience and designs inconvenient but beneficial systems. In this case, introducing quantitative evaluations is difficult. For example, providing users a space to develop their skills is one element of fuben-eki. The space can be objectively evaluated, but its value depends on the users that cannot be objectively measured.

In this way, the designs of systems, based on principles other than efficiency, are difficult to quantify and manage in a systematic way. Nevertheless, it is desirable to manage design processes on mathematical bases. As such a base, this paper introduces Channel Theory [4], which is also called qualitative information theory and provides a mathematical framework to Situation Semantics, to the ideation of design concept.

2 Designing Systems Based on Fuben-eki

2.1 Outline

In this paper, fuben-eki denotes the benefits of inconvenience. Fuben-eki examples stand for existing examples that provide fuben-eki, and fuben-eki systems stand for newly designed systems that give users the space to experience fuben-eki. This paper supports the ideation processes for designing fuben-eki systems. Fuben-eki examples include:

- a cell production system instead of a line production system,
- barrier-*aree*[1] instead of barrier-free.

Compared with a line production system that is convenient for workers because they are only required to be skilled at one specific task, a cell production system is superficially inconvenient because it requires that workers have the skills to assemble such complex productions as automobiles in small groups. But the latter allows workers to be skilled and encourages them to understand production that in turn motivates them. Compared with barrier-free that eliminates inconvenient barriers from living spaces, barrier-aree introduces such minor barriers as differences in level on floor on purpose to enjoy maintaining physical abilities.

As a result of analyzing almost 100 examples from the point of view of the relationship between inconvenience and benefit, inconvenience-oriented benefits are classified into eight categories [5], as shown on the right of Fig. 1.

2.2 Premises of Fuben-eki Examples

The following three factors are the premises of fuben-eki examples:

factors of inconvenience: fuben-eki must originate from inconvenience,

[1] *Aree* is Japanese for existence.

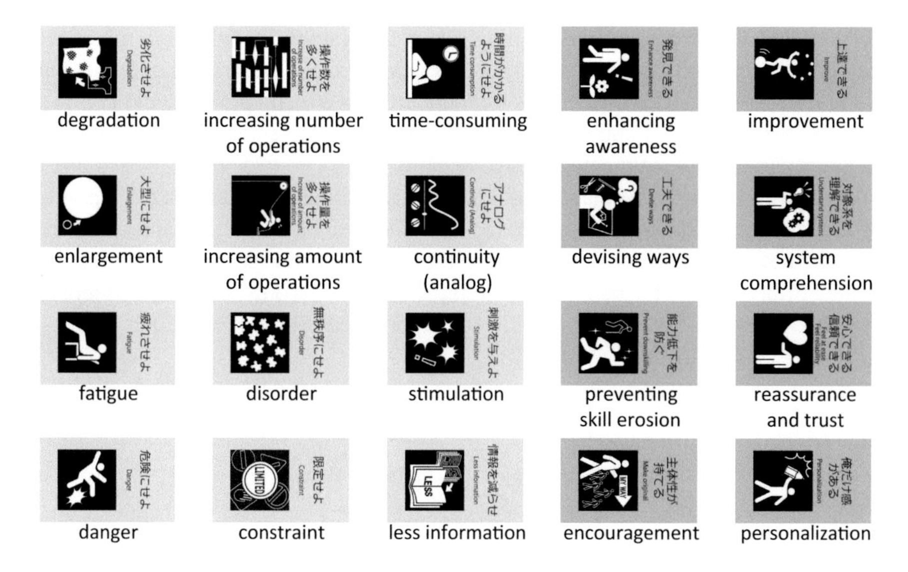

Fig. 1 Fuben-eki principles and categories displayed on cards [6]

main tasks: fuben-eki need to be collateral effects of achieving main tasks,
comparison examples: conveniece/inconvenience are comparative.

2.3 Subjective or Objective View

As described in Section 1, an example of fuben-eki, which provides a space for skill development, is objectively observable, but its value is subjective; appreciating the space is left to users. On the contrary, some users negatively value the space and prefer to innocuously carry on routines.

The subject who reaped a benefit needs to be the person who suffered inconvenience. We do not want to encourage the exploitation of others for our personal benefit.

2.4 Design Support of Fuben-eki Systems

To demonstrate fuben-eki, we developed such fuben-eki systems as a prime number ruler, degraded navigation [7], and a gesture lock for smartphones [8]. They were designed by transforming normal systems into inconvenient ones that provide benefits.

For supporting such designs, we proposed an ideation (idea generation) support method [9, 10] based on TRIZ, which is a theory for inventive problem solving. It is based on patent analyses and provides 40 principles for technical innovations and a tool for searching for appropriate principles called the Contradiction Matrix.

We extracted 12 principles for designing fuben-eki systems by analyzing fuben-eki examples and created a tool called the Fuben-eki Matrix [5]. The principles are shown on the left of Fig. 1. The matrix was implemented on a web application [11] and a card-type tool [6]. Fig. 1 shows the cards. The yellow cards show 12 principles, and the green cards show eight categories of benefits.

We confirmed the performances of our web application and the cards for supporting the ideation of fuben-eki systems [6, 11]. They are based on principles that are derived from the abstractions of the examples. In contrast, this paper does not rely on abstract principles but on analogies to examples. In other words, our previous works supported rule-based reasonings; this work supports case-based reasonings.

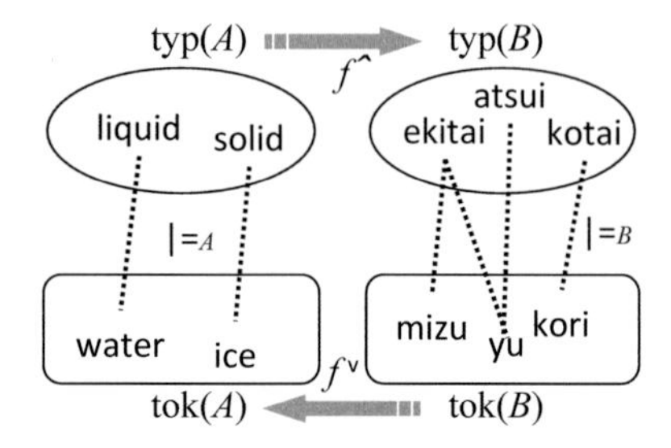

Fig. 2 Examples of classifications and infomorphism

3 Representational Scheme of Fuben-eki Examples Based on Channel Theory

While the standard Information Theory proposed by C. E. Shannon focuses on the quantity of information, Channel Theory [4], which provides a mathematical framework of Dretske's Situation Semantics, is called the qualitative theory of information.

Channel Theory is based on Category Theory and Boolean algebra and is closely related with Chu Space [12]. It is applied to Abstract Design Theory [13], Ontology Engineering [14], and so forth. Its ability to process semantics, which is always exclusive to quantitative operations, is suitable for at least representing fuben-eki, which cannot be evaluated by quantitative factors. Furthermore, its mathematical bases can be applied to support the design of fuben-eki systems.

3.1 Brief Introduction to Channel Theory

Channel Theory represents a local state by a *classification*, and the information flow between two classifications is represented by an *infomorphism*.

Classification $A = \langle \text{tok}(A), \text{typ}(A), \models_A \rangle$ consists of set $\text{tok}(A)$ of tokens, set $\text{typ}(A)$ of types, and binary relation $\models_A$ between $\text{tok}(A)$ and $\text{typ}(A)$. Generally, tokens are objects to be classified and types are their attributes. In this case, $a \models_A \alpha$ can be interpreted as "token a has attribute α," where $a \in \text{tok}(A)$ and $\alpha \in \text{typ}(A)$. Fig. 2 shows examples of classifications A of English words and B of Japanese words.

Infomorphism is a pair $\langle f^{\wedge}, f^{\vee} \rangle$ of functions. Given classifications A and B, an infomorphism from A to B, written as $f : A \rightleftarrows B$, satisfies

$$\forall \alpha \in \text{typ}(A),\ \forall b \in \text{tok}(B),\ b \models_B f^{\wedge}(\alpha) \Leftrightarrow f^{\vee}(b) \models_A \alpha. \tag{1}$$

For example, those who know both languages in Fig. 2 can imagine the following infomorphism:

$$f^{\wedge}(\text{liquid})=\text{ekitai},\ f^{\wedge}(\text{solid})=\text{kotai},$$
$$f^{\vee}(\text{mizu}) = f^{\vee}(\text{yu})=\text{water},\quad f^{\vee}(\text{kori})=\text{ice}.$$

This is semantically correct and syntactically satisfies Eq. (1). On the other hand, Eq. (1) does not always assume a unique and semantically correct morphism. For example, the following morphism satisfies Eq. (1) but is conventionally incorrect: ($f^{\wedge}$(liquid) $=$ kotai, $f^{\wedge}$(solid) $=$ ekitai, $f^{\vee}$(mizu) $=$ $f^{\vee}$(yu) $=$ ice, $f^{\vee}$(kori) $=$ water). In other words, infomorphism can break conventional assumptions.

Constraint $\Gamma \vdash_A \Delta$ satisfies

$$\forall a \in \text{tok}(A),\ (\forall \gamma \in \Gamma, a \models \gamma) \rightarrow (\exists \delta \in \Delta, a \models \delta),$$

where Γ and Δ are exclusive subsets of typ(A) of classification A. $\langle \Gamma, \Delta \rangle$ is called a sequent. The constraints of Channel Theory correspond to the implications of first-order predicate logic.

Local Logic $\mathfrak{L} = \langle A, \vdash_{\mathfrak{L}}, N_{\mathfrak{L}} \rangle$ consists of

- classification A,
- set $\vdash_{\mathfrak{L}}$ of sequents of A called the constraint of $\mathfrak{L}$,
- set $N_{\mathfrak{L}} \subseteq \text{tok}(A)$ of tokens that satisfy all elements of $\vdash_{\mathfrak{L}}$.

Each local logic of A establishes constraints by extracting a subset from tok(A) even if the constraints do not satisfy the whole set tok(A). Compared with first-order predicate logic, which quantifies tokens either $\forall$ (for all) or $\exists$ (it exists), the local logics of Channel Theory are for $N_{\mathfrak{L}}$ that are located between $\forall$ and $\exists$.

3.2 Representation and Analysis of Fuben-eki Examples Based on Channel Theory

This section represents fuben-eki examples by slightly extending the classification format of Channel Theory. Based on Channel Theory, infomorphism guides analogical reasoning, and the constraints derive beneficial attributes for supporting the ideation of fuben-eki systems.

Types and Tokens: By corresponding each token to an example and each type to an attribute of the examples, a set of fuben-eki examples is represented by a classification. As described in Section 2.2, examples must be represented by at least their main tasks, their comparison examples, and the factors that are inconvenient. As described in Section 2.3, we must also distinguish between subjective or objective viewpoints. Embracing these notions, each example is represented by the following three types:

common type (χ): common attributes of the fuben-eki example and its comparison example,
specific type (σ): objective functions specific to the fuben-eki example,
fuben-eki type (ϕ): subjective benefits felt by users who use the fuben-eki example.

Common types χ have such sub-categories as *is-used-for*, *is-a*, *has-a*, and *be-related-to*. Specific types σ are the factors that make examples inconvenient and derive fuben-eki (benefits of inconvenience).

Each token represents an fuben-eki example. Each token is labeled by the name of a concept.

Classifications *F* and *T*: A classification is defined by determining the binary relations between sets of tokens and types. The set of known fuben-eki examples is encoded into a classification called F. Tok(F) consists of the names of examples. Typ(F) consists of common, specific, and fuben-eki types. $\models_F$ indicates the types of each example.

The theme of ideation is also encoded into a classification that we call T. Its token is the target of ideation that we call t. Assuming a situation where users are faced with generating ideas for target t, $\text{tok}(T) = \{t\}$ and $|\text{tok}(T)| = 1$.

Informorphism $F \rightleftharpoons T$: The infomorphism between F and T represents the similarity between known examples and the target of ideation. Both constituents of infomorphism $\langle f^\wedge, f^\vee \rangle$ are functions where each element of their domain needs to be mapped uniquely to an element of their codomain. If $f^\vee : \text{tok}(F) \rightarrow \text{tok}(T)$, then all the known examples are mapped to a single target t. This mapping is semantically abnormal. Normally, a target is associated not with all the known examples but some similar examples. Therefore, $f^\vee : \text{tok}(T) \rightarrow \text{tok}(F)$. By remapping $f^\vee(t)$, similar examples are explored in tok(F). In this case, the definition of infomorphism derives $f^\wedge : \text{typ}(F) \rightarrow \text{typ}(T)$.

Constraint $\Gamma \vdash_F \Delta$: The constraints of F and its local logics support analogical reasoning. When generating ideas by analogies to known examples, specific types σ of the examples are beneficial for transforming target t into an inconvenient one. Therefore, beneficial constraints can be selected from those of F and its local logics by checking Δ, where all the constituents are specific types σ. Furthermore, for practical use, the constraints with $\neg(\exists f \in \text{tok}(F),\ \forall \gamma \in \Gamma,\ f \models_F \gamma)$ or $\Delta = \emptyset$ are eliminated even if they are logically correct.

On the other hand, even it is logically incorrect to derive $\Gamma \vdash \delta_i$ from $\Gamma \vdash \delta_1 \vee \cdots \delta_i \vee \cdots \delta_n$ $(n \geq 2)$, we employ Γ as the possibility of δ_i when Γ consists of common χ or fuben-eki types ϕ. When Γ consists of specific types σ, $(\{a\} \vdash \{b\}) \wedge (\{b\} \nvdash \{a\})$ gives an order relation between a and b, and $(\{a\} \vdash \{b\}) \wedge (\{b\} \vdash \{a\})$ gives an equivalent relation to a and b. The order and the equivalent relations make a partially ordered set of specific types σ.

Local Logic: The constraints of F are too strong for the ideation processes because they embrace all the tokens of F. Some are irrelevant to target t. Constraints $\vdash_{\mathfrak{L}}$ of local logic $\mathfrak{L}$ for a set of relevant tokens $N_{\mathfrak{L}} \subseteq \mathrm{tok}(F)$ are appropriate. The variation of selecting $N_{\mathfrak{L}}$ supports the diverseness of ideation.

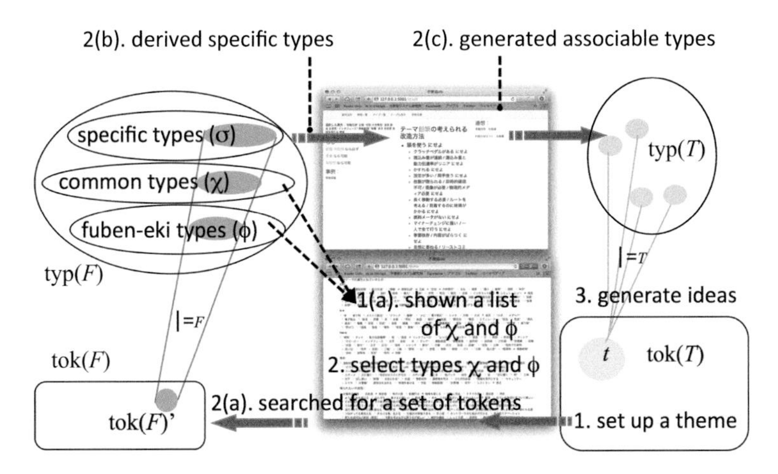

Fig. 3 Flow of ideation support

4 Implementation and Validation of a System for Ideation Support

4.1 Ideation Support Process

Based on the settings described in Section 3.2, this section proposes the ideation support process as shown in Fig. 3. In the figure, the process flow is superposed on the infomorphism flow.

1. The system user sets up a target of ideation. In other words, token t is put into classification T.

   (a) Lists of common types (χ) and fuben-eki types (ϕ) is shown.

2. From the lists, the user selects common types (χ) that are related to t and desirable fuben-eki types (ϕ). Consequently, the domain of $f^{\wedge}$ is determined.
   (a) Set of tokens $\text{tok}(F)'$ is extracted from $\text{tok}(F)$, s.t. each constituent of $\text{tok}(F)'$ is at least one of the selected types. The codomain of $f^{\vee}$ is $\text{tok}(F)'$, so each constituent of $\text{tok}(F)'$ can be $f^{\vee}(t)$, and the possible number of infomorphisms is equivalent to $|\text{tok}(F)'|$.
   (b) Local logic $\mathfrak{L}$ is defined where $N_{\mathfrak{L}} = \text{tok}(F)'$. Following $\vdash_{\mathfrak{L}}$, a set of specific types σ is derived from selected χ and ϕ.
   (c) The associable types of derived σ are generated in $\text{typ}(T)$ as $f^{\wedge}(\sigma)$.
3. The user generates ideas by referring to σ and $f^{\wedge}(\sigma)$. Consequently, binary relations $t \models_T f^{\wedge}(\sigma)$ are established.

In Step 3 (c), associable types $f^{\wedge}(\sigma)$ are generated for translating derived σ in Step 3 (b) into new types that fit t.

4.2 Specifications of Experimental System

The experimental system is implemented by Python as a web application. The system's database consists of four tables: $\text{tok}(F)$, $\text{typ}(F)$, $\models_F$, and associable types. The data sizes are as follows:

$$\begin{aligned} \text{token} &\quad 81, \\ \text{common type} &\quad 195, \\ \text{specific type} &\quad 191, \\ \text{fuben-eki type} &\quad 37, \\ \text{binary relation } (\models_F) &\quad 1{,}150, \\ \text{candidates of associable type} &\quad 206{,}432. \end{aligned}$$

The system's web site consists of pages for registering $\text{tok}(T)$, for registering $\text{typ}(T)$, for browsing $\models_T$, for ideation support, and for showing results. The result page shows ten associable types for every specific type.

4.3 Experimental Settings and Procedure

To validate the performance of our proposed process, we conducted intra-subject experiments using our experimental system. The seven participants had almost equal levels of understanding of fuben-eki. Written informed consents were obtained from them. Each participant did three sessions: (1) normal idea generation, (2) idea generation with a partial support system, and (3) idea generation with a full support system.

Instruction (10 min.): The participants were instructed about the concept of fuben-eki.

(1) Normal idea generation (15 min.) The participants generated ideas about transforming a wristwatch into fuben-eki systems and then wrote them down.

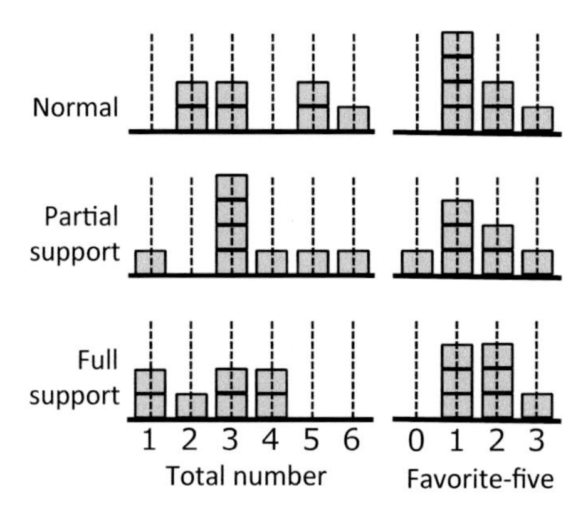

Fig. 4 Numbers of generated ideas

Table 1 Number of ideas generated by each participant

	Total number								Favorite-five								
Sessions	A	B	C	D	E	F	G	sum	A	B	C	D	E	F	G	sum	rate
(1) Normal	3	5	5	2	2	6	3	26	2	1	1	2	1	1	3	11	11/26=0.42
(2) Partial support	3	3	6	3	3	4	1	23	2	2	3	0	1	1	1	10	10/23=0.44
(3) Full support	1	2	3	3	4	4	1	18	1	2	1	2	3	2	1	12	12/18=0.67

They mentioned as many kind of benefits as possible that are derived from the transformation.

Practice (10 min.): The participants used the support system. They were given a minute in which to repeat the selection of relevant types for a given theme.

(2) Idea generation with partial support system (15 min.): The participants generated ideas of transforming a wristwatch into fuben-eki systems. They were supported by specific types σ, constraints $\vdash_{\mathfrak{L}}$, and examples $\mathrm{tok}(F)'$. They wrote down their ideas with relevant σ.

(3) Idea generation with full support system (15 min.): The participants generated ideas of transforming a wristwatch into fuben-eki systems. They were supported by σ, $\vdash_{\mathfrak{L}}$, $\mathrm{tok}(F)'$, and associable types $f^{\wedge}(\sigma_i)$. They wrote down their ideas with relevant $f^{\wedge}(\sigma_i)$.

Selection of favorite five ideas (5 min.): The participants selected their favorite five ideas from their own ideas generated in sessions (1), (2), and (3).

Answering questionnaires (10 min.)

4.4 *Experimental Results and Discussions*

Number of Ideas: The left parts of Table 1 and Fig. 4 show the numbers of generated ideas by each participant in each session. All sessions lasted 15 minutes, so the results

Table 2 Questionnaire scores

Participant	A	B	C	D	E	F	G	average
Q II	3.5	3	4	2	3	2	2	2.8
Q III	3.2	4	5	3	4	5	3	3.9
Q IV-1	150	200	200	120	105	200	130	151.9
Q IV-2	110	160	250	110	150	50	80	130.0
Q V	4.3	5	5	3	5	5	4	4.5
Q VI	3.8	5	5	4	5	5	2	4.3

Table 3 Questionnaire items

Q II	: I made full use of the system. (no: 0, yes: 5)
Q III	: The system helped me conceive ideas. (no: 0, yes: 5)
Q IV-1	: How much did the system help you in session (2)?
Q IV-2	: How much did the system help you in session (3)?
Q V	: Some ideas could not been conceived without this system. (no: 0, yes: 5)
Q VI	: I want to use this system in the future. (no: 0, yes: 5)

show that the support systems reduced the number of ideas per unit time. We got the following related comments from participants:

order effect: "I squeezed out most of my ideas in the first session."
time resource: "I spent most of the time of sessions using the system."

In spite of these setbacks, participants D and E generated more ideas in sessions (2) and (3) than in session (1). According to the result of session (1), they are poor at ideation. These facts support our hypothesis that the support system is effective for those who are not good at ideation.

The right parts of Table 1 and Fig. 4 show the numbers of favorite ideas. The support system reduced the total number of ideas but did not reduce the number of favorite ideas. More than half of the generated ideas in session (3) were favorite. These facts support our hypothesis that associable types $f^{\wedge}(\sigma_i)$ improve the quality of ideas.

The favorite-five include such ideas as a "game watch" that shows the time only after users win a simple game, and a "gesture watch" that does not show the time unless users shake it in a pre-registered gesture.

Questionnaires: Table 2 shows the questionnaire answers, and Table 3 shows the questions.

Questions V and VI earned full marks from more than half of the participants and both average scores are high. The system did help participants conceive ideas.

The scores of question IV were normalized so that the score of session (1) is 100. Both sessions (2) and (3) got more than 100 on average. Most participants preferred session (2) to session (3). We got the following comments: "the associable types were too numerous;" "most of the associable types were wasteful;" "they confused me;" and "most were badly written Japanese sentences." On the other hand, the participants who gave session (3) high marks said: "generating ideas was fun;" and "the associable types were concrete, and some were almost self-completed."

5 Conclusions

This paper introduced Channel Theory as the framework of an ideation (idea generation) support process based on case-based reasoning. We encoded a set of examples into a *classification*, and the theme of ideation is also encoded into another *classification*. The *infomorphism* between two classifications relates the target of ideation to known examples. The *local logics* are established by relevant examples of the target, and the *constraints* of the local logics derive *types* that are beneficial for ideations. The proposed processes were implemented in a web application and applied to the ideation of fuben-eki systems to give users beneficial inconvenience. Our experimental results elucidated the effect of the proposed process and got positive feedback from the experiment participants.

Even though the scale of the experiments was inadequate for firmly stating a conclusion about the performance of our proposed process, our results support the following hypotheses:

- The support system is effective for persons who are poor at ideation,
- The associable types improve the quality of ideas.

The research reported in this paper can be developed in the following directions:

Improving the Case Base: The case base of the web application is a set of fuben-eki examples. Enriching them would enhance the ability of ideation support.

Improving Evaluation Method: In our reported experiments, ideas were subjectively evaluated by the participants. Introducing evaluations by bystanders would improve the objectivity of the experiments.

Generalization: In this paper, our proposed process was applied to the ideation of fuben-eki systems, but it was not restricted to fuben-eki. By replacing the case base, our proposed process can be applied to the ideation of other fields where qualitative evaluations are difficult to introduce.

References

1. Norman, D.A.: Human centered design considered harmful. Interactions **12**(4), 14–19 (2005)
2. Norman, D.A.: Emotional design. Basic Books (2005)
3. Kawakami, H.: Inconvenience utilizing design. Kagaku-Dojin Publishing Company Inc., (2011) (in Japanese)

4. Barwise, J., Seligman, J.: Information Flow: The Logic of Distributed Systems. Cambridge University Press (1997)
5. Naito, K., Kawakami, H., Hiraoka, T.: Design support method for implementing benefit of inconvenience inspired by TRIZ. In: 12th ETRIA TRIZ Future Conference, pp. 351–356, October 2012
6. Hasebe, Y., Kawakami, H., Hiraoka, T., Naito, K.: Card-type tool to support divergent thinking for embodying benefits of inconvenience. Web Intelligence **13**(2), 93–102 (2015)
7. Kitagawa, H., et. al: Implementing a degrading navigation system as an explanatory example of "benefit of inconvenience. In: Proc. of SICE Annual Conference, pp. 1738–1742 (2010)
8. Araki, T., Kawakami, H., Hiraoka, T.: Making systems inconvenient to stimulate motivation of competent users. In: Proc. of IEEE SMC 2015 (2015). (in print)
9. Altshuller, G.: 40 Principles: TRIZ Kyes to Technical Innovation. Technical Innovation Center, Inc. (1997)
10. Mann, D.: Hands-on systematic innovation. CREAX Press (2002)
11. Naito, K., Kawakami, H., Hiraoka, T., Inui, K.: Support system for embodying benefit of inconvenience by referring to TRIZ. In: Proc. of 2012 Int. Conf. on Humanized Systems, pp. 175–180 (2012)
12. Chu Spaces. http://chu.stanford.edu
13. Kakuda, Y.: Design and Information Flow. J. of the Japan Assoc. Phil. Sci. **29**(1), 1–6 (2001). (in Japanese)
14. Kalfoglou, Y., Schorlemmer, M.: IF-map: an ontology-mapping method based on information-flow theory. In: Journal on Data Semantics, vol. 2800, pp. 98–127. LNCS (2003)

A Solution to the Cold-Start Problem in Recommender Systems Based on Social Choice Theory

Li Li and Xiao-jia Tang

Abstract Recommender systems are a popular approach for dealing with the problem of product overload. Collaborative Filtering (CF), probably the best known technique for recommender systems, is based on the idea of determining and locating like-minded users. However, CF suffers from a common phenomenon known as the cold-start problem, which prevents the technique from effectively locating suggestions for new users. In this paper we will investigate how to provide a recommendation to a new user, based on a previous group of users opinions, by utilizing techniques from social choice theory. Social choice theory has developed models for aggregating individual preferences and judgments, so as to reach a collective decision. We then determined how these can best be utilized to establish a collective decision as a recommendation for new users; hence, a solution to the cold start problem. This solution not only solves the cold-start problem, but can also be used to give existing users more accurate suggestions. We focused on models of preference aggregation and judgment aggregation; specifically, by using the judgment aggregation model to solve the cold-start problem, which is a novel approach.

Keywords Collaborative filtering algorithm · Social choice theory · Preference aggregation model · Judgment aggregation model · Cold-start problem

1 Introduction

Recommender systems (RSs) [1] technology was designed to help agents solve the product overload challenge, as it can suggest items to users based on their preferences. In general, recommendations are relied on items related to the characteristics or users'

L. Li(✉) · X.-j. Tang(✉)
Center for the Study of Logic and Intelligence, Southwest University,
1st TianSheng Road, Beibei District, Chongqing 400715, China
e-mail: lilylea220@gmail.com, txj@swu.edu.cn

K. Lavangnananda et al. (eds.), *Intelligent and Evolutionary Systems*,
Proceedings in Adaptation, Learning and Optimization 5,
DOI: 10.1007/978-3-319-27000-5_22

features, while some of the recommendations are reliant on other users who have similar interests or tastes. The outcome of the process is determined by combining the information of users and items; thus, being able to more adequately ascertain the recommendation results. Recommendation systems algorithms attempt to explore the connection between users and items and allows the users more personalized recommendations. Almost all websites use recommender systems to relate some useful and appropriate advice to their customers.

Social choice theory focuses on group decision processes and procedures, and a growing literature research in the aggregation of individual inputs (e.g. votes, preferences, judgments, welfare) into collective outputs (e.g, collective decisions, preferences, judgments, welfare). The preference aggregation model [2]is designed to aggregate several individuals' preference orders into a collective preference order over alternatives set. Judgment aggregation [3, 4] is a new branch of social choice theory, which is based on research related to how individual judgments can be aggregated into a collective decision, which is based on deductive logic and reason. More specifically, judgment aggregation considers how multiple sets of logical formula can be combined into a single consistent collective set. There are some academic literary publications related to judgment aggregation rules [5, 6], which are used to determine the consistent collective judgment set. Compared with preference aggregation model, judgment aggregation model is built on the basis of logic, which is more general and accurate. Judgment aggregation model always is applied in aritficial intelligence and computer science.

The Cold-start problem is one of the most difficult problems to solve in RSs, and it has a close relationship with data sparsity in recommendation algorithms. Since the algorithms cannot obtain previous knowledge of the users, the recommender systems are inhibited in the function of transmitting a recommendation to the new user. In this paper we will focus on solving the cold-start problem for new users. Because people easily adopt the group of previous users' opinions, we proposed an algorithm to employ a preference aggregation model and judgment aggregation model of social choice theory to locate a group decision as a recommendation to a new user. There are three phases we designed, so as to better attain the recommendation outcome for a new user. The first phase was to select a group of previous users, who had sufficient data for appropriate application. The second phase was to aggregate the individuals' preferences and judgments to be a group decision set as a recommendation result. The last phase should present the recommendation result to the new user.

When it comes to related work, some solutions to cold-start problem have been proposed recently. S.Shaghayegh et al.(2011) [7] presented a solution based on homophily in a social network, which use social networks' information in order to fill the gap and detect similarities between users. S.Part et al. (2009) [8] proposed predictive feature-based regression models that leverage all available information of users and items to alleviate cold-start problems. A.I.Schein et al.(2002) [9]developed a new performance metric named CORC curve and demonstrated empirically that the various components of their testing strategy combined, to obtain a deeper understanding of the performance characteristics of recommender systems, and their testing is based on cold-start recommendations. In this paper we attempt to solve the recommendations on existing items for new users, which is a type of cold-start problem.

2 Collaborative Filtering Algorithm Based on Social Choice Theory

There are numerous literatures describing the traditional collaborative filtering algorithm; in this section we want to explore a new and unique method to gather the group's user information, so as to analyze and obtain the group decision, then recommend the collective decision result to a new user. In this case, we will have an initial set of users U which we assume is a finite set consisting of at least two elements: $U = \{u_1, \ldots, u_n\}$, where $n \geq 2$. Also, each systems will have a fixed set of items $I = \{I_1, \ldots, I_m\}$,where $m \geq 3$, and we suppose that a new user will have made little or nothing known about her views regarding what she feels about the items, the system cannot give a recommendation to the new user. In this paper, we will solve this type of cold-start problem and we will focus on the recommender systems based on preference aggregation model and judgment aggregation model which assume that newcomers often like to take the suggestions from previous users' opinions about some items in a total new website. So we will describe a function which, on the basis of some attitudes by the users concerning the items, we will extraplate the collective attitude of the group on the set of items, and give the new user this collective attitude. In this section,we plan to change the collaborative filtering procedure for recommendation, based on preference and the judgment aggregation model in the following graph, as compared with the traditional user-based collaborative filtering algorithm.

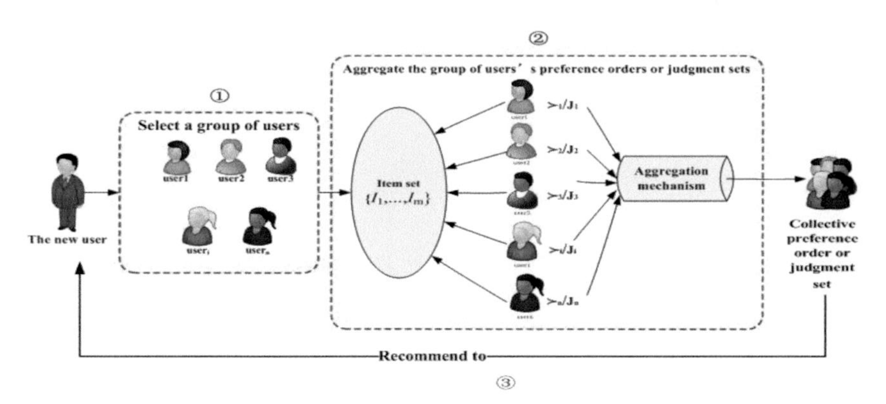

Fig. 1 Collaborative filtering algorithm based on social choice theory

Phase 1: Select a Group of Users

From Figure 1, we can clearly determine that the first step of our algorithm was to select a group of users. At the beginning we had an active user, who had no purchase history; only little data about his/her registration information or location IP address etc. were available. For example, her location, gender, E-mail address and many website preferences were available through logging in with the users' Facebook , Twitter and Linkedin accounts, which were included in the registration information database, and the other information, such as which kind of course the user had chosen in Coursera, etc. The social network information of users is quite important, as it can be a significant tool in solving the cold-start problem in recommender systems. If the database has more information about the active user, then the recommender systems can more easily select some people who are his/her relatives, friends from Facebook or Linkedin, followed users on Twitter, and classmates from Coursera. These people, most likely, share a significant part of the active user's interests or preferences. If we aggregate the ratings that were provided by the user's friends, the system could predict the active user's interest for some select items.

However, if we cannot find the active user's friends from Facebook or Twitter etc, we will select some users in the database by random selection. The only condition related to these users is that the users must assign selected item scores. If the users have enough information to utilize for aggregation, these users can be in a group temporarily, as it is a reasonable method to form a group, as well as determining the group decision.

Therefore, the first step of our algorithm was to select some users as a group; the number of the members should be more than two and odd in number, in order to keep the procedure proceeding successfully and avoid a tie-breaking case result. After forming a group, the user-item ratings matrix of the group members was collated and inputted into the aggregating mechanism.

Example 1. Given the active user is John, who had no previous data for use in this website, and we also selected five other users in a group $U = \{u_1, \ldots, u_5\}$, five items $I = \{I_1, ..., I_5\}$ and the user-item ratings matrix information as input in the following table.

Table 1 User-item ratings matrix

	I_1	I_2	I_3	I_4	I_5
$user_1$	5	3	4	1	2
$user_2$	5	4	3	2	1
$user_3$	1	4	3	2	5
$user_4$	2	4	3	5	1
$user_5$	5	3	4	2	1

Phrase 2: Aggregate the Group of Users' Preferences or Judgments

It was an important step for our algorithm for the group aggregation; this step should aggregate individuals' preference orders or judgments to a group preference order or a collective judgment set. There are two types of aggregation approaches in social choice theory, which are preference aggregation and judgment aggregation respectively. In this paper we used both the preference and judgment aggregation model to attain the group decision. We will present them respectively in the following subsections. Each record of user-item ratings matrix expresses the user's preference for each item, and the system takes it as a vote and computes the winner, based on these records.

2.1 Group Decision Based on Preference Model

When the system has received the data, such as the above user-item ratings matrix, this step should obtain the group decision. A very simple and effective mechanism for aggregating preferences of multiple users is to apply preference rules.

The preference model is defined as follows. Let a set of individuals or users $U = \{u_1, \ldots, u_n\}(n \geq 2)$, $I = \{I_1, \ldots, I_m\}(m \geq 3)$ be nonempty set of items. Each individual $u_i \in U$ has a preference order $\succ_i$ over items set I: a linear order (that is, irreflexive, transitive, and complete)on I. Let $\mathcal{L}(I)$ denote the set of all linear orders on I. A *profile* $V = (\succ_1, \ldots, \succ_n) \in \mathcal{L}(I)^{|U|}$ is a vector of preferences. A preference rule F is a function that assigns of preference orders to a collective order, that is $F : \mathcal{L}(I)^{|U|} \rightarrow \mathcal{L}(I)$.

Example 2. Given a group of users $\{u_1, u_2, u_3, u_4, u_5\}$, the items set $I = \{I_1, I_2, I_3, I_4, I_5\}$ and user-item ratings matrix in table 1 as input, the system can transform the user-item ratings matrix to the preference orders, based on the preference model from the above definitions. From table 1, we can see that every user has assigned a score to each item, for example, $user_1$ assigned 5 to $item_1$, 3 to $item_2$, 4 to $item_3$, 1 to $item_4$, and 2 to $item_5$. It indicates that $user_1$ likes $item_1$ the most, $item_3$ is in the second place, $item_2$ is in the third place, $item_5$ is in the fourth place, while $item_4$ is the worst one in her mind. So her preference order should be $I_1 \succ I_3 \succ I_2 \succ I_5 \succ I_4$. According to the preference model definition, we can transform the user-item ratings matrix to the corresponding preference order for all users in table 2.

Table 2 The corresponding preference profile for every user

	Preference orders
$user_1$	$I_1 \succ I_3 \succ I_2 \succ I_5 \succ I_4$
$user_2$	$I_1 \succ I_2 \succ I_3 \succ I_4 \succ I_5$
$user_3$	$I_5 \succ I_2 \succ I_3 \succ I_4 \succ I_1$
$user_4$	$I_4 \succ I_2 \succ I_3 \succ I_1 \succ I_5$
$user_5$	$I_1 \succ I_3 \succ I_2 \succ I_4 \succ I_5$

A. Linear Borda Count Rule

Definition 1 ***Linear Borda Count.*** *Let* $I = \{I_1, ..., I_m\}(m \geq 3)$ *be a set of items,* $U = \{u_1, \ldots, u_n\}(n \geq 2)$ *be a set of users, and* $B(u)$*be a linearly ordered ballot*

on I of each user. Let I_i be the ith most preferred item among the n total items in B(u). The Linear Borda Score of I_i in ballot B(u) is defined as $BS_{I_i}(u) = n - i$. Let $\mathcal{B} = \{B(u_1), ..., B(u_n)\}$ be a profile of I, $I_i \in I$ and $B(u_j) \in \mathcal{B}$. The Borda score of item I_i in ballot $B(u_j)$ is defined as $BS_{I_i}(u_j)$. The total Borda score of item I_i is defined as $\sum_{j=1}^{m} BS_{I_i}(u_j)$.

Each item gets 0 points for each last place vote received, 1 point for each next-to-last point vote, and so on, all the way up to n-1 point for each first place vote (where n is the number of items). According to the preference profile in table 2 and the definition of the Borda count method, we then have:

$$\sum_{j=1}^{5} BS_{I_1}(u_j) = 13; \sum_{j=1}^{5} BS_{I_2}(u_j) = 13; \sum_{j=1}^{5} BS_{I_3}(u_j) = 12; \sum_{j=1}^{5} BS_{I_4}(u_j) = 7; \sum_{j=1}^{5} BS_{I_5}(u_j) = 5.$$

Now we obtain the group preference ordering $I_1 \sim I_2 \succ I_3 \succ I_4 \succ I_5$. In this preference ordering, we could not see a difference between I_1 and I_2. Since they had the same scores, I_1 and I_2 should be the same best items; both of them were the winners under the linear Borda count.

B. Copeland Rule

Definition 2 ***Copeland Rule.*** *Let $I = \{I_1, \ldots, I_m\}$ be a set of items, the Copeland score can be defined as follows.*

$$C_{I_{ij}} = \begin{cases} 1, & \text{if } I_i \succ I_j \\ 0, & \text{if } I_i \sim I_j \\ -1, & \text{if } I_j \succ I_i \end{cases}$$

For every item I_i,its Copeland score is $\sum_{j=1}^{m} C_{I_{ij}}$, The Copeland rule choose the item with the highest Copeland score wins.

The Copeland rule chooses the item, which exceeds the highest number of other items in pairwise preference order. Therefore, we needed to transform the preference order presented in table 2 to pairwise preference order profile in table 3. For instance, $user_1$'s preference order was $I_1 \succ I_3 \succ I_2 \succ I_5 \succ I_4$, in fact, she agreed that $I_1 \succ I_2$, $I_1 \succ I_3$, $I_1 \succ I_4$, $I_1 \succ I_5$, $I_2 \succ I_4$, $I_2 \succ I_5$, $I_3 \succ I_4$, and $I_3 \succ I_5$, which these pairwise preference orders were all true for $user_1$ and marked as 1, but she supported neither $I_2 \succ I_3$ nor $I_4 \succ I_5$ and should be marked as -1. We were then able to accurately calculate the Copeland score for each item in table 4, based on the pairwise preference order.

Table 3 Pairwise preference oder profile

	$I_1 \succ I_2$	$I_1 \succ I_3$	$I_1 \succ I_4$	$I_1 \succ I_5$	$I_2 \succ I_3$	$I_2 \succ I_4$	$I_2 \succ I_5$	$I_3 \succ I_4$	$I_3 \succ I_5$	$I_4 \succ I_5$
$user_1$	1	1	1	1	-1	1	1	1	1	-1
$user_2$	1	1	1	1	1	1	1	1	1	1
$user_3$	-1	-1	-1	-1	1	1	-1	1	-1	-1
$user_4$	-1	-1	-1	1	1	-1	1	-1	1	1
$user_5$	1	1	1	1	-1	1	1	1	1	1

Table 4 Copeland scores

	I_1	I_2	I_3	I_4	I_5	Copeland scores
I_1	0	1	1	1	3	6
I_2	-1	0	1	3	3	6
I_3	-1	-1	0	3	3	4
I_4	-1	-3	-3	0	1	-6
I_5	-3	-3	-3	-1	0	-10

The highest Copeland score was I_1 and I_2 from calculating the Copeland scores, then I_3, I_4 and I_5; therefore, the collective preference ordering $I_1 \sim I_2 \succ I_3 \succ I_4 \succ I_5$, the I_1 and I_2 have same Copeland score, both of them are winners under Copeland rule. The collective preference order could be a recommendation to the new user.

2.2 Group Decision Based on Judgment Aggregation Model

Judgment aggregation is a new area of research, as related to social choice theory; the main research problem is discovering a collective judgment set that represent group members' judgments on logical issues [6]. The interest in combining judgment aggregation theory with recommender systems was sparked by the literature on judgment aggregation rules based on minimization [5]; perhaps the consistent collective judgments, which are obtained by judgment aggregation rules can provide a new user with a solution to buy some items or read some news, books, videos, etc. In this section we will transform the user-item rating matrix to a judgment aggregation profile, while utilizing the judgment aggregation rule to determine the collective judgment set; thus, the collective judgment set will be a recommendation result which can help a new user solve the cold-start problem.

Judgment aggregation model is defined as follows. Let $U = \{u_1, \ldots, u_n\}(n \geq 2)$, $I = \{I_1, \ldots, I_m\}(m \geq 3)$ be a nonempty set of items. Let $\mathcal{L}$ be a set of well-formed propositional logic formulas, which not only include $\bot$ and $\top$, but also some logical connectives, such as $\neg$, $\wedge$, $\vee$, $\rightarrow$ and $\leftrightarrow$.

Issues issue is a proposition of $\mathcal{L}$, it means $item_i$is assigned j by a user, which can be denoted by $P_{I_i j}$, for ease of programming, we shall abbreviate "$P_{I_i j}$" to "P_{ij}" in the following paper.

Agenda The set $\mathcal{A}$ of propositions under consideration is called agenda, and it subdivided some issues; especially in this paper. Agenda $\mathcal{A}$ =$\{P_{ij}|i \in I, 1 \leq j \leq$

$|I|\} \cup \{\neg P_{ij} | i \in I, 1 \le j \le |I|\}$ means that every item should be assigned a natural number as the score of it is according to the users' preferences. For instance, if an agenda is $\mathcal{A}$ =$\{P_{1,1}, P_{1,2}, P_{1,3}, P_{1,4}, P_{1,5}, \ldots, P_{i,j}\}$, the issues of the agenda would be $P_{1,1}, \ldots, P_{i,j}$.

Constraint $\Gamma \in \mathcal{L}$ is a formula of $\mathcal{L}$. In this paper, we wanted to obtain a consistent and complete collective judgment set, so the constraint would be that one score can be assigned to one item, and only one item can get one score. $\Gamma = \bigwedge\{\neg(P_{ij} \wedge P_{i,j'}) | i \in I, j \neq j', 1 \le j, j' \le |I|\} \cup \bigwedge\{\neg(P_{ij} \wedge P_{i'}j) | i, i' \in I, i \neq i', 1 \le j \le |I|\}$. The constraint means that the system cannot accept two scores assigned to one item, and two items cannot have the same score.

It is clear that the individual judgment set depends on the agenda $\mathcal{A}$ and constraint Γ, and we denoted the individual judgment set as $(\mathcal{A}, \Gamma)$-judgment set. An individual judgment set is a set of formulas; $J \subset \mathcal{A}$ such that, J is consistent, if and only if, $J \cup \{\Gamma\}$ is consistent and J is complete if for every $P_{ij} \in \mathcal{A}$, J contains a member of each issue $\{P_{ij}, \neg P_{ij}\}$. Therefore, a fully rational user's judgment set would be consistent and complete; the set of all rational judgment set is denoted by $\mathcal{D}(\mathcal{A}, \Gamma)$.

Thus, users express their opinions through the scores for items, and then the scores reflect the users' opinions on the issues of the agenda, which are the judgments of individual users. We were then able to collect the users' judgment sets, which looked like a function $J \colon \mathcal{A} \to \{1, 0\}$; the 1 standing for "acceptance" and 0 representing "rejection". Since the judgment set is a consistent and complete subset of the agenda, the individual judgment set contains the issues which the user accepted. According to the users' accepted set, we easily obtained the complement of the agenda, which was the users' rejected set, because the consistent and complete set could guarantee that every user was rational. Users had to provide their opinions for every proposition of the agenda in completion, and could not be internally contradictory (consistency) of their judgment set, while the judgment set was produced by every user assigning a score to each item.

Judgment profile n-user judgment profile $\mathcal{P}$ was based on $\mathcal{A}$ and Γ, and a collection of n$(\mathcal{A}, \Gamma)$-judgment sets profile $\mathcal{P} = \langle J_1, \ldots, J_n \rangle$. For example, we can see in table 1 that the set of users was $U = \{u_1, \ldots, u_5\}$ and the set of items was $I = \{I_1, \ldots, I_5\}$, the matrix indicating the users' judgments for each item, and the agenda could be $\mathcal{A}$ =$\{P_{1,1}, \ldots, P_{5,5}\}$. The profile was $\mathcal{P} = \langle \{P_{1,5}, P_{2,3}, P_{3,4}, P_{4,1}, P_{5,2}\}, \{P_{1,5}, P_{2,4}, P_{3,3}, P_{4,2}, P_{5,1}\}, \{P_{1,1}, P_{2,4}, P_{3,3}, P_{4,2}, P_{5,5}\}, \{P_{1,2}, P_{2,4}, P_{3,3}, P_{4,5}, P_{5,1}\}, \{P_{1,5}, P_{2,3}, P_{3,4}, P_{4,2}, P_{5,1}\} \rangle$. The set of users in $\mathcal{P}$ with judgment sets that contain P_{ij} is $N(\mathcal{P}, P_{ij}) = \#\{k | P_{ij} \in J_k, J_k \in \mathcal{P}\}$.

Definition 3 ***Judgment Aggregation Rule.*** *A judgment aggregation rule is a function F, which is mapping every judgment aggregation profile $\mathcal{P}$ to a nonempty judgment set, based on $\mathcal{A}$ and Γ.*

Judgment aggregation rules require a function to map the judgment aggregation profile as input and get a subset of the agenda as output. It is a decision-making procedure, that the subset of the agenda is the collective judgment set, which will be the recommendation result.

Based on the above, we have defined agenda $\mathcal{A}$ and constraint Γ, when the system has read the user-item ratings matrix as input, it then needs to convert the data format to a judgment profile data type. The judgment profile is a $|U| \times |I|^2$ table where each row indicates a user, and each column represents a specific issue of agenda $\mathcal{A}$. The intersection of a row and a column reflects the user's judgment, "1" stands for true, otherwise, "0" is false. If the $user_1$ gives 1 to item 1, then the intersection between $user_1$ and $P_{1,1}$ can be marked as 1; otherwise, the other intersection about $item_1$ $P_{1,2}$, $P_{1,3}$, $P_{1,4}$, and $P_{1,5}$ can only be 0, and so on. The following table is an example, which has been transformed from the standard user-item ratings matrix in table 1 to a judgment profile on agenda $\mathcal{A}$.

Example 3. Consider the agenda $\mathcal{A} = \{P_{1,1} \dots P_{5,5}\}$, $U = \{u_1, \dots, u_5\}$, let $\Gamma = \{\neg(P_{ij} \wedge P_{ij'}) | i \in I, j \neq j', 1 \leq j, j' \leq |I|\} \cup \{\neg(P_{ij} \wedge P_{i'}j) | i, i' \in I, i \neq i', 1 \leq j \leq |I|\}$ be the constraint under $\mathcal{A}$. For $user_1$ in table 1, we observed that her scores for each item were $\{5, 4, 3, 2, 1\}$; therefore, she had to prefer $item_1$ the most and dislike $item_5$. We can now establish her judgment set $J_1 = \{P_{1,5}, P_{2,3}, P_{3,4}, P_{4,1}, P_{5,2}\}$, and the set is satisfied with the constraint Γ. Table 5 shows the process of the user-item ratings matrix transforming to the judgment aggregation profile under $\mathcal{A}$ and Γ.

Table 5 Judgment profile $\mathcal{P}$

	$P_{1,1}$	$P_{1,2}$	$P_{1,3}$	$P_{1,4}$	$P_{1,5}$	$P_{2,1}$	$P_{2,2}$	$P_{2,3}$	$P_{2,4}$	$P_{2,5}$	$P_{3,1}$	$P_{3,2}$	$P_{3,3}$	$P_{3,4}$	$P_{3,5}$	$P_{4,1}$	$P_{4,2}$
$user_1$	0	0	0	0	1	0	0	1	0	0	0	0	0	1	0	1	0
$user_2$	0	0	0	0	1	0	0	0	1	0	0	0	1	0	0	0	1
$user_3$	1	0	0	0	0	0	0	0	1	0	0	0	1	0	0	0	1
$user_4$	0	1	0	0	0	0	0	0	1	0	0	0	1	0	0	0	0
$user_5$	0	0	0	0	1	0	0	1	0	0	0	0	0	1	0	0	1

	$P_{4,3}$	$P_{4,4}$	$P_{4,5}$	$P_{5,1}$	$P_{5,2}$	$P_{5,3}$	$P_{5,4}$	$P_{5,5}$
$user_1$	0	0	0	0	1	0	0	0
$user_2$	0	0	0	1	0	0	0	0
$user_3$	0	0	0	0	0	0	0	1
$user_4$	0	0	1	1	0	0	0	0
$user_5$	0	0	0	1	0	0	0	0

The system then required a judgment aggregation rule, so as to aggregate the individual judgment set to a collective judgment set. The aggregation rule F maps the judgment set profile $\mathcal{P} = \langle J_1, ..., J_n \rangle$ to a collective judgment set. There are many judgment aggregation rules which have been explored, for instance, some rules are based on the majoritarian judgment set, some rules are based on the weighted majoritarian judgment set, and some rules are based on the removal or change of individual judgments,and scoring rules for judgment aggregation, etc. Here we selected the maxweight sub-agenda rule [6].

Definition 4 ***Maxweight Sub-agenda Rule.*** *For every agenda $\mathcal{A}$, for every $\Gamma \in I$, for every$(\mathcal{A}, \Gamma)$-profile $\mathcal{P}$, this maxweight sub-agenda rule (MWA) is defined as follows:*

$$MWA_{\mathcal{A},\Gamma}(\mathcal{P}) = \underset{J \in \mathcal{D}(\mathcal{A},\Gamma)}{argmax} W_{\mathcal{P}}(J), where W_{\mathcal{P}}(J) = \sum_{P_{ij} \in J} N(\mathcal{P}, P_{ij})$$

Example 4 Consider judgment profile in table 5, as we utilize Maxweight sub-agenda rule to aggregate individual judgment set to collective judgment set, we have:

$N(\mathcal{P}, P_{1,1})=1$, $N(\mathcal{P}, P_{1,2})=1$, $N(\mathcal{P}, P_{1,3})=0$, $N(\mathcal{P}, P_{1,4})=0$, $N(\mathcal{P}, P_{1,5})=3$;
$N(\mathcal{P}, P_{2,1})=0$, $N(\mathcal{P}, P_{2,2})=0$, $N(\mathcal{P}, P_{2,3})=2$, $N(\mathcal{P}, P_{2,4})=3$, $N(\mathcal{P}, P_{2,5})=0$;
$N(\mathcal{P}, P_{3,1})=0$, $N(\mathcal{P}, P_{3,2})=0$, $N(\mathcal{P}, P_{3,3})=3$, $N(\mathcal{P}, P_{3,4})=2$, $N(\mathcal{P}, P_{3,5})=0$;
$N(\mathcal{P}, P_{4,1})=1$, $N(\mathcal{P}, P_{4,2})=3$, $N(\mathcal{P}, P_{4,3})=0$, $N(\mathcal{P}, P_{4,4})=0$, $N(\mathcal{P}, P_{4,5})=1$;
$N(\mathcal{P}, P_{5,1})=3$, $N(\mathcal{P}, P_{5,2})=1$, $N(\mathcal{P}, P_{5,3})=0$, $N(\mathcal{P}, P_{5,4})=0$, $N(\mathcal{P}, P_{5,5})=1$.

$MWA_{\mathcal{A},\Gamma}(\mathcal{P}) = \{\{P_{1,5}, P_{2,4}, P_{3,3}, P_{4,2}, P_{5,1}\}\}$, since $W_{\mathcal{P}}(\{P_{1,5}, P_{2,4}, P_{3,3}, P_{4,2}, P_{5,1}\}) = 15$ is the maximal with respect to all $J \in \mathcal{D}(\mathcal{A}, \Gamma)$, the collective judgment set is $\{P_{1,5}, P_{2,4}, P_{3,3}, P_{4,2}, P_{5,1}\}$. This set is complete and consistent, it can represents a collective outcome; in the next step it will be used as a recommendation to a new user.

Phrase 3: Producing a Recommendation

The last step is producing a recommendation. In this step the system will output the recommendation to the new user. If the new user doesn't have any previous rating for the items, the group decision could give him/her a choice; for the preference model, our algorithm could give the user a group preference order or just conveying to the new user which item is the best one from the previous users' preference orders. For the judgment aggregation model, our algorithm could give the new user a collective judgment set. According to the group preference order or collective judgment set, the user can decide if it is worth purchasing or not. For example, if the group preference order is $I_1 \sim I_2 \succ I_3 \succ I_4 \succ I_5$ under linear Borda rule from example 3, the active user John might buy $item_1$ and $item_2$; however, he may hesitate to buy $item_5$, because $item_1$ and $item_2$ were better than the other items from the group preference order, but $item_5$ was the worst one at the same time. Furthermore, if the collective judgment set was $\{P_{1,5}, P_{2,4}, P_{3,3}, P_{4,2}, P_{5,1}\}$, John could see that $item_1$ was assigned the highest score by the group of people. He could have made the decision to buy $item_1$ as well; however, he would have considered more carefully regarding $item_5$, since it was not supported by most of people in the group. Compared with preference aggregation procedure, the judgment aggregation process is related to logic reasoning, which can be more specifically analyzed the aggregation procedure, and could give the new user a recommendation not only includes the information about which kind of items the goup of users like or dislike, but includes how much they like or dislike the item.

3 Demo

We used .NET to develop a demo to present our algorithm. There are three tables in the program, the first one is to input the individuals' user-item ratings matrix; the second one is to transform the user-item ratings matrix to preference order profile or judgment profile, based on the preference aggregation model or the judgment aggregation model; the third one is to output the recommendation. The three methods will be presented in the following program summary.

1. The Linear Borda Count Method Based on Preference Model
The user-item ratings matrix was given in example 1 as input; however, we transformed the matrix to preference order in the second table, as the 1st place was the best and the 5th place the worst.The third table contains all the items Borda counts.The third table also shows the group preference order,with the best one having the largest Borda count,which can be the recommendation for the new user. This method can be implemented for the cold-start problem, as is shown in the Fig. 2.

	Item1	Item2	Item3	Item4	Item5
User1	5	2	4	3	1
User2	1	2	3	4	5
User3	2	3	4	5	1
User4	1	5	3	4	2
User5	3	5	1	4	2

Run

Copeland Rule Borda Count MWA Rule

	1st	2nd	3rd	4th	5th
User1	I1	I3	I4	I2	I5
User2	I5	I4	I3	I2	I1
User3	I4	I3	I2	I1	I5
User4	I2	I4	I3	I5	I1
User5	I2	I4	I1	I5	I3
Borda	I4	I2	I3	I1	I5

	Borda Count
Item1	7
Item2	12
Item3	10
Item4	15
Item5	6

Fig. 2 Linear Borda rule based on preference model

2. The Copeland Rule Based on Preference Model
The first table is the user-item matrix depicted in example 1. We were able to change the user-item ratings matrix to the pairwise preference order; then used the pairwise preference order to obtain the Copeland scores in the second table, with the third table presenting the result where the most popular one is gaining the highest Copeland score.So the best item can be ranked in the 1st place from the Copeland score; the group decision can be recommended to a new user as is reflected in Fig. 3.

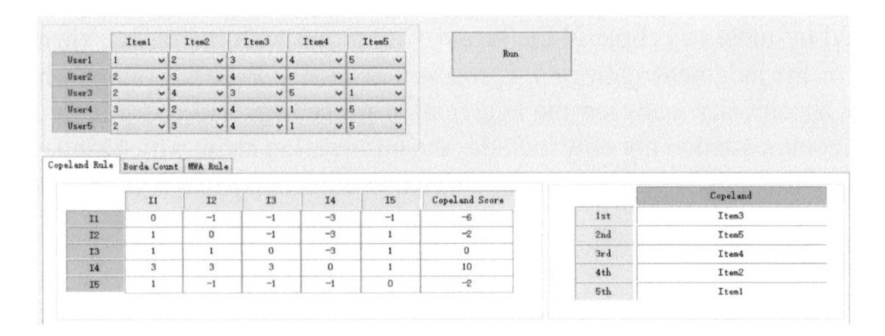

Fig. 3 Copeland rule based on preference model

3. Maxweight Sub-agenda Rule Based on the Judgment Aggregation Model
The first table also contains the user-item matrix shown in example 1. The second part was to transform the matrix to a judgment profile, based on the judgment aggregation model. The third table shows the outstanding options from the calculation by MSA rule in the second table, and were marked in red in the corresponding boxes, which could be a recommendation result for a new user, as it shows in Fig. 4.

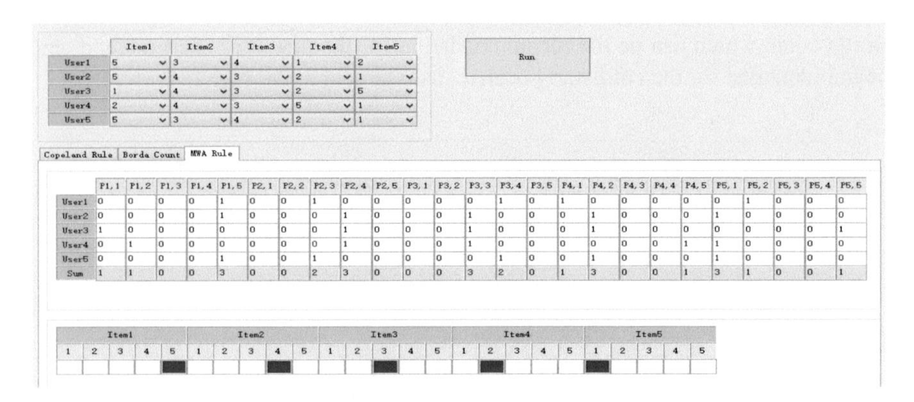

Fig. 4 Maxweight sub-agenda rule based on judgment aggregation model

4 Conclusions

Recommender systems can assist people in locating various items they really like and prefer as soon as possible from a vast amount of products or items in the systems, which have also proved useful for our online shopping. Collaborative filtering recommendation algorithm is the most successful method of utilizing the recommender systems in recent years. The traditional collaborative filtering algorithm focuses on calculating the similarity between users or items; then predicts the item which the active user likes or dislikes. It is very difficult to solve the cold-start problem when the new user does not have enough data to be available for collection. One contribution of this paper has been to introduce the group decision mechanism as a protocol to more effectively offer an individual user a new recommendation. It applies preference aggregation and judgment aggregation techniques in social choice theory to change the collaborative filtering algorithm, so as to obtain a recommendation for the new user. It is a relatively new method to provide a solution for the cold-start problem in recommender systems.

Acknowledgements This work was partially supported by the Philosophy Social Science Fund of Southwest University under Grant No. (12XDSKZ003). I would like to express my sincere appreciation to the China Scholarship Committee for providing financial support for me while I studied at the University of Bergen as a visiting PhD student. I sincerely appreciate the assistance of professor Thomas Ågotnes, for all his advice and encouragement in helping me to select my project and his patient guidance in helping me complete this paper. I also wish to thank Marija Slavkovik, who assisted me in my solution when I was somewhat confused and anxious. I would also like the thank Truls Pedersen, who greatly inspired and motivated me with discussions relating to this solution. Furthermore,we would like to thank all the anonymous judges, who so generously contributed their time and effort.

References

1. Jannach, D., Zanker, M., Felfernig, A., Friedrich, G.: Recommender systems: an introduction. Cambridge University Press (2010)
2. Arrow, K.J., Sen, A., Suzumura, K.: Handbook of Social Choice and Welfare, vol 2. Elsevier (2010)
3. Dietrich, F.: Judgment aggregation:(im) possibility theorems. Journal of Economic Theory **126**(1), 286–298 (2006)
4. List, C., Pettit, P.: Aggregating sets of judgments: An impossibility result. Economics and Philosophy **18**(01), 89–110 (2002)
5. Lang, J., Pigozzi, G., Slavkovik, M., van der Torre, L.: Judgment aggregation rules based on minimization. In: Proceedings of the 13th Conference on Theoretical Aspects of Rationality and Knowledge, pp. 238–246. ACM (2011)
6. Lang, J., Slavkovik, M.: Judgment aggregation rules and voting rules. In: Algorithmic Decision Theory, pp. 230–243. Springer (2013)
7. Sahebi, S., Cohen, W.W.: Community-based recommendations: a solution to the cold start problem. In: Workshop on Recommender Systems and the Social Web, RSWEB (2011)
8. Park, S.T., Chu, W.: Pairwise preference regression for cold-start recommendation. In: Proceedings of the Third ACM Conference on Recommender Systems, pp. 21–28. ACM (2009)
9. Schein, A.I., Popescul, A., Ungar, L.H., Pennock, D.M.: Methods and metrics for cold-start recommendations. In: Proceedings of the 25th Annual International ACM SIGIR Conference on Research and Development in Information Retrieval, pp. 253–260. ACM (2002)

Named Entity Recognition Through Learning from Experts

Martin Andrews

Abstract Named Entity Recognition (NER) is a foundational technology for systems designed to process Natural Language documents. However, many existing state-of-the-art systems are difficult to integrate into commercial settings (due their monolithic construction, licensing constraints, or need for corpuses, for example). In this work, a new NER system is described that uses the output of existing systems over large corpuses as its training set, ultimately enabling labelling with (i)better F1 scores; (ii)higher labelling speeds; and (iii)no further dependence on the external software.

Keywords Named Entity Recognition · NER · Natural Language Processing · NLP · Recurrent Neural Network · RNN · Unstructured data

1 Introduction

One key capability required of natural language processing (NLP) systems is to be able to identify the people, organisations and locations mentioned in a given text. These labels (plus further categories that include times, dates, and numeric quantities, for instance) are essential for understanding the facts described, yet they do not *per se* add much to the linguistic structure of the text. Therefore, building systems that can reliably perform this Named Entity Recognition (NER) has been a focus of NLP research, since it is an essential stepping-stone to exploring the other linguistic content in unstructured text.

M. Andrews(✉)
Red Cat Labs, Singapore, Singapore
e-mail: Martin.Andrews@RedCatLabs.com
http://www.RedCatLabs.com

K. Lavangnananda et al. (eds.), *Intelligent and Evolutionary Systems*,
Proceedings in Adaptation, Learning and Optimization 5,
DOI: 10.1007/978-3-319-27000-5_23

Unfortunately, while the NER task might be considered largely conquered from a linguistic research viewpoint, building an effective system is still a challenge in a commercial setting:

1. Licenses for many existing academic systems are not conducive to being embedded within commercial systems
2. Often, existing codebases focus on 'tweaks' rather than solid engineering
3. Commercial systems may have particular task-specific requirements that are difficult to implement on a pre-built system
4. Training corpuses can be a limiting factor, since commercial uses focus on specific domains of interest, rather than domains that have well understood corpuses already available

This work describes an NER system that can be trained from the output of 'known good' systems. Since the system developed here only requires large volumes of (machine) annotated text, it essentially sidesteps several of the problems that these existing systems have in commercial settings.

Moreover, the experiments show that the new system can learn to be better than its teachers - both in the test scores obtained and labelling speed.

Importantly, the results obtained during training and testing are described here in full - the models have not been cherry-picked and tweaked for publication - which illustrates the robustness of this type of model and training process.

2 Model

2.1 Vocabulary Building

As described below, the CoNLL-2003 [1] NER datasets were chosen as the test-bed for this work, and the unlabelled 'Large Corpus' was used to build the vocabulary and word-embedding features.

A vocabulary was built from the contents of the whole Large Corpus (there were 484k distinct tokens in the 1.0Gb corpus) with the following additional tokenization steps taken prior to insertion into the dictionary:

1. Convert to lower case
2. Replace each string of digits within the token with `NUMBER` (so that, for instance, '12.3456' becomes '`NUMBER.NUMBER`')

2.2 Word Embedding Layer

Skip-gram embeddings of size 100 were pre-trained over the whole large corpus and vocabulary using `word2vec` [2] as provided by the Python package `gensim` [3] (this required only 15 minutes of wall-clock time).

The token embedding was filtered so that only tokens with 10 mentions or more were included, yielding an effective vocabulary size of 118,695 distinct tokens. To cope with words not present in the embedding, a special token `<UNK>` was added to the embedding space, with a vector that corresponded to the mean vector over the rest of the known dictionary.

2.3 Additional Features

The only feature added to the vector representation of each token was an indicator $\{0,1\}$ as to whether that token/word had originally contained upper-case characters. Therefore, for each token the extended vector given to the next stage was 101 elements in length.

2.4 Bi-Directional Recurrent Neural Network (RNN)

Having mapped each token to a numerical input vector, a bi-directional recurrent neural network was used to map the token embeddings to hidden states. Since each timestep corresponded to exactly one output label, it was not necessary to separate ingestion and output RNNs: a lock-step arrangement was sufficient.

The model was built using Theano using the recently announced `blocks` [4] framework, which provides many useful primatives, and is currently under active development. The sizes of the embedded parameters are given in Table 1.

In the interests of initially keeping the model as simple as possible, a very basic recurrent network was used:

$$\begin{aligned} \mathbf{h}_t^F &= \tanh(\mathbf{W}^F \mathbf{h}_{t-1}^F + \mathbf{x}_t) \\ \mathbf{h}_t^B &= \tanh(\mathbf{W}^B \mathbf{h}_{t+1}^B + \mathbf{x}_t) \end{aligned}$$

where $\mathbf{h}_t^F$ and $\mathbf{h}_t^B$ refer the hidden states in the forward and backward chains respectively; $\mathbf{W}^F$ and $\mathbf{W}^B$ refer to independent weight matrices for each chain, and $\mathbf{x}_t$ is the extended token vector.

The initial conditions to the forward and backward chains, $\mathbf{h}_{-1}^F$ and $\mathbf{h}_{T+1}^B$ are set to values (that are also trainable inputs) at the beginning and end of each sentence respectively.

Labelling Output Layer. One feature of the CoNLL-2003 datasets was that in addition to the basic {`PER`, `ORG`, `LOC`, `MISC`} entity labels, there were also specific 'Beginning' labels to be used to separate two entities which abutted against each other without any other intervening token. However, situations in which this actually arose were very rare (respectively $\{0.0\%, 0.2\%, 0.1\%, 0.7\%\}$ of each token's occurrences). Therefore, to simplify the output stage logic, only 5 labels were learned (the entity labels, plus `O` for non-entity tokens).

The output stage consisted of a dense linear layer (with bias), with each of the 5 label outputs at a given timestep connected to all the RNN hidden units at the same timestep (both forwards and backwards chains), followed by softmax:

Table 1 Model Parameters

Parameter Set	Notation	Shape	# of Float32
Word Embedding (all tokens)	$\mathbf{x}$	(118695, 100)	11,869,500
Generated features	`token_ucase()`	(..., 1)	n/a
State transition matrices	$\mathbf{W}^F$ and $\mathbf{W}^B$	(101, 101) × 2	20,402
State initialisation vectors	$\mathbf{h}^F_{-1}$ and $\mathbf{h}^B_{T+1}$	(101, 1) × 2	202
RNN outputs to label matrix	$\mathbf{X}^F$ and $\mathbf{X}^B$	(202, 5)	1,010
RNN outputs to label biases	$\mathbf{b}_t$	(1, 5)	5
Total (RNN only)			21,619
Total Model			11,891,119

$$\mathbf{d}_t = \mathbf{X}^F \mathbf{h}_t^F + \mathbf{X}^B \mathbf{h}_t^B + \mathbf{b}_t$$

$$\mathbf{p}_t^i = \frac{e^{\mathbf{d}_t^i}}{\sum_k e^{\mathbf{d}_t^k}}$$

where $\mathbf{d}_t$ refers to the linear combination over the RNN outputs at that timestep; $\mathbf{X}^F$ and $\mathbf{X}^B$ refer to independent weight matrices for each chain; $\mathbf{b}_t$ is a bias term; and $\mathbf{p}_t$ is the softmax output for the assigned label.

This 'one-hot' representation was trained using Categorical Cross-Entropy for each label summed over batches of sentences as an objective function for gradient descent, which used an ADADELTA [5] step rule.

During the test phase, labels were simply read from the output stage, without post-processing.

2.5 External Models

As a basis for learning, the RNN was trained against the provided training set (3.3Mb) as well as the Large Corpus labelled by two external models. These models were chosen because they are both state-of-the-art, have acceptable licenses and were easiest to use off-the-shelf.

Please note that this paper's results are only possible because its RNN models are able to 'stand on the shoulders of giants': there is no intention here to detract from the fine work that went into creating these models in the first place.

Other potential candidate models are mentioned below (in Related Work), but studying the following was sufficient for the present experiments.

In all cases, care was taken to ensure that the models all treated the given tokenization in the same way, and that the results obtained from the models alone matched the reported results.

MITIE. According to its GitHub page (https://github.com/mit-nlp/MITIE), the MITIE project (built around `dblib` [6]) is a state-of-the-art information extraction tool, which performs named entity extraction and binary relation detection. It is available under a permissive Open Source license (which, interestingly, was one of the key objectives of the funding for the project provided by the DARPA XDATA program).

Model files specifically constructed for the CoNLL 2003 NER task are available for download. All MITIE output here was created using the 343Mb model file english_ner_model_just_conll.dat.bz2.

Stanford Named Entity Recognizer. According to its substantial documentation page (http://nlp.stanford.edu/software/CRF-NER.shtml) the Stanford Named Entity Recognizer provides a general implementation of (arbitrary order) linear chain Conditional Random Field (CRF) sequence models [7], and is included in the Stanford CoreNLP suite of NLP tools.

According to Stanford's NER benchmarks, the Stanford model was used to submit results in the original CoNLL-2003 competition, and performed well. The model file used here (v3.5.2 of english.conll.4class.distsim.crf.ser.gz) is close to (or an improvement on, it is unclear) the original CoNLL-tuned version. The compressed model size appears to be approximately 110Mb.

3 Experiments

3.1 CoNLL-2003

The experimental setting chosen was the same as given in CoNLL-2003 [1]. This provided several distinct datasets (statistics for which are given in Table 2), each of which were tokenised using the CoNLL-provided scripts:

"Large Corpus" This consists of 10 months of Reuters news stories, with no labelling provided;

Training Set This is a labelled set of data that models can be trained on - with the option also available (in 2003) of using external training data too;

Development Set This is a hold-out labelled test set ('`testa`') which was set aside for validation and/or hyper-parameter selection;

Test Set This is the labelled test set ('`testb`'), with scripts provided to calculate recall/precision/F1 scores both overall and for each category label.

As described earlier, no additional pre- or post- processing was applied to the data.

Table 2 Data set sizes

Data sizes	Bytes	Words	Sentences
"Large Corpus"	1.0Gb	184,717,139	11,869,032
Training Set	3.3Mb	204,567	14,987
Development Set	827Kb	51,578	3,467
Test Set	748Kb	46,666	3,685

3.2 Models and Training

Training. Initial training runs used 15 million labelled sentences (this figure was chosen to be approximately 1000 epochs on `train` - sufficient to fully learn the CoNLL-2003 provided data). For the more extensive runs, the number of labelled sentences was arbitrarily fixed at 100 million (this count does not include sentences that were excluded by the 'Consensus' technique below).

Expert Scores. Included in Table 3 are results for the base scores for the two expert models. These figures agree with their previously reported scores.

RNN Learning from Individual Experts. Test results are given for 15 and 100 million sentences of training (over the output of the respective expert labelling of the Large Corpus). These results are surprisingly close to the expert they are being trained from, despite having no knowledge of the internal workings (or tweaks, tricks, etc) being used.

In order to test the variability of models built, the 'RNN-MITIE' model was trained with 15 different random number seeds for the internal model initialisation (using, however, from the same initial word embedding data). The resulting set of `testb` F1 scores had mean 88.15% and standard deviation of 0.14%.

RNN Trained on Training Set Alone. Although the RNN has the benefit of a word embedding derived from the Large Corpus, the results show that solely learning the labelling task from the training data set (1000 epochs) was insufficient for good performance.

RNN 'Mixer'. This RNN was trained from a data source that took (in turn) one sentence from each of the Training Set, and Large Corpus sets as labelled by the MITIE and Stanford experts (i.e. 3 sources in equal measure - even though this implies considerably more epochs of Training Set data, since it is so much smaller in size).

RNN 'Consensus'. These RNNs were trained from a data source that took a fixed proportion α of sentences from the Training Set (given as a percentage in Table 3), and sentences whose labelling both the MITIE and Stanford experts agreed upon *in full*. The fixed proportion α, viewed as a hyper-parameter, was chosen according to the RNN performance on the Development Set (this was the only time the `testa` dataset was used).

Ensembling. Simple ensembles of the most promising 'Consensus' RNNs and the given experts were created (one of each type, doing a simple vote for each output label). In addition, RNN models trained on each expert were also tested as members of ensembles, to see whether ensembling gains could be made using solely RNN-trained models.

Table 3 F1 scores for individual and ensembled models

	Sentences (millions)	Training Set F1%	Dev. Set F1%	Test Set F1%
Individual Models				
Expert-MITIE	n/a	96.98	97.11	88.10
Expert-Stanford	n/a	97.66	91.79	88.19
RNN-MITIE	15	90.43	91.11	86.58
RNN-MITIE	100	93.08	93.25	88.08
RNN-Stanford	15	90.19	89.03	85.51
RNN-Stanford	100	91.93	90.26	86.24
RNN-TrainSet	15	**99.62**	84.47	79.50
RNN-Mixer	100	99.50	93.39	88.76
RNN-Consensus-00%	100	94.01	93.04	88.64
RNN-Consensus-05%	100	98.65	**93.66**	89.45
RNN-Consensus-10%	100	99.38	93.60	**89.51**
Ensemble Models (100 million sentences)				
Consensus-05 + RNN-MITIE + RNN-Stanford		95.85	93.64	89.52
Consensus-05 + Expert-MITIE + RNN-Stanford		97.77	94.69	89.68
Consensus-05 + RNN-MITIE + Expert-Stanford		98.22	94.08	89.92
Consensus-05 + Expert-MITIE + Expert-Stanford		98.72	95.34	**90.12**
Consensus-10 + Expert-MITIE + Expert-Stanford		99.00	95.38	**90.18**

4 Analysis

4.1 The CoNLL-2003 Task

One surprising aspect of the CoNLL-2003 task was that the `testb` data set (on which final F1s are measured) appears to be significantly different from the training data given. Several features stand out:

1. There are many sports scores in `testb` (presumably because Reuters news carried a lot of these articles during that end-of-summer time period);
2. Sports score summaries contain a lot of numeric tokens, with little in the way of other linguistic structure;
3. Sometimes labelling of teams can be problematic, with ‘China’ being both a location and a team (organisation) name.

The difficulty of `testb` is noticable specifically in the F1 scores for ‘RNN-TrainSet’, which completely ignores `testa` during training. Note, though, that all the other training runs may have implicit dependencies on `testa` simply because the MITIE and Stanford systems may have relied on hyper-parameter selection based on `testa` performance.

4.2 Model Complexity

As mentioned in the description of the models used, an attempt was made to keep the model becoming more complicated than necessary. The results obtained indicate that the Simple Recurrent setup used is sufficient for the NER task.

However, for more complex tasks, it seems likely that Gated Recurrent Units (GRU [8]) or their highly parameterized predecessor Long Short-Term Memory (LSTM [9]) may have more expressive power (particularly since these are now commonly being stacked in layers). Fortunately, the `blocks` framework chosen here is flexible enough to accommodate these enhancements.

In the context of the NER task, it is possible that the Stanford model incorporates processes that are difficult to learn for the Simple Bi-Directional RNN used - as evidenced by the F1 scores converging more slowly during training than is the case for the MITIE model. In addition, during ensembling, using RNN-Stanford was significantly less impactful than RNN-MITIE, which is a pity, since the Stanford model is heavier computationally, as can be seen from Table 4.

Table 4 System labelling speed

	Sentences per second	Comment
Expert-MITIE	1,646	OpenBLAS / Lapack found during compilation, but the system appeared to run single-threaded
Expert-Stanford	48	This was invoked through `Stanford CoreNLP`, but only stages relevant to NER were run
RNN (all)	3,123	This implementation was GPU-based, and timings were taken during backprop training (simply labelling requires fewer operations)

4.3 Implementation Speed

The RNN implementation benefited significantly from using a consumer-grade GPU. One feature of the Theano/`blocks` framework is that the model description is coded independent of the target computing device, since Theano is capable of dynamically creating `C++`, `OpenCL`, and `CUDA` code as required.

By choosing an appropriate batch size for the training (so that multiple sentences to be trained in parallel), a speed-up of 35x was realized over the initial choice of parameters used by example code online (see Table 5).

A GPU blocksize of 256 was chosen, since higher blocksizes appeared to cause significant delays in saving `Checkpoint` data to disk (the 1Gb files saved for the 256 blocksize were deemed acceptable).

Overall, the training time on each 100 million sentence experiment was 7-8 hours. The Consensus experiments took approximately 50% longer, solely because much of the data ingested was immediately discarded (and not learned).

Table 5 Training time in seconds on 150k sentences (lower is better)

	CPU	GPU
	i7-4770 CPU	GTX 760
batchsize	@ 3.40GHz	(2Gb)
8	1030	455
64	254	70
256	211	29
512	n/a	23

4.4 Consensus Methods

Evidently, training a model solely on the data upon which experts agree is an effective approach. What is surprising is that it still works in the cases where the experts would disagree, because the model would not have received training from either expert in these circumstances.

Comparing the Consensus models with 'Mixer' (which is very similar in design, except that no filtering is taking place: the three sources of training data are used on a round-robin basis), it is clear that filtering the training examples is actually beneficial to learning.

There is also a sense in which the Consensus models are performing an ensembling-together of three different training datasets - with the ensemble voting taking place during the ingestion phase, rather than on the final output labels. Interestingly, this puts a heavy burden on the generalization ability of the RNN model to cases in which its supposed teachers disagree. Apparently, this is something these models are capable of doing.

4.5 Ensembling

The best results obtained in this paper were (unsurprisingly) from ensembles of models. Indeed, some of these results broke through the apparent 90% F1 score barrier. However, it was somewhat disappointing that ensembles of pure RNN models didn't reach the same levels of performance of RNN models ensembled with the original experts. This is particularly true of the Stanford model, which is the more desirable of the two models chosen to eliminate (due to speed and licensing considerations).

On the other hand, from a practical point of view, optimising out the last ounce of performance is probably less important than the overall lessons to be learned: Ensembling does work between models, but the implicit ensembling provided by the training of the Consensus models may be both more robust and easy to implement.

4.6 Further Enhancements

The commercial setting in which this work takes place is particularly focused on English-language documents sourced from the ASEAN region.

Given the variability of names in the region (specifically names of people), and their 'obvious' differences in spelling from English words, one further enhancement to the system is the creating of additional word features using a letter-based RNN trained on databases of English prose and ASEAN names (these corpuses have already been curated).

Thus, instead of embedding concrete gazetteers (as is common for more traditional systems), the plan is to train an RNN on the NER task on a character-by-character basis. The trained RNN can then be 'cut off at the output stage' so that its internal pre-output state (a 20 element vector) can be used as additional features for each token for the RNN described in this paper (name tokens that might otherwise all be assigned to `UNK`). This scheme may also offer the opportunity to further characterize names by country-of-origin, for instance.

5 Related Work

Surprisingly, an approach that used LSTM neural networks was previously undertaken for the CoNLL task in 2003 [10]. However, this was published well before importance of word-embedding was understood, so the results reported there (<75% F1 overall) are essentially from a different era.

Work by Collobert *et al.* [11] published in 2011, demonstrated that a pure data-driven neural network approach to language tasks can be very effective. They made use of extensively trained word-embeddings, but did not make use of Recursive Neural Networks (their 'sentence scoring' element was performed using a max-pooling approach over a convolutional layer on top of the word embeddings). Their sofware `SENNA` is published under a No Commercial Usage license, and achieves approximately the same performance as the Consensus models created here.

Presented at ICLR (in May 2015), Oriol Vinyals et al. [12] essentially repurposed Google's LSTM translation framework to learn 'Grammar as a foreign language'. This task is more difficult than the step-wise labelling performed herein, and required considerably larger computation resources. For example, their network needed to produce 100 different labels, and they made use of a 512-dimensional embedding, and large multi-layered LSTM networks with a 4000-dimensional internal state. Overall, their model included 34 million trainable parameters. That being said, their approach strongly influenced the direction of this work.

5.1 Other External Models Considered

Berkeley Entity Resolution System. The Berkeley NER [13] system is also a state-of-the-art NER system, and is part of the suite of software used in the 'Grammar as a Foreign Language' work cited above. It is GPL3+ licensed, which would be acceptable for the current work, however it was not used here purely for time reasons.

Illinois Named Entity Tagger. This NER system [14], created by the Cognitive Computation Group from the University of Illinois at Urbana-Champaign, reports scoring 90.8% `testb` F1 on the CoNLL-2003 task, which makes it an attractive candidate system to learn from.

However, despite the Illinois NER system being available under a broadly copyleft license to a Licensee for "its own academic and research purposes", the license includes the following explicit non-commercial usage clause:

> "No license is granted herein that would permit Licensee to incorporate the Software into a commercial product, or to otherwise commercially exploit the Software. "

This current work illustrates the type of legal questions that learning systems bring into focus: If the software is solely used to create a corpus annotation, and a model is trained from that corpus, has the Software been commercially exploited? Is the Licensor asserting come kind of usage rights over all output of the Software? This is surely a new set of challenges to be faced by software license writers, similar to how the GPL has evolved to avoid the 'Tivoization' problem.

6 Conclusions

This work has shown that it is possible to build a near state-of-the-art NER system based solely on the output of externally created software systems.

Even without ensembling (from which even better results were obtained), the resulting system was shown to have learned to exceed the capabilities of its teachers, while being significantly more amenable to usage within a commercial environment.

Acknowledgments The author thanks DC Frontiers, a Singapore-based company that has created the data-centric service 'Handshakes' (http://www.handshakes.com.sg/), for their willingness to believe the system created herein was feasible.

DC Frontiers is the recipient of a Technology Enterprise Commercialisation Scheme grant from SPRING Singapore, under which this work took place.

References

1. Tjong Kim Sang, E.F., De Meulder, F.: Introduction to the conll-2003 shared task: language-independent named entity recognition. In: Proceedings of the Seventh Conference on Natural Language Learning at HLT-NAACL 2003 - Volume 4. CONLL 2003, pp. 142–147. Association for Computational Linguistics, Stroudsburg (2003)
2. Mikolov, T., Chen, K., Corrado, G., Dean, J.: Efficient estimation of word representations in vector space. CoRR **abs/1301.3781** (2013)
3. Řehůřek, R., Sojka, P.: Software framework for topic modelling with large corpora. In: Proceedings of the LREC 2010 Workshop on New Challenges for NLP Frameworks, pp. 45–50. ELRA, Valletta, May 2010. http://is.muni.cz/publication/884893/en

4. van Merriënboer, B., Bahdanau, D., Dumoulin, V., Serdyuk, D., Warde-Farley, D., Chorowski, J., Bengio, Y.: Blocks and fuel: Frameworks for deep learning. CoRR **abs/1506.00619** (2015)
5. Zeiler, M.D.: ADADELTA: an adaptive learning rate method. CoRR **abs/1212.5701** (2012)
6. King, D.E.: Dlib-ml: A machine learning toolkit. Journal of Machine Learning Research **10**, 1755–1758 (2009)
7. Finkel, J.R., Grenager, T., Manning, C.: Incorporating non-local information into information extraction systems by gibbs sampling. In: Proceedings of the 43rd Annual Meeting on Association for Computational Linguistics. ACL 2005, pp. 363–370. Association for Computational Linguistics, Stroudsburg (2005)
8. Cho, K., van Merrienboer, B., Gülçehre, Ç., Bougares, F., Schwenk, H., Bengio, Y.: Learning phrase representations using RNN encoder-decoder for statistical machine translation. CoRR **abs/1406.1078** (2014)
9. Graves, A.: Supervised sequence labelling with recurrent neural networks. Vol. 385. Springer (2012)
10. Hammerton, J.: Named entity recognition with long short-term memory. In: Proceedings of the Seventh Conference on Natural Language Learning at HLT-NAACL 2003 - Volume 4. CONLL 2003, pp. 172–175. Association for Computational Linguistics, Stroudsburg (2003)
11. Collobert, R., Weston, J., Bottou, L., Karlen, M., Kavukcuo-glu, K., Kuksa, P.P.: Natural language processing (almost) from scratch. CoRR **abs/1103.0398** (2011)
12. Vinyals, O., Kaiser, L., Koo, T., Petrov, S., Sutskever, I., Hinton, G.E.: Grammar as a foreign language. CoRR **abs/1412.7449** (2014)
13. Durrett, G., Klein, D.: A joint model for entity analysis: coreference, typing, and linking. In: Proceedings of the Transactions of the Association for Computational Linguistics (2014)
14. Ratinov, L., Roth, D.: Design challenges and misconceptions in named entity recognition. In: Proceedings of the Thirteenth Conference on Computational Natural Language Learning. CoNLL 2009, pp. 147–155. Association for Computational Linguistics, Stroudsburg (2009)

Appendix

Working code to implement the RNN scheme outlined in this paper is available through links on: https://github.com/mdda

A Decision-Support Tool for Humanitarian Logistics

Takushi Ashinaka, Masao Kubo and Akira Namatame

Abstract Humanitarian missions are complex operations that require emergency resources to be delivered in a timely fashion to a disaster area. This article describes the development of a decision-support tool to improve the effectiveness of humanitarian operations through efficient inventory management and quick distribution of emergent resources for disaster areas. Such humanitarian logistics necessitate better coordination and planning. Unlike commercial logistics, humanitarian logistics demands from disaster areas cannot be predicted. Thus, to support a quick and efficient relief operations is important by developing ICT-based decision aids. A decision-support tool is just such an attempt, which allows the design of logistics networks for effective disaster responses. Such a decision-support tool may require the following two key decisions: determining temporary warehouse locations and deciding the means of transportation to points of destination (POD). The design of the humanitarian logistic networks includes: (a) a supply chain network that consists of inventory and distribution management, and (b) a logistic network that includes multimodal transportation of different scales for transportation times.

Keywords Supply chain management · Multi-echelon · Integrated distribution and inventory management · Disaster relief operations · Last-mile logistics

1 Introduction

With the establishment of the ASEAN Economic Community, the Mekong River inland area will be improved as a transportation infrastructure. As a result many countries will be connected to an economic corridor and considerable growth is

T. Ashinaka(✉) · M. Kubo · A. Namatame
Department of Computer Science, National Defense Academy of Japan,
Yokosuka, Japan
e-mail: em53041@nda.ac.jp

K. Lavangnananda et al. (eds.), *Intelligent and Evolutionary Systems*,
Proceedings in Adaptation, Learning and Optimization 5,
DOI: 10.1007/978-3-319-27000-5_24

expected in the area as it is a large economic zone. The improved efficiency and the increased resilience of sea lines that connect Singapore, Shanghai, Hong Kong, Pusan, and the main ports in Japan is not only important for further developments of the ASEAN Economic Community, but also for supporting global economic development through logistics. In the meantime, industrial zones are developing in Thailand and Indonesia, with Thailand being the largest global industrial establishment after China, and Indonesia having its geographic advantage as a maritime nation.

However, Southeast Asia has experienced major natural disasters quite often in the last decades. The 2011 catastrophic flood in Thailand not only damaged the industrial complex directly but caused a great financial loss to ASEAN nations, which have a strong and close economic connection to Thailand. Manufacturing of electronic products and automobile parts stopped for an extended time, and took several months to recover. This affected numerous global corporations for a long period of time. It is recognized that natural disasters such as the Great East Japan Earthquake and the floods in Thailand in 2011 not only impact the locals, but also damage the global economy. The science of logistics and supply chain management has become critically important for private sector logisticians. Logistics has started to be recognized as integral to any relief operation. A natural disaster can occur anywhere at any time, and all countries in the world must be prepared in order to minimize the eventual damage of the catastrophe. The globalization of production and optimization of supply chains have increased systemic efficiencies in the global economy but have exacerbated the speed and scope of contagion in the event of shocks. They pose particular threats to the just-in-time business model. In an increasingly connected global economy and society, more people are affected by shocks [1].

There is an urgent need to rescue lives and support the livelihood of people in the affected areas. As a result of recent disaster, disaster management was proved to be important to reduce damage. When natural disasters hit industrial areas, production activities will be interrupted for an extended time due to damages in the infrastructure and logistic pathways in the locality. Because the extent of damage of natural disasters in industrial areas has national and global economic impacts, it is necessary to create a systematic approach by which local and national government and international organizations can collaborate.

Although all types of catastrophes are different, they share similar consequences: destruction of infrastructure, destruction of roads, and health diseases. This is why an effective coordination of the supply chain and a constant flow of information are the keys to improving the delivery and the performance of humanitarian operations. The effective management of aid determines the number of lives that can be saved. According to the type of disaster whether earthquake, flood, or hurricane, the specific requirements of the affected population vary. For example, when an earthquake occurs, shelter is one of the immediate requirements, but when a flood occurs, lack of food is the major concern. The specific conditions of the place where the disaster occurs also determine the kind of assistance needed.

We should also recognize the value of utilizing integrated ICT-based systems to capture and analyze information; these can result in a more effective and efficient relief effort. Humanitarian logistics information systems improve information flows and provide better feedback to donors, thus performing effective operations [2]. Humanitarian logistics activities are generated across the disaster management cycle of preparedness, response, recovery and mitigation. Humanitarian logistics information systems improve the continuity of humanitarian operations by sharing information throughout the transition of different phases and logistics activities.

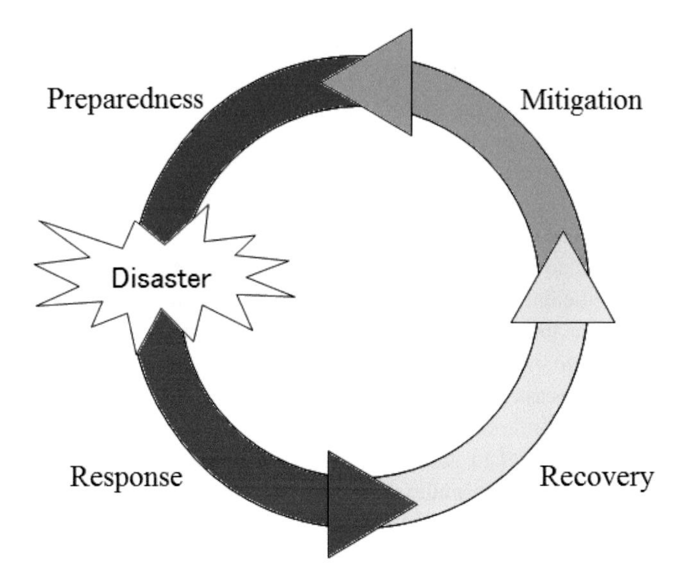

Fig. 1 Disaster management cycle

This paper is concerned with developing an ICT-based decision-support tool that supports the planning and decision-making of disaster responses in the region, at the ASEAN, national, and local levels. This paper emphasizes the importance of building emergency supply chains for assistance to the needy, considering how different actors interact across the humanitarian logistics to face the needs of people in a disaster area. The supply chain of humanitarian aid is a complex and interlinked network in which different types of decisions and information are mixed to serve the needs of the victims in a catastrophe. Globalization, outsourcing and interdependencies have increased the complexity in the structures of global maritime transportation networks.

2 Problem Context

Disaster management includes the care of the survivors' needs. To deliver necessary goods to a disaster area is effective disaster supply chain and one of essential function of disaster management [3, 4].

Humanitarian logistics is a word that also includes the operations concerning distribution and the flow of relief goods needed for speeding the recovery from disaster. Quick restoration needs managing of the supply chain in disaster relief operations. Therefore between commercial logistics and humanitarian logistics exists different processes and mechanisms. In point of loss, the delay of supplies in a humanitarian supply chain is more severe than a commercial supply chain [5].

The problem of last-mile logistics is also a fundamental issue in which aid supplies are piled up at the distribution center near a disaster area but not delivered to the people who need them; this occurred in recent disasters such as the Great East Japan Earthquake. In a disaster, delivery of necessary goods becomes a difficult task that requires the mobilization of a great number of resources under many conditions of uncertainty, including transportation disruptions. The supply chain is therefore stretched to its limits and that is why it is so important to understand the complex operations of humanitarian logistics.

The process of delivering supplies to a community that has suffered a natural disaster implies the coordination and execution of multiple organizations. Humanitarian aid is handled by national governments, local governments, international and local non-governmental organizations (NGOs), and community-based local partners. The challenge is to bring about a smooth, speedy, and coordinated response and create systemic change by helping multiple organizations overcome the blurring of boundaries among the roles assigned to different organizations.

The recent world conference on disaster reduction called for better preparedness for disaster relief in natural disasters, but being better prepared can also mitigate the effects of man-made disasters [4]. The establishment of pre-positioning warehouse is strategy which a humanitarian organization generally takes. The pre-positioning strategy establish a warehouse in advance where the regions need supplies. Therefore, pre-positioning can respond quickly to a demand for the opportunity of relief goods even if the disaster happened. The strategy of pre-positioning utilizes a facility location model to identify the warehouse locations for relief items to be stored by considering changes to the natural disaster trends observed.

Specific models have been proposed to determine the locations and the number of pre-positioning warehouses needed to maximize the total expected demand covered given a set of scenarios [1]. The humanitarian logistics begin once a disaster takes place. The national entity in charge needs to elaborate a detailed evaluation of the disaster and suggest the most suitable actions according to the stocks of available products and logistic teams in place. Then it verifies if the local teams request for assistance can be received, processed, and delivered according to their logistical resources. If the local unit is unable to respond to these necessities, the second level of assistance is activated, and the regional and strategic units are contacted to provide aid. If stock levels in these units are still not enough to provide for the needs of the affected population, the principal unit of assistance (i.e., the national unit) is alerted. Depending on the inventory and the amount of food requested by the regional unit, the national unit determines if it is necessary to provide supplies.

3 The Architecture of FEMA for Disaster Relief Operations

The structure and coordination of the supply chain conditions the success or failure of the effort. Therefore we need to standardize the process of delivering humanitarian aid. However, extreme standardization might be a mistake, because all disasters are different, not only in their causes, but also in the particular characteristics of the region in which they occur.

Federal Emergency Management Agency of the United States (FEMA) has four main stocks at local, regional, strategic, and national warehouses for its supply chain. There are seven main components in the supply chain that provide relief commodities for disaster victims [6].

4 The Architecture of Decision-Support System

There are several commodities that need to be distributed among the disaster victims. The commodity changes by many reason. Victims may strike random zones within the region therefore resulting in stochastic demands of different types of emergency supplies. FEMA suggests a list of required items and the amount per day per survivor so a total of about 20 ft^3 of relief items per day is required [6].

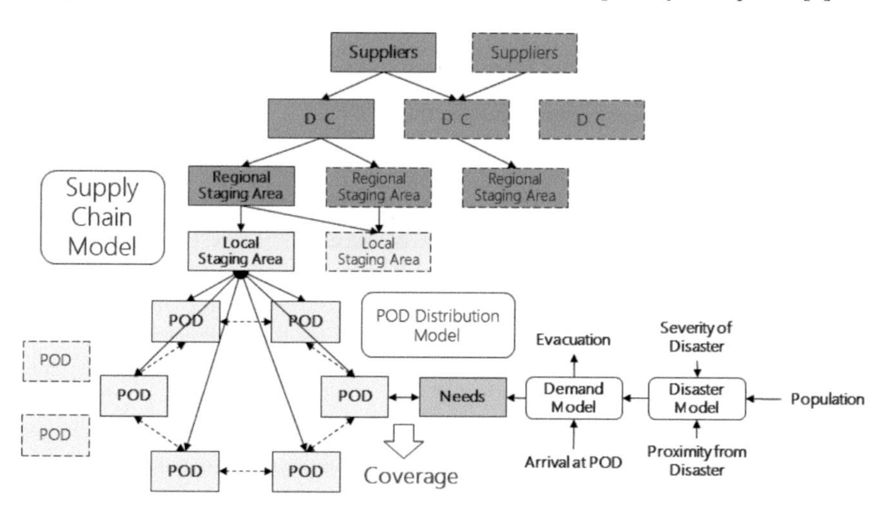

Fig. 2 Architecture of model

In our model, we assess the delivery of aid kits, and specifically model the flow of emergency goods. Of course, extreme standardization may be a mistake, because all disasters are different, not only with respect to their causes, but also in terms of the particular geographical or cultural characteristics of the region in which they occur; this should be taken into account at the time of delivering aid. First, we evaluate the damages and needs of the population, and then determine how many emergency kits we will deliver to the population.

A humanitarian supply chain is characterized by the presence of different kind of delays. These occur during three stages in the aid process: in the evaluation of the disaster and the measurement of the numbers of people affected; during the process of delivering aid, including the logistics needed to transport the aid from the place where donations are received to the location of the disaster; and in the process of relief and distribution of aid to the recipients.

The main operation center is located in the country, where the strategic warehouses are also located. The regional warehouses as well as local warehouses are located in the towns. The strategic warehouses are used when the aid in regional centers is insufficient in the face of an emergency. The main functions of the model are:

- To determine the requirement of emergency goods by victims
- The delivery of goods available in regional and local warehouses
- The delivery of goods in the strategic and national warehouses

The humanitarian supply chain assistance begins once a disaster takes place. The national entity in charge elaborates a detailed evaluation of the disaster, and suggests the most suitable actions according to the stocks of available goods and logistic units in place. Then it verifies if the local teams request for assistance can be received, processed, and delivered according to their logistic resources. If the local unit is unable to respond to these necessities, the second level of assistance is "activated," and the regional and strategic units are contacted to provide aid. If stock levels in these units are still not enough to provide for the needs of the affected population, the national unit is alerted.

The issue is how to connect regional networks. The supply chain structure at the national level offers the interface between a flow of commodities in national level facilities and the designated region level facilities. We specify and select the transfer terminals between the two levels of networks. How the FOSA nodes connect two sub-networks with different time steps is shown in Table 1. For this numerical study, the time step chosen for the basic time unit at the national network level is one day, and the time step selected for the region's network level is one hour. The travel times are calculated based on the distance and a fixed average travel speed.

At the local network level, we assume that demands are known in advance, and in the stochastic model we assume that we know the possible scenarios that could realize uncertain demands.

The local center compares the requirements measured in emergency goods with the resources stocked in several warehouses. This proposition determines the number of emergency kits that should be requested. The deficit is the amount that the local centers ask from the regional centers. This value is adjusted by certain percentages: It is possible that the regional center would be unable to commit to delivering the full amount of aid because the resources must also be used to cover other emergencies. The orders received by the regional centers are compared with the amount of goods available in the regional warehouse, and the deficit is also calculated. This deficit represents the amount that is ordered for the next stage in the national center.

A framework is suggested for modeling the humanitarian supply chain. The main characteristics of the modeling approach can be summarized as follows:

1. Time-Space Network
2. Facility Location
3. Facility Capacity
4. Demand
5. Supply
6. Multi-modal
7. Vehicle Routing
8. Capacity Constraints
9. Integrated Model

The model assumes that the initial requirements are given at the beginning of the simulation. This means that only an initial requirement is given, and then the supply chains answers to that initial requirement, which represents the result obtained through the evaluation of damages and needs. Initially, some resources are pre-positioned in different warehouses because many emergencies occur in specific seasons and these can be foreseen although not predicted.

5 Implementation with Realistic Data and Simulation Results

Routing models need the latest and high-quality data for humanitarian settings. Betterment in the availability and adoption of latest information in disaster bring the chance to provide the data necessary for credible decision-making.

We used ARTISOC, the multi-agent simulation. The necessary data for tool making are shown below:

1. Position of each POD
2. The number of victims
3. Harbor, airport, position of the station that are to become the supply point
4. Capacity for transportation and supply of the supply point
5. The number of vehicles that can be managed
6. Loading capacity of vehicles
7. The number of prior warehouses

We input the static data, which could change these initial settings in this report and maintain greater availability.

We assumed the cluster ring [7] by the k-means method, the center position of the prior warehouse is made according to the number of the placements of the PODs, in which the relief supplies were previously placed.

A supply chain network, including the prior warehouses is created, which produces on demand the amount needed for the number of victims. The supply delivered from each supply point at the end of the simulation is a stage beyond the quantity of total demand that totaled the quantity of demand for each refuge. Supplies of sufficient quantity are deployed in the prior warehouse beforehand and can originally include the difference at the time of the simulation's end for a quantity of necessary relief supplies.

We assumed each Thai traffic infrastructure to be a supply point and assumed the central location of municipalities' PODs. The supply point to use adopted an airport, a harbor, the first-class station, and ft3 to convert the transport volume needed in a year was assumed.

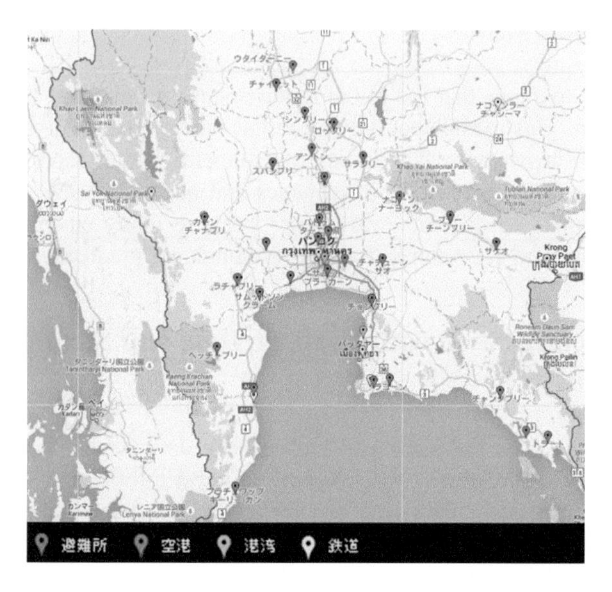

Fig. 3 Positions of POD and Supply points in Thailand

The number of victims was derived from the population data [8, 9] and was included without passing through the decrease by the disaster model. Thus, we modeled the disaster to scale with corresponding damages when we assumed that suffering of probability 1 had been generated. We assumed a general container transportation truck for the vehicle, and a vehicle speed of 40km/h; the transportation capability was assumed to be 1 TEU, $20ft^3$, and the number of vehicles was assumed to be 2,000.

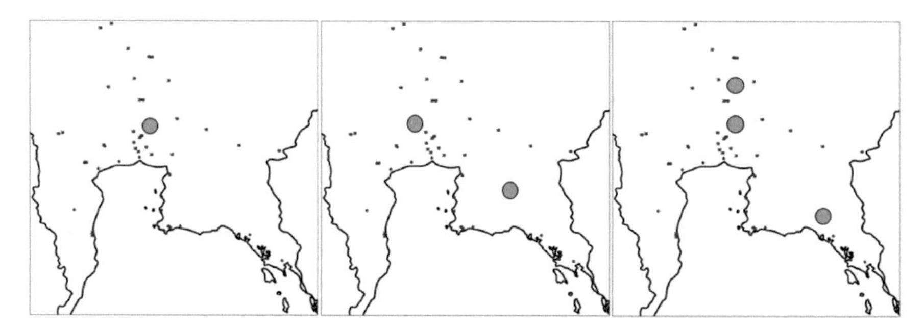

Fig. 4 Positions of prior warehouses

We considered the case that placed the number of prior warehouses at 1–3 each, and the position—the link situations—were decided from the aggregate by the Hartigan statistic of the k-means method.

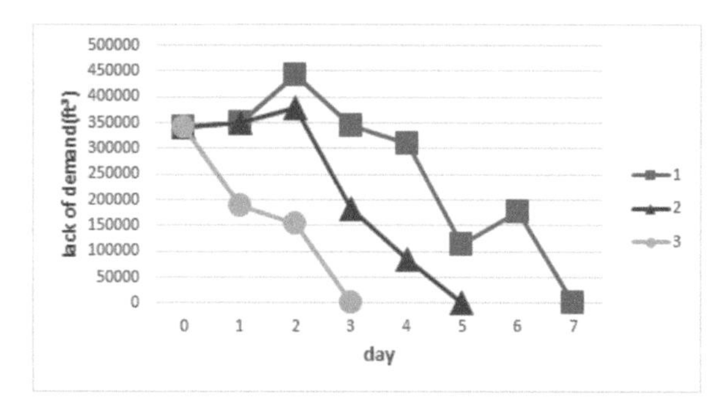

Fig. 5 Lack of demand over days as the function of the number of prior warehouses

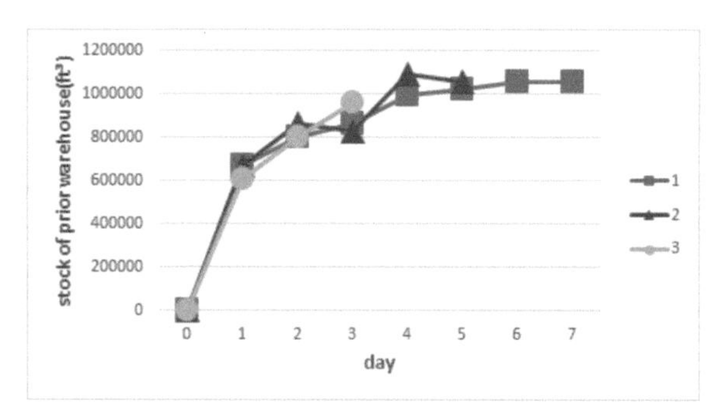

Fig. 6 Necessary stock of prior warehouses as the function of the number of prior warehouses

Table 1 Summary of simulation results

The number of prior warehouses	1	2	3
The days until lack of supplies cancellation	7	5	3
Necessary stock of prior warehouses (10^3 ft^3)	1050	1050	960

As a result of mentioning it specifically, a difference emerged in the quantity of insufficient demand of the first day. Here, the quantity of insufficient demand was located at 1 or 2 whereas it was about 350,000 ft^3, and the lack of demand quantity on the first day was located at 3 where it was about 190,000 ft^3. In this respect, the distance from each prior warehouse to each POD is shortened by an increase in the number of warehouses and their placement in a more effective position; it is thought that the time to extend beyond the delivery is a shortened result. In addition, we obtained approximately a 10% reduction of the results about the stock quantities, which should be prepared beforehand, for a similar reason. Based on the above mentioned result, we found that:

When the three prior warehouses are located about 20 million victims of the region in central Thailand are assumed, and the required relief supplies total 960,000 ft^3, emergency goods would be supplied smoothly within 3 days.

6 Future Works

Here, we make recommendations for further developments to this study:

i. The inflection of dynamic data
We built population, the number of victims, the transport volume, haul distance, and the coordinates of the POD by static data into this report. Therefore we cannot refer to the latest information when there is a turn of events and updates occur. The inflection of the dynamic data require consistent updates to solve this problem is necessary. Network services extend further and further, and future emergencies can specialize if the acquisition about a rescuer is enabled, according to the number and the position of the POD.

ii. Optimizing placement by the population ratio
We used only the position of the POD for the placement of the prior warehouse in this report. As a result, consideration by distance, and the number of victims in the refuge are not considered although it is done.

The refuge where support supplies are not delivered may equally not exist, but it is unequal with respect to the quantity distributed to each person. Based on the foregoing characteristics, in a cluster ring, the technique that assumed the number and weight of the victims is expected.

iii. Application of the disaster model
We assumed the suffering probability in this report by disaster 1, and derived the number of the victims from population data. However, in the case of a natural disaster such as a flood, the damage increases if a river water system is nearby, or there may be extremely little damage depending on the actual place. In addition, with regard to the placement position of the prior warehouse, the warehouse itself may suffer. Therefore examination by the disaster model, including risk management, is necessary.

iv. Integration with the real-time mapping technology
Feedback of clear information in the disaster location is necessary for expansion of available and reliable decision-making. Given images including road information, mapping information, and real-time data provided by crowdsourcing, decision-making becomes a more useful tool for practitioners.

v. Expansion of the scale
We considered a potential disaster at the center of Thailand in this report and studied the prior placement of relief supplies and delivery by a domestic network. However, a delay can originally occur in the delivery of relief supplies external to the territory and negotiations about the preparedness and readiness to receive must occur. In addition, placement within a large framework, such as the ASEAN countries, and numerical examinations are necessary as are discussions in a single country about the prior warehouses.

7 Conclusion

A result of this study is that we developed a decision-aiding tool for ICT-based humanitarian support logistics and were able to solve the issue of final last-mile logistics problem using the prior warehouses. In this article we modeled an experimental area in central Thailand and identified the implementation using realistic data such as the quantity of relief supplies, the number of prior warehouses, and the positions that should be involved in making preparations for an emergency.

Acknowledgements This work was supported by KAKENHI Grant number 25330277 and 26282089.

References

1. Minic, S.M.: Three-Echelon Supply Chain Management for Disaster Relief Operations. CIRRELT-2014-28 (2014)
2. Crowley, J.: Connecting Grassroots and Government for Disaster Response. Commons Lab of the Woodrow Wilson International Center (2013)
3. ManMohan, S.S.: Buttressing Supply Chains against Floods in Asia for Humanitarian Relief and Economic Recovery. Production and Operations Management **23**(6), 938–950 (2013)
4. Manopiniwes, W., Nagasawa, K., Irohara, T.: Humanitarian Relief Logistics with Time Restriction, Thai Flooding Case Study. Industrial Engineering & Management Systems **13**(4), 398–407 (2014)
5. Tabbara, N.L.: Emergency Relief Logistics: Evaluation of Disaster Response Models. Oxford Brookes University, a project report (2008)
6. Haghani, A.: Supply Chain Management in Disaster Response. Grant-DTRT07-G-0003, Mid-Atlantic Universities Transportation Center (2009)
7. Hartigan, A.J.: A K-Means Clustering Algorithm. Journal of the Royal Statistical Society **28**(1), 100–108 (2012)
8. Lee, M.Y.: Simulating distribution of emergency relief supplies for disaster response operations. IBM, Winter Simulation Conference (2009)
9. Barahona, F.: Agile logistics simulation and optimization for managing disaster responses. IBM, Winter Simulation Conference (2013)

B-Spline Curve Knot Estimation by Using Niched Pareto Genetic Algorithm (NPGA)

Vahit Tongur and Erkan Ülker

Abstract In this paper, estimated curve Knot points are found for B- Spline Curve by using Niched (Celled) Pareto Genetic Algorithm which is one of the multi objective genetic algorithms. It is necessary to know degree of the curve, control points and knot vector for drawing B-Spline curve. Some knot points are of very few or no effect at all on the drawing of B-Spline curve drawing. Omitting such points will not effect the shape of curve in curve drawing. In this study, it is aimed to find and omit these ineffective curve points from drove of curve. Performance of proposed method are compared with selected studies from literature.

Keywords NPGA · Genetic algorithm · B-Spline

1 Introduction

B-Spline curves is defined by corner points polygon which are called as control points. As well as not crossing control points, the form of the curve and surfaces,which are acquired by using these points, are shaped according to the position of these points. The polygon which is created by these control points is called as control polygon. These points make the curve to follow control polygon by acting like a magnet and ultimately a characteristical and smooth curve is acquired which is within the borders of control polygon. But as the amount of control points increase, the amount of

V. Tongur
Department of Computer Engineering, Faculty of Engineering and Architecture,
Necmettin Erbakan University, Konya, Turkey
e-mail: vtongur@konya.edu.tr

E. Ülker(✉)
Department of Computer Engineering, Faculty of Engineering,
Selçuk University, Konya, Turkey
e-mail: eulker@selcuk.edu.tr

K. Lavangnananda et al. (eds.), *Intelligent and Evolutionary Systems*,
Proceedings in Adaptation, Learning and Optimization 5,
DOI: 10.1007/978-3-319-27000-5_25

curve effecting control points decrease. Most of the control points which are given in very frequent intervals barely effect the curve. In this situation, such points can be determined and omitted, so the curve can be drawn again without damage. Amount of the points which are choosen as knot directly effects the amount of control point. The smallest number of knot point should be chosen for decreasing control points. Omitting unnecessary knot points can provide a faster drawing in very complex studies. It is important to determine the right knot points in eliminating the points. Otherwise the shape of the curve can be deformed. It is not possible to determine unnecessary points by just looking. In such situation, optimization algorithms are used. Especially heuristic artificial intelligence techniques give good results in problem solving. It will be objective to use multi purpose genetic algorithms in this kind of optimization processes. Because the shape of the curve will be deformed if too much points are omitted. On the other hand if less than necessary points are omitted than the shape will not be deformed but targeted point elimination will be insufficient.

Many problems consist of concurrent optimization of more than one confounding targets. While one of the targets moves through the best the other starts to moves away. Accepting both targets in considerable values means to bring multi objective problems to the optimum level.

Estimation of knot points by artificial intelligence techniques are studied in literature. Yuan et al. have used two staged Knot selection method in their study for the structure of the curve. In the first stage a subset of the basic functions is chosen from previously determined multi resolution basic set by using a statistical variance selection method (Lasso). In the second stage a vector space is constructed for it is determined that it is enough to characterise vector space for conveying the vector space to basic function [1]. Li et al. have proposed adaptive gradient choice method in B-Spline curve estimation [2]. Jacobson and Murphy have developed automatic Knot placement algorithm for activating the use of NonUniform Rational B-Splines (NURBS) [3]. Valenzuela et al. have presented a new methodology for the optimal selection and position of the curves for curve creating and simulation [4]. Ülker proposed the use of PESA algorithm in Knot estimation of B-Spline curves [5]. Ülker and Arslan used artificial immune system for overcoming negativities sourced from Knot points in B-Spline drawing and finding a smaller amount of estimated Knot points [6]. Gülcü and Ülker offered Strength Pareto Evolutionary Algorithm 2 algorithm in their study for decreasing Knot points of the B-Spline curves in their study [7]. Galvez and Iglesias have presented a new method in their study for creating a B-Spline curve creating from a noisy data points set. This method calculates all parameters of B-Spline curve creating. This multimodel is necessary for solving the optimization problem of unlinear smallest squares . They have solved this optimization problem by applying firefly algorithm in approaches [8]. Ülker and Değer offered Strength Pareto Optimization Algorihm (SPEA) method for Knot point estimation for B-Spline curves [9]. Weishi et al. have developed an adjustable Knot placing algorithm for busy and noisy points in B-Spline curves [2]. Zhaoa et al. have tried to find the best Knot points of B-Spline curve drawing by GMM-based continuous optimization algorithm [10]. They acquired starting positions of each Knot in algorithm by using Monte Carlo method. Ma and Kruth have applied the projection of measured points to the ground surface for attaining parameter value to the randomly measured points for the smallest placement of the B-Spline surfaces [11]. Park, Lee have presented a new approach for B-Spline curve creating which is called as repressed point [12]. Pham have developed a simple and efficient method for creating an offset curve for a uniform cubic B-Spline by using a set of control Knots [13]. Galvez and Iglesias have offered again using Particle Swarm Optimization algorithm for Knot optimization [14]. There are another studies in literature on Bezier and NURBS surfaces but we have not mention it in this section as the article is about Knot optimization in B-spline curves.

In this study, the minimum change of curve shape is aimed by the minimum Knot points. NPGA algorithm is used in Knot estimation which is not studied in literature. Performance of the algorithm is shown in experimental studies.

2 Problem Definition

It is necessary to produce approximate points while decreasing the number of Knots in B-Spline curves. Normal three basic parameter are needed while drawing a B-Spline curve. These are order (degree+1) Knot vector and control points of the curve. When it is desired to reduce the number of Knots in a curve, new Knot points should be calculated and Knot vector should be determined again. The parameter of the degree of curve does not change. The first thing to do in this situation is to produce new control points basing real coordinate points.

Some of the real coordinates are chosen for this and the points are checked if they are the best or not. In this step, Euclid distance total between the produced approximate points and real points is calculated. If there are previously tried points the error rates are checked for these sets and the best point sets are tried to be determined. This stage includes below steps;

a. Some of the real point coordinates are chosen as Knot and centripetal vector is calculated by Eq. (1).

b.

$$\overline{u_0} = 0, \overline{u_m} = 1, \qquad \overline{u_i} = \overline{u_{i-1}} + \frac{\sqrt{|Q_i - Q_{i-1}|}}{\sum_{j=0}^{m} \sqrt{|Q_j - Q_{j-1}|}} \tag{1}$$

where, Q_i factor shows the i. point which is one of the points chosen from real coordinate points. And the m parameter is the quantity of the Knots found in point cloud

c.

$U = \{0, 0, \ldots, 0, u_{d+1}, \ldots, u_m, 1, 1, \ldots, 1\}$,

$$u_{j+d} = \frac{1}{d} \sum_{i=j}^{j+d-1} \overline{u_i} \qquad j = 1, \ldots, m-d \tag{2}$$

Where, d is the degree of the curve. Eq. (2) is used for the real points which are chosen as Knot for calculating U Knot vector for B-Spline curve. The first d element and the last d element are chosen as zero and one in order while calculating U Knot vector. Intermediate elements are found by the help of Eq. (2).

d. According to B-spline curve formula $Q = PxR$ matrix is accepted. In here, R matrix is the matrix which produces B-spline blending function according to N values. Q matrix is the set of points which is chosen as Knots. P is the estimated control points. Eq. (3) can be solved by what is known for calculating P control points.

$$P = QxR^{-1} \tag{3}$$

R matrix is acquired from recursive cox-de Boor function.

$$N_{i,1}(u) = \begin{cases} 1 & if \quad t_i \leq u \leq t_{i+1} \\ 0 & else \end{cases}$$

$$N_{i,k}(u) = \frac{u - t_i}{t_{i+k-1} - t_i} N_{i,k-1}(u) + \frac{t_{i+k} - u}{t_{i+k} - t_{i+1}} N_{i+1,k-1}(u) \tag{4}$$

Where $u \in [0, 1]$, $t \in U$ are the Knot vector elements, i is the existing point coordinate and k represents the order.

After R matrix elements are calculated in Eq. (4), reverse of the acquired matrix is taken and $R - 1$ is acquired. P points are acquired from Eq. (3). Calculating the difference between the estimated points (S) and real points (F) is depended upon F point amount which is the increasing amount of u parameter in B–spline curve formula. If the F point amount is named as c, increasing amount of u is found as $1/c$. So, while drawing P point curve of the m quantity estimation, as much coordinate as the number of F line is acquired.

$$S(u) = \sum P_i N_{i,d}(u) \tag{5}$$

In this situation the error between the F and S can be calculated by the Euclid equation;

$$Error = \sqrt{\sum_{i=1}^{M_u} |S_i - F_i|^2} \tag{6}$$

3 Genetic Algorithms

Genetic algorithms are heuristic research method which base producing new sets of chromosomes for solving complex regular problems. Each chromosome in genetic algorithm represents a solution in problem space. Chromosomes coding can be done in various ways. The most commonly used coding is the binary coding. Chromosomes are subjected to selection process after coded for crossover. There are also different ways for chromosome selection. A selection process is applied according to the individual's chromosome. Selected individuals are intercrossed as couples. Single point or multi point crossover methods are used in crossover. A part of the chromosomes is marked in single point crossover method and crossover is done by replacing the genes on the left and right sides. In multi point crossover chromosomes seperates the genes in at least two points and make the gene transfer between the couples between determined areas. Mutation happens as a result of changing a gene in a chromosome. For example, if chromosome has dual code and zero gene value, this gene is mutated by valuing it as 1. Termination criteria is checked after this step. Termination process is done as soon as proper value is reached or a specific generation is created. If the termination criteria is not met, crossover and mutation steps are repeated until the termination criteria is met.

Pseudo code of the genetic algorithm is as below;

1. *Create the beginning population*
2. *Calculate fitness values for each individual*
3. *Use selection operator for selecting the individuals for crossover*
4. *Crossover the selected individuals*
5. *Apply mutation operator the created new individuals*
6. *Repeat the step 2-5 until the desired generation is reached*
7. *Give the best individual of all generations as the result*

4 Implementing of the Problem to Multi Objective Genetic Algorithm

There should be at least two aim function in multi objective genetic algorithms. One of the objective of this study is to minimize the number of the Knots and the other one is to minimize the error rate. The error rate will increase while minimizing the number of predictable Knot points.

On the other hand, the chosen Knot numbers should be equal to real Knot numbers for minimizing the error rate. Multi objective genetic algorithms may not be the best single solution for all purposes. In this situation, decision maker is asked to choose a solution among a determined set. Proper solution should provide an acceptable performance for all purposes.

It is necessary to create chromosome structures for solving the problem with genetic algorithm. A random population is created. Chromosomes are coded. Coding for the subject problem is done in binary system as Figure 1.

The length (number of genes) of the chromosome is as much as the number of real points. Genes of the chromosomes are filled with 0 and 1 values randomly. the value of 1 in the gene i means that the point is chosen as the Knot. So random points are chosen among the real points. All chromosomes of the population are coded in the same way. Coded chromosomes are chosen for crossover. In this study niched pareto method is used for selection.

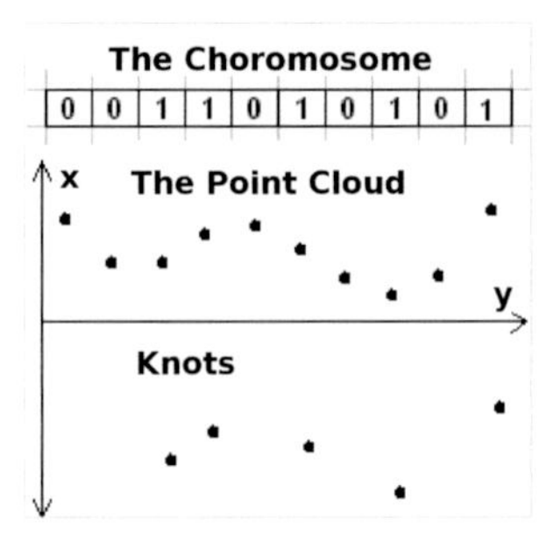

Fig. 1 Selection of Random Knots and representation of chromosome [6]

5 Niched Pareto Genetic Algorithm

Niched pareto Genetic Algorithm (NPGA) has been suggested firstly by Horn et al. (1994) for the problems about hydro systems [15]. Tournament selection has been changed in two ways for preventing the convergence in Niched pareto genetic algorithm and acquiring Pareto-Optimal solutions. These are; Adding Pareto Domination Tournament and using sharing for the winner when there is an unsuppressed tournament. When two randomly selected individual are compared according to their pareto superiorities, naturally they are subjected to a dual tournament. If one of them suppress the other one , it is the winner. Two canditate individuals and their comparision set (crossover pool) is selected randomly from existing population. After that each of the candidates are compared to each candidates in comparison set (crossover pool). If a candidate is suppressed by comparison set and the other is not, the other one is chosen for reproduction.

In this study, equivalent kind sharing method is used. Sharing is applied in the situations that both candidates of the crossover pool are suppressed or unsuppressed. Each candidate checks the number of the candidates that set diameter is given. The set in which there are less canditates is chosen and the other one is dismissed.

In literature, there are studies with niched pareto genetic algorithm in different areas. Zhang et al. (2009), have used niched pareto genetic algorithm for optimizing a micro grooved heating pipe [16]. In their study, they have suggested NPGA for a lot of multi objective optimization problems in which heat transfer capacity and total heat resistance are inversely proportional. Ozugur et al. (2001) have suggested four layer 2G and 3G for UMTS coverage zone and NPGA method for hierarchical mobility management optimization problem [17]. Baradli et al. (2009) have used NPGA algorithm for propert selection in temporary classifying process [18]. In Figure 2, equivalent class sharing is seen. Number of individuals is checked according to the given diameter value. Candidate 2 is sent to crossover pool as it has less individual and Candidate 1 is dismissed.

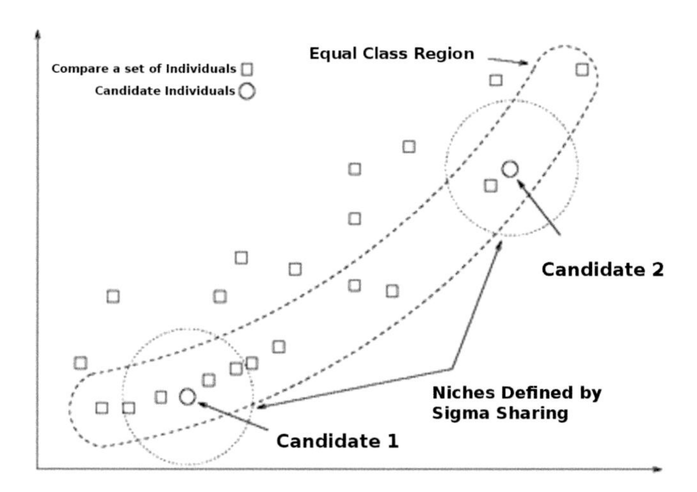

Fig. 2 Equivalent Classifying Sharing [15]

Pareto Domination Tournament pseudo code as follow;

1. Select 2 random candidates from population
2. Compare each of the candidates with a candidate in comparison set
3. Send the unsuppressed candidate to crossover pool
4. Apply sharing if both of the candidates are suppressed or unsuppressed
5. Repeat step 2-5 until crossover pool is full

6 Experimental Results

In Figure 3, B-Spline curve parametization example for accessing NPGA algorithm in estimating of the B-Spline curve Knots is given. 10% noise is added to the clean points in Figure 3.a and 200 points are acquired in total. Modelled draft is a non uniform B-spline cubic curve which has {0.0, 0.0, 0.0, 0.0, 0.25, 0.5, 0.75, 1.0, 1.0, 1.0, 1.0} Knot vector and 7 control points. NPGA algorithm is repeated as 50,100,150 individuals and for each experiment #1, #2 and #3 numbers are given. Chromosome length and crossover rate is standard in each experiments and it is 0,3 and 100, respectively. All experiments are executed as 100 iteration. Average Error and Average AIC values are calculated for each population. AIC is an information retrieval method used in literature [5].

Figure 4.a, 4.b and 5 shows the best results for all generations in Experiment #1, Experiment #2, and Experiment #3, respectively. Table 2, 3 and 4 show average error and AIC values according to generations. Best average value for AIC is 3600 when examined results of the study in [9]. According to this, proposed method are given better results than SPEA algorithm for same sample.

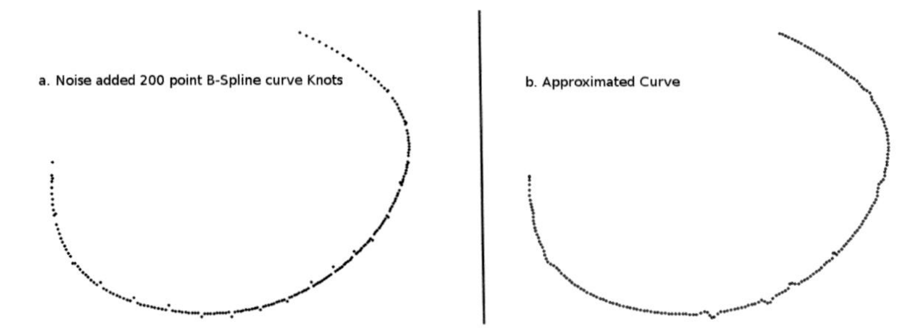

Fig. 3 B-Spline curve knots and approximated curve using 200 points

Table 1 Parameter values used in NPGA for Experiment #1, #2, #3

Parameter	Exp#1	Exp#2	Exp#3
Population	50	100	150
Chromosome Length	200	200	200
Crossover Rate	0.3	0.3	0.3
Iteration	100	100	100

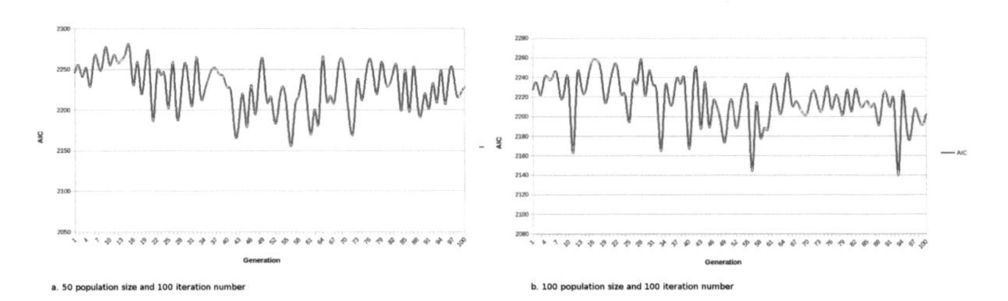

Fig. 4 AIC values according to generations for Experiment #1, #2

Table 2 Average error and AIC values according to generations for Experiment #1

Generation	Error (Average)	AIC (Average)	AIC (Average) [9]
5	5292.14	2331.32	3580
10	5301.64	2336.64	3563
25	5201.22	*2320.34*	3470
50	5180.66	2326.82	3390
75	5189.36	2331.06	3372
100	*5144.36*	2325.84	3298

According to experimental results, Considering the average AIC values, 2319.19 and population size 100 gives the best result. Studying average error values, 5052.68 and population size 150 experiments give the best results. The minimum Knot number is 73 as a result of the experiments.

Table 3 Average error and AIC values according to generations for Experiment #2

Generation	Error (Average)	AIC (Average)
5	5273.84	2332.65
10	5261.64	2330.48
25	5211.92	2328.05
50	5169.12	*2319.19*
75	5163.52	2323.72
100	*5119.09*	2320.44

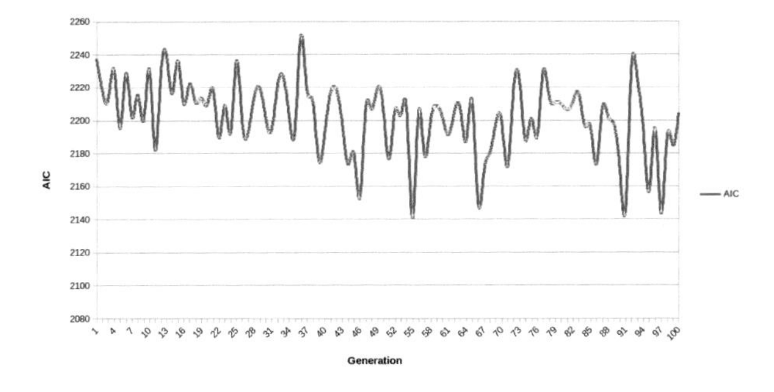

Fig. 5 AIC values according to generations for Experiment #3

Table 4 Average error and AIC values according to generations for Experiment #3

Generation	Error (Average)	AIC (Average)
5	5321.89	2330.74
10	5301.69	2332.54
25	5271.34	2329.16
50	5214.69	2328.31
75	5112.98	2326.48
100	*5052.68*	*2324.38*

Two data set in [1] are used for second and third case studies. First data set and second data set consists of 500 points that obtained from functions in Eq. (7) and Eq. (8), respectively.

$$g(t) = \frac{1}{2.3935}\left(1.5exp\left(-\frac{(t-0.1)^2}{0.3}\right) + 0.1exp\left(-\frac{(t-0.5)^2}{2}\right) + 2exp\left(-\frac{(t-0.8)^2}{0.02}\right)\right), t \in [0, 1] \quad (7)$$

$$g(t) = \sin(4t-2) + 2exp(-30(4t-2)^2), t \in [0, 1] \quad (8)$$

When examined of experiment results from #4 to #9, it is shown that obtained curves from proposed method and real curves are matched with each other. Also, fitting errors of proposed method are equivalent to results in [1].

Table 5 Parameter values used in NPGA for Experiment #4, #5, #6, #7, #8, #9

	Eq.(7)			Eq.(8)		
Parameter	Exp#4	Exp#5	Exp#6	Exp#7	Exp#8	Exp#9
Population	50	100	150	50	100	150
Chromosome Length	500	500	500	500	500	500
Crossover Rate	0.3	0.3	0.3	0.3	0.3	0.3
Iteration	100	100	100	100	100	100

Table 6 Average error and AIC values according to generations for Experiment #4, #5, #6

	Exp#4		Exp#5		Exp#6	
Generation	Error (Avg.)	AIC (Avg.)	Error (Avg.)	AIC (Avg.)	Error (Avg.)	AIC (Avg.)
5	160.08	4067.44	157.86	4046.12	158.64	4057.09
10	159.64	4060.92	158.34	4043.01	159.03	4053.09
25	159.30	4057.08	158.29	4038.90	158.47	4053.48
50	159.84	4059.26	157.64	4041.64	158.44	4054.49
75	159.92	4054.98	156.47	4033.57	158.83	4056.32
100	160.18	4064.76	157.90	4039.46	159.34	4060.97

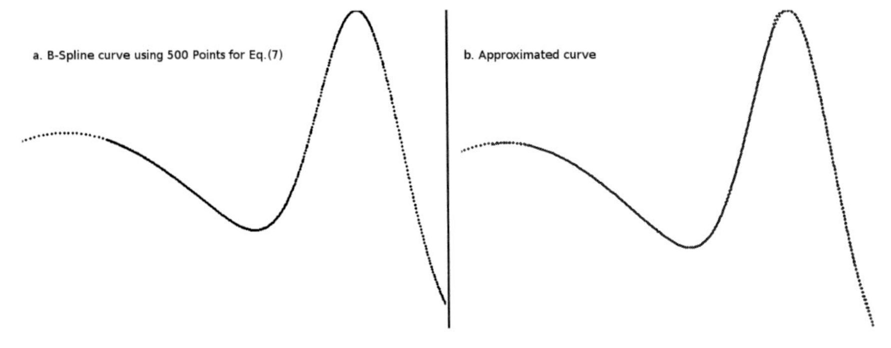

Fig. 6 B-Spline curve knots and approximated curve using 500 points for Eq. (7)

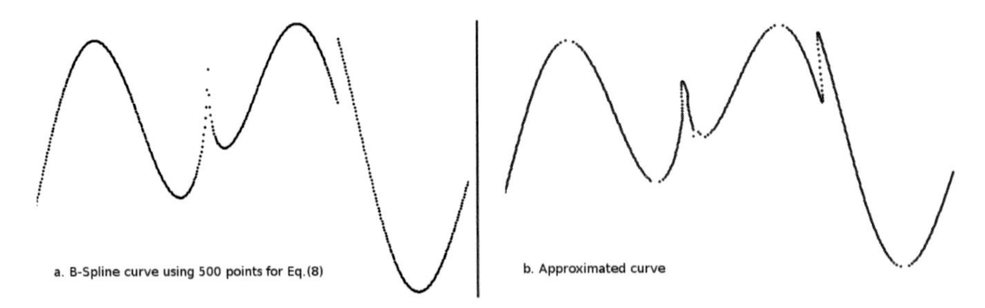

Fig. 7 B-Spline curve knots and approximated curve using 500 points for Eq. (8)

Table 7 Average error and AIC values according to generations for Experiment #7, #8, #9

	Exp#7		Exp#8		Exp#9	
Generation	Error (Avg.)	AIC (Avg.)	Error (Avg.)	AIC (Avg.)	Error (Avg.)	AIC (Avg.)
5	195.68	4172.72	195.61	4152.42	196.30	4153.14
10	196.46	4169.62	195.61	4155.53	196.08	4159.55
25	195.58	4160.14	195.89	4151.15	196.17	4157.10
50	196.78	4153.86	196.19	4147.92	197.18	4162.12
75	196.02	4158.84	195.62	4149.51	197.50	4157.14
100	196.26	4160.28	197.13	4155.62	197.60	4152.18

7 Conclusion

The curve reconstruction by choosing optimal number of Knots from digitized points data causes calculative complex optimization problems. As the problem is of great complexity and an extremely non-linear optimization problem, non-deterministic optimization strategies should be practised for acquiring optimal interpolation results. Evolutionary algorithms show great flexibility and sustainability in such problems. Even in complex problem spaces' of evolutionary algorithms, it has provided certain solutions and this is one of the factors increased its popularity. In this paper, when a large points set is given for a widely placed plane, NPGA algorithm is used for conforming parametric B-Spline curve for irregularly placed points. Performance of proposed method are compared with selected studies from literature. Points are dimensions taken from a physical fact or geometrical model and the aim is to be able to rebuild the B–spline curve with combination of Knot placements and numbers. The algorithm which is given here can be used in various applications especially in the places that the points are intense and noisy. Evolutionary algorithms represent high efficiency for complex (multi-model, combinational and calculative difficult) optimization problems. However proper parametrization of an evolutionary algorithm can be pretty difficult. As B-spline curve estimation problem is highly combinational complex, convergence speed decreases according to the speed of the points. For this reason optimization of many points can be slow to convergence to number of points although evolutionary algorithms are close to local optimum. Generally efficient optimization strategies and quality criteria should be considered for producing a more efficient CAD surface samples and rare numeric point data. The most important part of our method is that it can define concurrent placement of the Knots and enough numbers automatically. If the method can have the ability of recognising the weights of control points, it can be developed to Non-Uniform B-spline curves.

Acknowledgements This work is supported by Scientific Research Project of Selçuk University.

References

1. Yuan, Y., Chen, N., Zhou, S.: Adaptive B-Spline Knot Selection Using Multi-Resolution Basis Set. IIE Trans. **45**(12), 1263–1277 (2013)
2. Li, W., Xu, S., Zhao, G., Goh, L.P.: Adaptive Knot Placement in B-Spline Curve Approximation. Computer-Aided Design **37**(8), 791–797 (2005)
3. Jacobson, T.J., Murphy, M.J.: Optimized Knot Placement For B-Splines in Deformable Image Registration. Medical Physics **38**(8), 4579–4592 (2011)
4. Valenzuela, O., Delgado-Marquez, B., Pasadas, M.: Evolutionary Computation for Optimal Knots Allocation in Smoothing Splines. Applied Mathematical Modelling **37**(8), 5851–5863 (2013)
5. Ülker, E.: B-Spline Curve Approximation Using Pareto Envelope-Based Selection Algorithm- PESA. International Journal of Computer and Communication Engineering **2**(1), 60–63 (2013)
6. Ülker, E., Arslan, A.: Automatic Knot Adjustment Using an Artificial Immune System for B-Spline Curve Approximation. Information Sciences **179**(10), 1483–1494 (2009)
7. Gülcü, Ş., Ülker, E.: Knot estimation of the B-spline curve with strength pareto evolutionary algorithm 2 (SPEA2). In: Proc. of the 16th WSEAS Int. Conf. on Computers (part of CSCC 2012), Kos Island, Greece , pp. 308–314 (2012)
8. Galvez, A., Iglesias, A.: Firefly Algorithm for Explicit B-spline Curve Fitting to Data Points. Mathematical Problems in Engineering 2013 (2013)
9. Ülker, E., Değer, S.: Optimization of knot points in B-spline curve fitting with strength pareto optimization algorithm (SPEA). In: Proc. of 6th International Conference of Advanced Computer Systems and Networks: Design and Application (ACSN 2013), pp. 90–97. Lviv (2013)
10. Zhao, X., Zhang, C., Yang, B., Li, P.: Adaptive Knot Placement Using A Gmm-Based Continuous Optimization Algorithm in B-Spline Curve Approximation. Computer-Aided Design **43**(6), 598–604 (2011)
11. Ma, W., Kruth, J.: Parameterization of Randomly Measured Points for Least Squares Fitting of B-Spline Curves and Surfaces. Computer-Aided Design **27**(6), 663–675 (1995)
12. Park, H., Lee, J.H.: B-Spline Curve Fitting Based on Adaptive Curve Refinement Using Dominant Points. Computer-Aided Design **39**(6), 439–451 (2007)
13. Pham, B.: Offset Approximation of Uniform B-Splines. Computer-Aided Design **20**(8), 471–474 (1988)
14. Galvez, A., Iglesias, A.: Efficient Particle Swarm Optimization Approach for Data Fitting with Free Knot B-Splines. Computer-Aided Design **43**(12), 1683–1692 (2011)
15. Horn, J., Nafpliotis, N., Goldberg, D.E.: A niched pareto genetic algorithm for multiobjective optimization. In: Proceedings of the First IEEE Conference on Evolutionary Computation: IEEE World Congress on Computational Intelligence, pp. 82-87 (1994)
16. Zhang, C., Chen, Y., Shi, M., Peterson, G.P.: Optimization of Heat Pipe with Axial Ω-shaped Micro Grooves Based On A Niched Pareto Genetic Algorithm (NPGA). Applied Thermal Engineering **29**(16), 3340–3345 (2009)
17. Ozugur, T., Bellary, A., Sarkar, F.: Multiobjective hierarchical 2G/3G mobility management optimization: niched pareto genetic algorithm. In: IEEE Global Telecommunications Conference (Cat. No.01CH37270), GLOBECOM 2001, vol. 6, pp. 3681–3685 (2001)
18. Baraldi, P., Pedroni, N., Zio, E.: Application of A Niched Pareto Genetic Algorithm for Selecting Features for Nuclear Transients Classification. International Journal of Intelligent Systems **24**(2), 118–151 (2009)

Part V
Nature Inspired Creative Computing

Computational Red Teaming in a Sudoku Solving Context: Neural Network Based Skill Representation and Acquisition

George Leu and Hussein Abbass

Abstract In this paper we provide an insight into the skill representation, where skill representation is seen as an essential part of the skill assessment stage in the Computational Red Teaming process. Skill representation is demonstrated in the context of Sudoku puzzle, for which the real human skills used in Sudoku solving, along with their acquisition, are represented computationally in a cognitively plausible manner, by using feed-forward neural networks with back-propagation, and supervised learning. The neural network based skills are then coupled with a hard-coded constraint propagation computational Sudoku solver, in which the solving sequence is kept hard-coded, and the skills are represented through neural networks. The paper demonstrates that the modified solver can achieve different levels of proficiency, depending on the amount of skills acquired through the neural networks. Results are encouraging for developing more complex skill and skill acquisition models usable in general frameworks related to the skill assessment aspect of Computational Red Teaming.

Keywords Neural network · Domain propagation · Skill acquisition · Supervised learning

1 Introduction

In Computational Red Teaming (CRT) a Red agent takes actions to challenge a Blue agent, with a variety of purposes. In the cognitive domain, one of these purposes, which generated an intense interest in the scientific community in recent years, is to force a human Blue agent to improve its skills. This process involves two major aspects. First, the Red must find the proper ways of action for challenging the Blue;

G. Leu(✉) · H. Abbass
School of Engineering and Information Technology,
University of New South Wales, Canberra Campus, Campbell, Australia
e-mail: {G.Leu,H.Abbass}@adfa.edu.au

K. Lavangnananda et al. (eds.), *Intelligent and Evolutionary Systems*,
Proceedings in Adaptation, Learning and Optimization 5,
DOI: 10.1007/978-3-319-27000-5_26

this is the *task probing*. Second, in order to find those ways of action, the Red must first assess Blue's skills for finding its weaknesses and hence, potential directions of improvement. This second aspect is the *skill assessment* aspect, in which the representation of Blue's skills is essential.

In this paper we apply the CRT to Sudoku puzzle and we focus on the representation of the skills used for solving a Sudoku game. We investigate the Sudoku literature in order to establish what are the skills that humans apply to solve the puzzles, and then we create their computational representation, in a manner that is cognitively plausible. We use feed-forward neural networks (NN) to represent the skills, and we model the skill acquisition process through supervised learning and back propagation. The NN-based skills are then embedded into a classic hard-coded constraint propagation Sudoku solver, endowing it with the ability to learn Sudoku skills through training. While the Sudoku solving sequence remains hard-coded, the computational solver uses at each predefined step the pattern recognition capability of the neural networks, and thus, its proficiency varies based on the skills embedded in its structure. In order to demonstrate this we use two skill setups: a first one in which the neural networks can only detect the existence of a favourable pattern on te Sudoku board, and a second one in which the pattern can be both detected and localised. Simulation results show how the realistic skill-based solver can achieve different levels of proficiency in solving Sudoku in the two setups, with a higher level of proficiency reached for the first skill setup.

The paper is organised as follows. The second section presents the existing computational approaches on solving Sudoku and draws a conclusion on the lack of skill-based computational solvers. The third section shows how we choose from the range of human skills used in Sukodu solving, in order to transfer them into the proposed skill representation and acquisition model. The fourth section describes the methodology used for modelling the skills and the NN-based skill acquisition process. The fifth section presents and discusses the results of the experiments. Last section concludes the study and summarises the main findings.

2 Background on Computational Sudoku Solving

The existing computational Sudoku solvers focus mostly on reducing the implementation and computational complexity, and on solving the puzzle as a search/optimisation problem, without Sudoku domain-specific knowledge or concerns about the cognitive plausibility.

From a computational perspective several Sudoku solvers have been reported in the literature. The simplest, but also the least effective is the backtracking solver, a brute force method that uses the full space of possible grids and performs a backtracking-based depth-first search through the resultant solution tree [1]. Another simple solver is the "pencil and paper" algorithm [2] which visits cells in the grid and generates on the fly a search tree.

In a strict mathematical view, the general $n \times n$ formulation of Sudoku is considered a non-deterministic polynomial time (NP) problem. An open question still exists in the literature on whether Sudoku belongs or not to the subclass of NP-complete problems, however more authors seem to be on the NP-completeness side [3–7]. Yato [3] and Yato and Seta [4] first demonstrated that the generalised $n \times n$ Sudoku problem can be solved in polynomial time. Later, another approach [6] converted through reduction a Sudoku problem into a "Boolean Satisfiability" problem, also known as SAT. The approach allowed not only the solving, but also the analysis of a Sudoku puzzle difficulty from the polynomial computation time perspective. A similar SAT-based solver was also proposed in [5], where the author describes a straightforward translation of a Sudoku grid into a propositional formula. The translation, combined with a general purpose SAT solver was able to solve 9×9 puzzles within milliseconds. In addition, the author suggests that the algorithm can be extended to enumerate all possible solutions for Sudoku grids that are beyond the unique solution grids posed for usual commercial puzzles.

A distinct class of computational solvers is based on stochastic techniques. A solver based on swarm robotics was proposed in [8]. The solver uses an artificial bee colony (ABC) for a guided exploration of the Sudoku grid search space. The algorithm mimics the behaviour of bees when foraging, behaviour which is further used for building partial (local) solutions in problem domain. The purpose of the algorithm is to minimise the number of duplicate digits found on each row and column. The authors compare the ABC algorithm with a Sudoku solver based on a classic genetic algorithm proposed by Mantere [9], and demonstrate that the ABC solver outperforms the GA solver significantly (i.e., on average 6243 processing cycles for ABC, versus 1238749 cycles for GA). In a different study Perez and Marwala [10] proposed and compared four stochastic optimisation solvers: a Cultural Genetic Algorithm (CGA), a Repulsive Particle Swarm Optimisation (RPSO) algorithm, a Quantum Simulated Annealing (QSA) algorithm, and a Hybrid method combining a Genetic Algorithm with Simulated Annealing (HGASA). The authors found that the CGA, QSA and HGASA were successful with runtimes of 28, 65 and 1.447 seconds respectively, while the RPSO failed to solve any puzzle. The authors concluded that the very low runtime of HGASA was due to combining the parallel searching of GA with the flexibility of SA. In the same time, they suggested that RPSO was not able to solve the puzzles because the search operations could not be naturally adapted to generating better solutions.

Another class of computational solvers is based on neural networks [11, 12]; however, these solvers are not emphasising on the cognitive plausibility of the neural networks, but rather on their mathematical mechanism. In [11] the authors propose a Sudoku solver based on the Q'tron energy-driven neural-network model. They map the Sudoku constraints in Qtron's energy function, which is then minimised ensuring the local minimums are avoided through a noise-injection mechanism. The authors show that the algorithm is totally unsuccessful in the absence of noise, while with the noise the success rate is 100% and the runtime is within 1 second. Also they demonstrate that the algorithm can be used not only for solving, but also for generating puzzles with unique solution. In a different approach, Hopfield [12] considers that neural networks do not work well when applied to Sudoku, because they tend to make errors on the way. While [11] treats this problem by injecting noise in the

Q'tron, Hopfield assumes that the search space during a Sudoku game can be mapped into an associative memory which can be used for recognising the inherent errors and reposition the NN representation of the Sudoku grid on the proper search path.

One particular class of computational Sudoku solvers, which is of major interest for our study, is the Constraint Propagation (CP) solvers. Several studies considered that Sudoku puzzle can be treated as a Constraint Satisfaction Problem [13, 14], and hence, can be solved using constraint programming techniques. Constraint Propagation solvers are purely computational methods, and the studies that proposed them followed the same purpose as the rest of the computational approaches, i.e. to produce proficient Sudoku solvers with minimal computational complexity and no domain knowledge. However, the constraint propagation processes described in both [13] and [14] are considered to be similar to the steps undertaken by human players when solving Sudoku. In his study [14] Norvig emphasises that the major task performed by humans when playing Sudoku is not to fill an empty cell, but to eliminate the multiple candidates for it, as a result of applying and propagating the Sudoku constraints. Yet, Norvig does not mention in which way the propagation of constraints resembles the human thinking. Instead, Simonis [13] does, and states that the various Sudoku-related skills used by the human players when trying to eliminate redundant candidates from cells are actually propagation schemes that participate to a constraint propagation process which eventually solves the constraint satisfaction problem. Simonis considers that "*they [human players] apply complex propagation schemes with names like **X-Wing** and **Swordfish** to find solutions of a rather simple looking puzzle called Sudoku. Unfortunately, they are not aware that this is constraint programming*". An even more advanced step towards demonstrating this concept is taken in [1] where the authors implement the constraint propagation based on a set of Sudoku skills (e.g. naked candidates, hidden candidates, Nishio-guess). The authors do not relate their algorithm to the constraint propagation formalism, and refer to it as "rule-based", but they emphasise it "*consists of testing a puzzle for certain rules that [...] eliminate candidate numbers. This algorithm is similar to the one human solver uses*".

In this study we build on the concepts proposed in the last class of computational Sudoku solvers, and we consider the skill-based approach on constraint propagation problem as central for the skill representation aspect of CRT applied to Sudoku. Thus, in the following section we describe in detail the Sudoku constraints and some of the skills used by human players in solving the game.

3 The Sudoku Game and Skills

Sudoku is a number puzzle which in its most known form consists of 81 cells contained in a 9×9 square grid that is further divided into 9 boxes of 3×3 cells. The aim of the game is to fill all cells in the grid with single digits between 1 and 9, so that a given number appears no more than once in the *unit* it belongs to, where the unit can be a row, a column or a box. These are the Sudoku rules or the constraints. In general the Sudoku problem can be seen as a $n \times n$ grid with n subsequent boxes of $\sqrt{n} \times \sqrt{n}$ cells. The constraints for a grid G can be then expressed in general as follows:

1. **Cell.** A cell $C_{ij} \in G$ must be filled with exactly one digit d_{ij} with value between 1 and n
2. **Row.** All values in a row i must be unique: $d_{ij} \neq d_{ik}, \forall i = 1, n$ and $\forall j, k = 1, n$ with $j \neq k$.
3. **Column.** All values in a row j must be unique: $d_{ij} \neq d_{kj}, \forall j = 1, n$ and $\forall i, k = 1, n$ with $i \neq k$.
4. **Box.** All values in a box $B_i \in G$ must be unique: $d_{jk} \neq d_{pq}, \forall d \in B_i$, with $i = 1, n$.

3.1 Playing a Game

In this study we consider the 9×9 version of Sudoku. A player applies the Sudoku constraints to empty cells and generates lists of candidates for the visited cells. This process is displayed in Figure 1, where Figure 1(a) shows the application of rules to cell C_{G4}, and Figure 1(b) shows the lists of candidates for all empty cells in the grid.

(a) Sudoku constraints

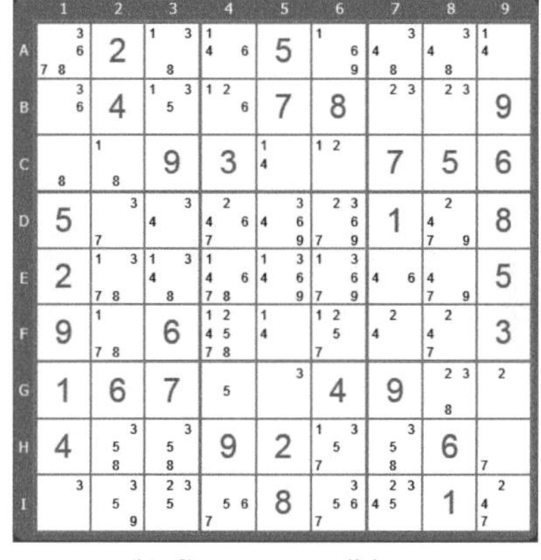

(b) Generate candidates

Fig. 1 The game knowledge.

The purpose of the game is to apply Sudoku skills and propagate the Sudoku constraints in order to reduce these lists of candidates to unique candidates [13] for all empty cells in the grid, which equals to filling the grid and, thus, solving the puzzle.

Performing Sudoku Skills. In order to reduce the lists of candidates, players use various skills which propagate the domain. In [15–17] the authors note that players choose the skills based on the perceived context at the current move. The skills considered in this study belong to two categories, the naked and the hidden candidates, which allow the solving of a significant number of Sudoku games. More complex skills [18] can be involved for solving very difficult games, however it is outside the scope of this study to investigate an exhaustive list of skills.

The set of naked candidate skills consists of finding and propagating naked singles and doubles (Figure 2). Recognising and propagating a naked single is the simplest skill, where after the application of Sudoku constraints a cell has only one possible candidate. The value of this unique candidate solves the empty cell, and is propagated by removing the candidate value from the candidate lists of all other cells situated in the units the cell belongs to. A naked single is illustrated in Figure 2(a) in pink colour at C_{A1}. For the naked doubles, the lists of candidates are checked for a pair of cells in a Sudoku unit containing only the same two candidates. These candidates can only go in these cells, thus the propagation is done by removing them from the candidate lists of all other unsolved cells in that unit. In Figure 2(b) the cells coloured in pink, in column 3 at C_{F3} and C_{I3} show a naked double containing the candidate values (2,3).

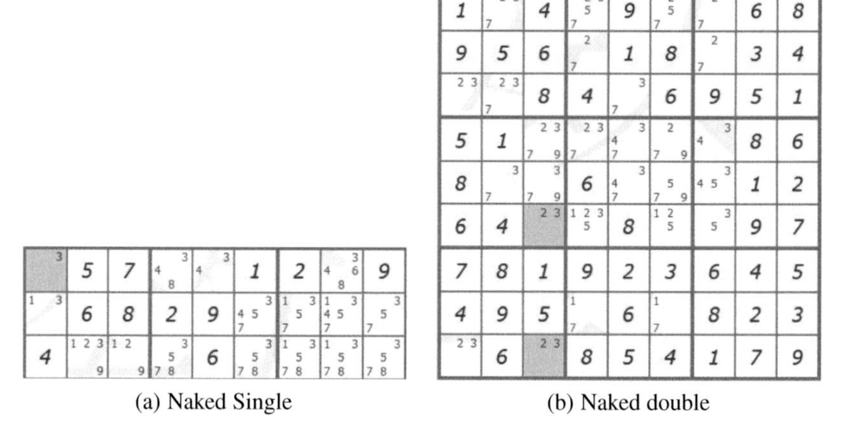

(a) Naked Single (b) Naked double

Fig. 2 Naked candidates

The set of hidden candidate skills consists of finding and propagating hidden singles and doubles (Figure 3). For the hidden single if a candidate value appears in only one cell in a Sudoku unit (row, column or box), the value becomes the unique candidate for that cell, the rest being removed. Thus, the candidate becomes a naked single and further propagates the domain as a naked single. Figure 3(a) shows value 3 as a hidden single in cell C_{D2}. For the hidden double, if a given pair of candidates appears in only two empty cells in a unit, then only these candidates must remain in these cells, the other candidates being removed. Thus, the hidden double becomes a naked double and further propagates the domain as a naked double. Figure 3(b) shows the hidden pair $(1, 8)$ in cells C_{E4} and C_{E6}.

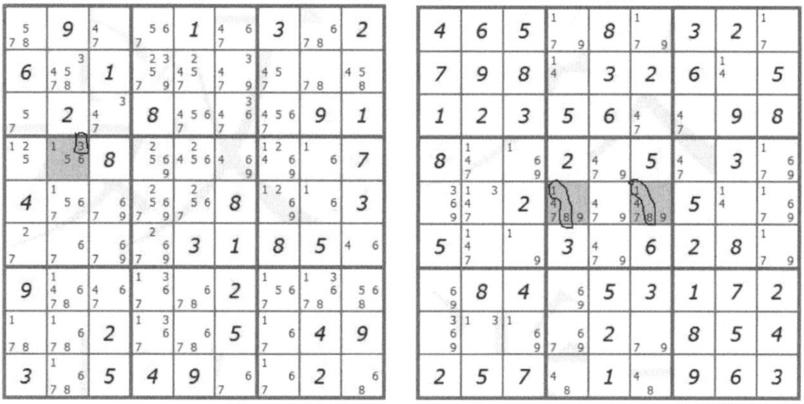

(a) Hidden Single

(b) Hidden double

Fig. 3 Hidden candidates

4 Methodology

In this study we consider that performing skills is subject to pattern recognition, where the player must recognise the pattern of a skill in the lists of candidates in a unit, in order to be able to apply that skill. We model the acquisition of skills through supervised training of feed-forward neural networks with back-propagation mechanism, one network for each skill. We treat two possible situations in skill acquisition. First, we train the ability to recognise the existence of a skill pattern in a Sudoku unit (cell, column or box) and we call this case "skill detection". Second, we train the ability to recognise not only the existence of a skill pattern, but also the cells in the unit which the skills is applicable to. We call this case "skill localisation". In the two cases the resultant neural networks have similar number of neurons in the input and hidden layer, and similar training sets x for learning the skills, but they have different number of neurons in the output layer and, consequently, different target sets t.

Figure 4 shows the encoding of candidate list information into the input layer of neural networks. In a Sudoku unit, each of the nine cells can have a maximum of nine potential candidates, i.e. the digits from 1 to 9. However, at a certain step in the game the current candidate lists usually contain less than nine digits; the lists can be depicted as in the third row of the table. We encode the decimal values of the candidates into binary values as presented in the third row. The total length of the binary encoded lists of candidates is 81, thus, we use neural networks with 81 neurons in the input layer.

Sudoku unit (row/col/box)	Cell 1	...	Cell 9
All possible candidates	123456789	...	123456789
Current candidate list	2 4 7 9	...	3 5 89
Encoding	010100101	...	001010011

NN input: 81 neurons

Fig. 4 Neural network input

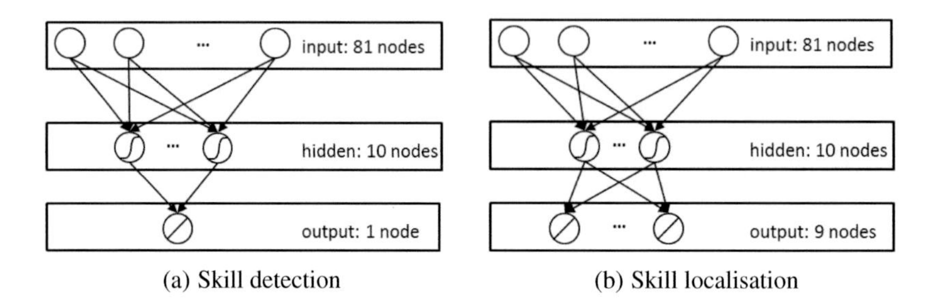

(a) Skill detection (b) Skill localisation

Fig. 5 Neural networks for skill pattern recognition

4.1 *Skill Detection*

For detecting the states we adopt the network structure presented in Figure 5(a), with one node in the output layer which shows if the pattern of a skill S_i is present in a Sudoku unit. For each skill we use artificially generated training and target sets, as presented in Algorithm 1. For the skills treated in this study a training sample x_i is a binary vector with 81 elements corresponding to the 81 nodes in the input layer.

Single Candidates. The training dataset is a binary matrix consisting of 162 samples $(X_{ij}, i = 1 : 162, j = 1 : 81)$. 81 samples correspond to all possible appearances of a naked or hidden single in a unit (i.e. there can be 81 naked single situations in a Sudoku column, row or box), for which the values of the target set $t(1 : 1296) = 1$. The other 81 samples do not contain the single candidate pattern, hence the values of the target set are $t(82 : 162) = -1$.

Double Candidates. The training dataset for double candidate skills (naked and hidden doubles) is a binary matrix consisting of 2592 samples $(X_{ij}, i = 1 : 2592, j = 1 : 81)$. 1296 samples correspond to all possible appearances of a skill in a unit (i.e. there can be 1296 naked double situations in a Sudoku column, row or box), for which the values of the target set $t(1 : 1296) = 1$. The other 1296 samples do not contain the skill pattern, hence the values of the target set are $t(1297 : 2592) = -1$.

Algorithm 1. Skill detection: training and target sets for a skill $S_i \in SkillSet$

1: {Input: Skill S_i}
2: **for** i = 1 **to** No. of S_i patterns in a Sudoku unit **do**
3: **for all** cells in the Sudoku unit **do**
4: trainingSet: $x(i, allcells) = S_i\,pattern$
5: **end for**
6: targetSet: $t(i) = 1$
7: **end for**
8: **for** j = 1 **to** No. of S_i patterns in a Sudoku unit **do**
9: **for all** cells in the Sudoku unit **do**
10: trainingSet: $x(i + j, allcells) = randompattern$
11: **end for**
12: targetSet: $t(i + j) = -1$
13: **end for**

4.2 Skill Localisation

For locating the patterns of skills we adopt the network structure presented in Figure 5(b), with nine nodes in the output layer. The nine nodes correspond to the nine cells in a Sudoku unit. Depending on which skill is subject to recognition, the cells in which the skill pattern exists will fire. The training dataset for this case is generated in a similar manner to the previous case. The generation is presented in Algorithm 2, where the training matrix X is similar to that from the skill detection case ($X_{ij}, i = 1 : 2592, j = 1 : 81$). The target set is a matrix $T(i, k)$ with ($i = 1 : TrainSetSize, k = 1 : 9$), defined as in Equation 1.

$$T(i,k) = \begin{cases} 1 & \text{if } X(i, 1:81) \text{ contains the skill pattern} \\ -1 & \text{otherwise} \end{cases} \quad (1)$$

Algorithm 2. Skill localisation: training and target sets for a skill $S_i \in SkillSet$

1: {Input: Skill S_i}
2: **for** i = 1 **to** No. of S_i patterns in a Sudoku unit **do**
3: **for all** cells in the Sudoku unit **do**
4: trainingSet: $x(i, allcells) = S_i\,pattern$
5: targetSet: $t(i, allOutputNodes) = 1$
6: **end for**
7: **end for**
8: **for** j = 1 **to** No. of S_i patterns in a Sudoku unit **do**
9: **for all** cells in the Sudoku unit **do**
10: trainingSet: $x(i + j, allcells) = randompattern$
11: targetSet: $t(i, allOutputNodes) = -1$
12: **end for**
13: **end for**

4.3 Network and Training Settings

We use the standard *tanh* for activation function of nodes in the networks and the mean square root error function (MSE) for the subsequent gradient minimisation.

The artificially generated training sets are split in ratios of 0.7, 0.15 and 0.15 for training, internal cross validation and generalisation testing, respectively.

4.4 Skill Aggregation - The Solver

The constraint propagation side of the Sudoku solving is hard-coded. However, the recognition of the patterns for each of the four skills considered in the study is implemented using the neural networks, and hence the ability to recognise either the existence of a skill pattern (detection) or its location (localisation) depends on the ability of the neural networks to produce the desired output. This implementation, with the hard-coded solving sequence, and the NN representation of the skills is error free from the Sudoku solving point of view, since it avoids situations when multiple states coexist in one board, i.e. a single candidate and double candidate simultaneously. Since the networks we propose are only meant to demonstrate the individual skills, they cannot treat cases where a combination of skills is present, or the player must choose from multiple skills. Since this aspect was outside of the scope of this study, we adopted a predetermined solving sequence implemented in the hard-coded constraint propagation module.

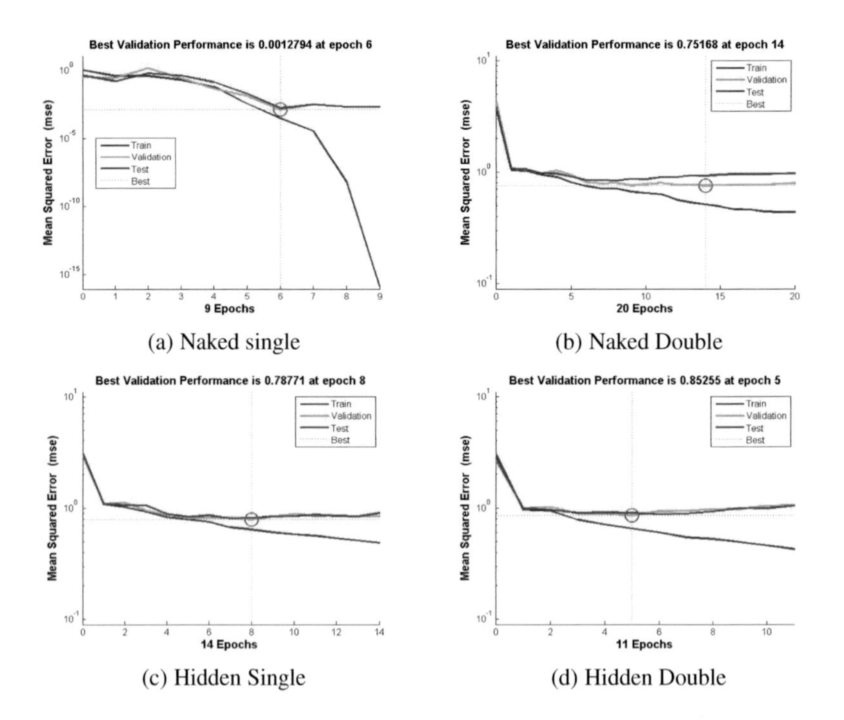

(a) Naked single (b) Naked Double

(c) Hidden Single (d) Hidden Double

Fig. 6 NN training for skill detection.

5 Results and Discussion

Figure 6 shows the results of the training process in the skill detection recognition case. The training of each of the four skills is considered finished when the best validation performance is reached. Table 1 and Figure 7 present the game solving results for both trained and untrained skills situations, where the untrained skills are the skills acquired after one epoch in the neural networks. Results demonstrate how the proficiency of the skill-based solver improves with the acquisition of skills. In the table the difference between the number of detected skill patterns is shown for the two cases, while in the figure the result of game solving is shown in terms of the degrees of freedom. We demonstrate that the NN-based skill detection training is able to solve the proposed Sudoku game, provided that the rest of the solving mechanism is hard-coded in the solver.

Table 1 Skill detection: Sudoku solving competency.

	Number of naked singles	Number of naked doubles	Number of hidden singles	Number of hidden doubles	Game result degree of freedom
untrained	2	0	17	1	153
trained	54	5	50	1	0

For the skill localisation case, results of the training process are shown in Figure 8. Similar to the skill detection case, the training of each of the four skills is considered finished when the best validation performance is reached. Table 2 and Figure 9 present the results of game solving for trained and untrained skills, where the untrained skills

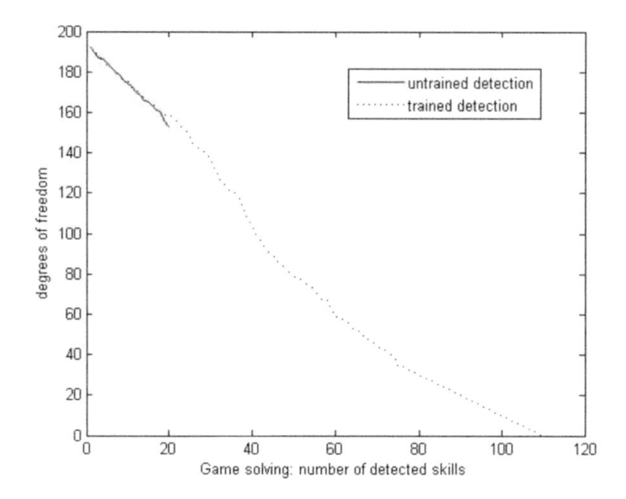

Fig. 7 Skill detection. Sudoku solving with trained and untrained skills.

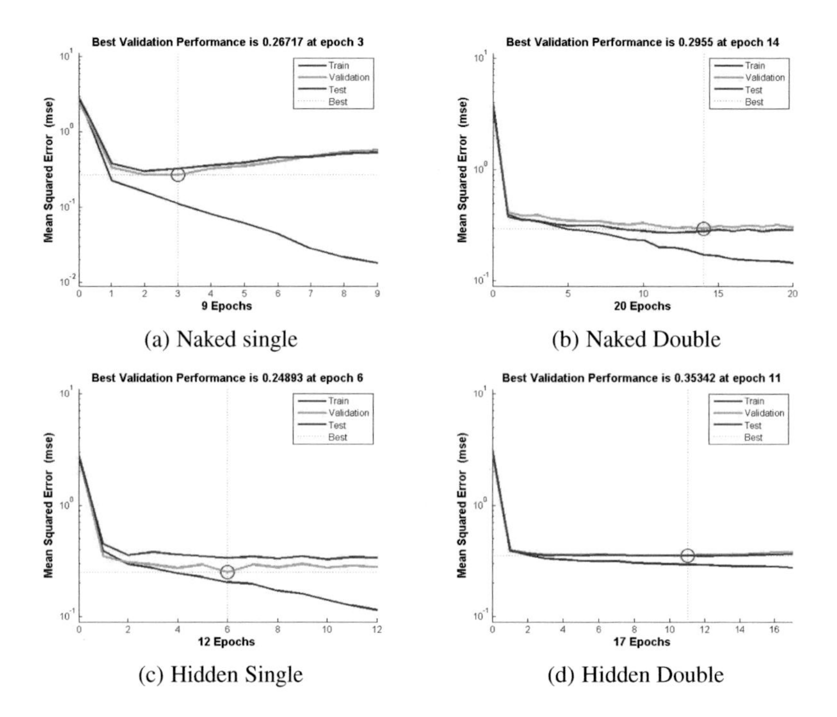

(a) Naked single (b) Naked Double

(c) Hidden Single (d) Hidden Double

Fig. 8 NN training for skill localisation.

Table 2 Skill localisation: Sudoku solving competency.

	Number of naked singles	Number of naked doubles	Number of hidden singles	Number of hidden doubles	Game result degree of freedom
untrained	0	1	6	1	181
trained	31	3	19	1	73

are the skills acquired after one epoch in the neural networks. Results demonstrate again that the proficiency of the skill-based solver improves with the acquisition of skills. The proficiency in this case is lower, a result which is expected given that the solver must recognise not only the existence of a skill in a unit, but also the skill pattern. Results show an improvement in the number of recognised skills, which subsequently leads to less degrees of freedom, but the solver still does not reach the end of the proposed Sudoku game. However, since the proficiency is not the purpose of this study, we emphasise on the improvement resulted from skill acquisition using NN training.

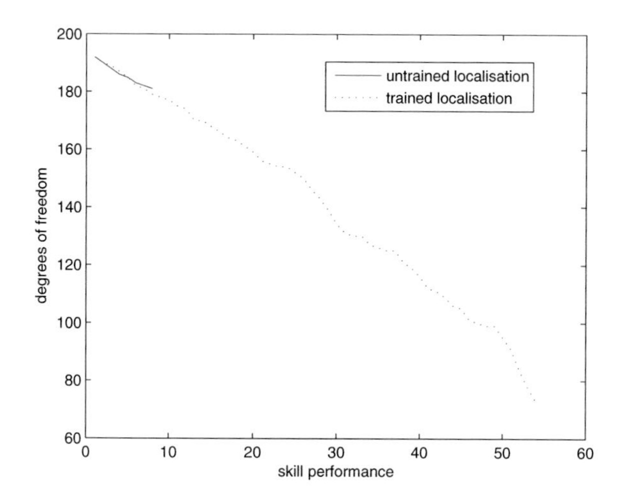

Fig. 9 Skill localisation. Sudoku solving with trained and untrained skills.

6 Conclusions

In this paper we focused on the skill assessment aspect of the CRT process, for which the representation of skills is a central and essential issue. We investigated this using the Soduku puzzle by introducing a plausible representation of Sudoku skills, and by modelling the process of acquiring these skills. We used feed-forward neural networks with back-propagation mechanism for training the skills and we tested the resultant skills in a cognitively plausible skill-based computational Sudoku solver.

The results of Sudoku game solving demonstrated the plausibility of using skills in computational Sudoku solvers, and also demonstrated the concept of skill acquisition in relation to the proficiency of this solver. We found that a skill-based computational Sudoku solver can achieve certain levels of proficiency by learning the Sudoku skills using neural networks. Results are encouraging for developing more complex skill and skill acquisition models usable in more general frameworks related to skill assessment stage of the Computational Red Teaming process.

Acknowledgements This project is supported by the Australian Research Council Discovery Grant DP140102590, entitled "Challenging systems to discover vulnerabilities using computational red teaming".

References

1. Berggren, P., Nilsson, D.: A study of sudoku solving algorithms. Master's thesis, Royal Institute of Technology, Stockholm (2012)
2. Crook, J.F.: A pencil-and-paper algorithm for solving sudoku puzzles. Notices of the AMS **56**(4), 460–468 (2009)
3. Yato, T.: Complexity and completeness of finding another solution and its application to puzzles. Master's thesis, Graduate SChool of Science, University of Tokyo (2003)
4. Yato, T., Seta, T.: Complexity and completeness of finding another solution and its application to puzzles. IEICE Transactions on Fundamentals of Electronics, Communications and Computer Sciences **86**(5), 1052–1060 (2003)
5. Weber, T.: A sat-based sudoku solver. In: The 12th International Conference on Logic for Programming Artificial Intelligence and Reasoning, pp. 11–15 (2005)
6. Ercsey, R.M., Toroczkai, Z.: The chaos within sudoku. Sci. Rep. **2** 10.1038/srep00725 (2012)
7. Goldberg, P.W.: Np-completness of sudoku, October 2015
8. Pacurib, J.A., Seno, G.M.M., Yusiong, J.P.T.: Solving sudoku puzzles using improved artificial bee colony algorithm. In: Fourth International Conference on Innovative Computing, Information and Control, pp. 885–888. IEEE (2009)
9. Mantere, T., Koljonen, J.: Solving, rating and generating sudoku puzzles with ga. In: IEEE Congress on Evolutionary Computation, pp. 1382–1389. IEEE (2007)
10. Perez, M., Marwala, T.: Stochastic optimization approaches for solving sudoku, May 2008. arXiv:0805.0697
11. Yue, T.W., Lee, Z.C.: 115. In: Sudoku Solver by Q'tron Neural Networks. Lecture Notes in Computer Science, vol. 4113, pp. 943–952. Springer, Heidelberg (2006)
12. Hopfield, J.J.: Searching for memories, sudoku, implicit check bits, and the iterative use of not-always-correct rapid neural computation. Neural Computation **20**(5), 1119–1164 (2008)
13. Simonis, H.: Sudoku as a constraint problem. In: CP Workshop on Modeling and Reformulating Constraint Satisfaction Problems. vol. 12, pp. 13–27. Citeseer (2005)
14. Norvig, P.: Solving every sudoku puzzle
15. Aslaksen, H.: The mathematics of sudoku (2014)
16. Davis, T.: The math of sudoku (2008)
17. Chadwick, S.B., Krieg, R.M., Granade, C.E.: Ease and toil: Analyzing sudoku. UMAP Journal **363** (2007)
18. Pitts, J.: Master Sudoku. Teach yourself. McGraw-Hill Companies, Inc. (2010)

Exploring Swarm-Based Visual Effects

Somnuk Phon-Amnuaisuk and Ramaswamy Palaniappan

Abstract In this paper, we explore the visual effects of animated 2D line strokes and 3D cubes. A given 2D image is segmented into either 2D line strokes or 3D cubes. Each segmented object (i.e., line stroke or each cube) is initialised with the position and the colour of the corresponding pixel in the image. The program animates these objects using the boid framework. This simulates a flocking behavior of line strokes in a 2D space and cubes in a 3D space. In this implementation the animation runs in a cycle from the disintegration of the original image to a swarm of line strokes or 3D cubes, then the swarm moves about and then integrates back into the original image.

Keywords Computer generated visual effects · Swarm-based VFX · Boid framework

1 Background

Early visual effects (VFX) in medias are mostly accomplished using non-digital techniques such as *stop-motion, optical printing, matte painting*, etc. Thanks to the advancement in digital image processing, VFX has now moved to a different level where the limitation is capped only by our imagination. Currently, most of the high-end post-production editing tools provide various VFX facilities. It is undeniable that VFX has become a crucial part of storytelling in games and films [1].

S. Phon-Amnuaisuk(✉)
Media Informatics Special Interest Group, School of Computing and Informatics,
Institut Teknologi Brunei, Gadong, Brunei
e-mail: somnuk.phonamnuaisuk@itb.edu.bn

R. Palaniappan
University of Kent, Canterbury, UK
e-mail: r.palani@kent.ac.uk

K. Lavangnananda et al. (eds.), *Intelligent and Evolutionary Systems*,
Proceedings in Adaptation, Learning and Optimization 5,
DOI: 10.1007/978-3-319-27000-5_27

In creative computing area, researchers have experimented with the idea of particles in many tasks e.g., NPCs' flocking formation in games [2], caricatures generation [3], music transcription [4]. The particle system has been extensively employed in computer generated VFXs where particles abstract objects in the physical world. The objects may have various physical properties (e.g., mass, shape, colour, velocity, viscosity, etc.). By carefully controlling those parameters, the particle system can emulate various natural events such as fire, smoke, clouds, explosion, etc. The particle approach also has successfully modelled stylistic swarm-like movements of animals such as bird flocking and fish schooling. The simulated swarm-like behaviours are emerging behaviours from the interactions of different individual particles. This kind of simulation is useful in investigating collective behaviours such as foraging, escaping, flocking, etc.

In this paper, we explore the potential of particle-based VFXs to control the swarming effect of small image segments. For example, a given 2D image could be segmented into many small tiles. This creates a mosaic rendering of an original image. Each mosaic tile can be programmed to move about in a 3D/2D space. To simulate the swarming effect, these tile particles move according to three forces: separation, alignment and cohesion [5]. Our system registered the initial location of each mosaic tile. Hence, the system could generate pleasing emerging behaviours of the tiles swarming out from a disintegrated image as well as swarming in to create an original image.

This paper is organised into the following sections: Section 2 gives an overview of related works; Section 3 discusses our proposed concept and gives the details of the techniques behind it; Section 4 provides the output of the proposed approach; and finally, the conclusion and further research are presented in Section 5.

2 Related Work

Arcimboldo produced many portrait paintings by compositing various objects such as vegetables, fruits, animals [6]. This activity creates new emerging meanings from basic primitives. This kind of concept has been widely explored by designers, artists and computer scientists e.g., emergent computing [7, 8]. Here we explore the idea of mosaics where an image emerges from a composition of various coloured tiles viewed from a distance.

Digital mosaics have been explored by many researchers in the past. Researchers have looked into various generative approach as considering the choices of tiles and how the tiles are placed. In its simplest form, an image can be easily converted into photomosaic of square tiles. In a more sophisticated manner, various components in an image can be segmented out first before generating mosaic patterns for each of them [9, 10]. The latter process produces a much more stylistic tile composition.

Mosaics are a static art form. The artefact is displayed when the tiles are fully composed. In this paper, we explore the movement of mosaic tiles and line stroke in a swarm-like movement in a 2D/3D space. In both 2D/3D animations, each mosaic tile or line stroke is represented as a particle in a particle swarm.

In [11], the author analysed empirical biological data and proposed a mathematical model to describe the *cohesiveness* of a fish school. Many reserachers have analysed swarm behaviours to gain more understanding of the emergent properties and the shape of the swarm [5, 12, 13].

Perhaps, the simulation by Reynolds [5] has the most impact in this area. He proposed the *bird-oid object* (boid) framework that describes swarm-like behaviours with simple intuitive rules. The boid framework emulates the flocking behaviour with the following basic individual behaviours: (i) avoid crowding particles by ensuring that particles cannot be too close to each other, (ii) steer itself toward the overall direction of the swarm, and (iii) steer itself toward the centre of the swarm. The original boid framework has been extended by other researchers to emulate other swarm behaviours such as splitting and uniting a flock.

Fig. 1 Left: the original image is segmented into many small cubes. Right: these cubes are animated using the boid framework.

3 Materials and Methods

Let $I(h, w)$ be a pixel from row h and column w of an image I having the width and height of h and w pixels. Let p_d be a particle d from a swarm $\mathbf{P}$ of size $|\mathbf{P}| = h \times w$. The image I can be represented using $h \times w$ particles where the particle p_1 is initialised with the information of the pixel at the location $I(1, 1)$, p_2 from $I(1, 2)$ and $p_{h \times w}$ from $I(h, w)$.

In this implementation, for a swarm-like visual effect, the size of each particle is bigger than a single pixel. Here, each particle occupies a space of size $m \times n$ pixels. Hence, $\frac{h}{m} \times \frac{w}{n}$ particles are employed to represent the original image. Each particle p_d is an instance of the object class particle, where $p_d.loc()$ returns the location information of the particle d in the 3D space and $p_d.vel()$ returns the velocity information accordingly.

Fig. 2 Left: the original image is segmented into many line stroke segments. Right: these line strokes are animated using the boid framework.

In this paper, the concept of particles was applied to create a visual effect of the swarm movements of many small image segments. Each segment represented a particle in a swarm where its position in the 3D space was controlled using the boid framework. Although these segments were rectangles of image components segmented from the original image, a swarm particle could take any geometrical shape and it would be initialised with the pixels' information of the corresponding segment. In this implementation, the swarm particles took two different types of shapes: cubes and lines. For a cube, all 6 faces were initialised with the pixels' information of the corresponding rectangle segment. The cube particles moved about in a 3D space and the line particles moved about in a 2D space.

3.1 The Boid Framework

The boid framework suggests three important heuristics that compute three velocities $\mathbf{v}_s$, $\mathbf{v}_a$, and $\mathbf{v}_c$ which are attributed to the following properties, respectively: separation, alignment and cohesion respectively. For each particle p_d with the velocity $\mathbf{v}_d$, the steering velocity $\mathbf{v}_d$ can be computed from $\mathbf{v}_d(t+1) = \mathbf{v}_s(t)+\mathbf{v}_c(t)+\mathbf{v}_a(t)-\mathbf{v}_d(t)$. The contributions from $\mathbf{v}_s$, $\mathbf{v}_a$, and $\mathbf{v}_c$ are computed as follow:

$$\mathbf{v}_s = \sum_{i=1}^{|\mathbf{P}|} \mathcal{N}_i k_s (p_d.loc() - p_i.loc())$$

where $\mathcal{N}_i = 1$ if p_i is within the desired neighborhood of p_d, else $\mathcal{N}_i = 0$. The neighborhood of p_d is a sphere of radius r (arbitrarily set by the users) and k_s is a normalisation factor that moderates the effect of the distance $\|p_d - p_i\|$. In other words, $\mathbf{v}_s$ is a vector formed from summation of all $(p_d.loc() - p_i.loc())$ vectors. $\mathbf{v}_a$ and $\mathbf{v}_c$ are computed in the same fashion but with a different normalisation condition:

$$\mathbf{v}_a = \sum_{i=1}^{|\mathbf{P}|} \mathcal{N}_i k_a p_i.vel(); \text{ where } i \neq d$$

and k_a is the normalisation factor.

$$\mathbf{v}_c = \sum_{i=1}^{|\mathbf{P}|} \mathcal{N}_i k_c (p_i.loc() - m_{\mathcal{N}}.loc())$$

where k_c is a normalisation factor that moderates the effect of the distance $\| p_i - m_{\mathcal{N}} \|$. Here $m_{\mathcal{N}}$ is the center of all particles in the neighborhood of p_d.

3.2 The Swarm

Given an input image, a swarm population is generated by segmenting the 2D image into many small segments. Each particle i in the swarm is initialised with a random velocity value $\mathbf{v}_i =< v_x, v_y, v_z >$. At each time step, the swarm behaviour emerges as a result from $\mathbf{v}_s$, $\mathbf{v}_a$, and $\mathbf{v}_c$ discussed in the previous section.

The boundaries front-back (z-axis), top-bottom (y-axis), and left-right (x-axis) are wrapped around. Hence, a particle that moves deep pass the bordering depth will enter the scene again at the frontmost position; a particle that moves out of the top border will emerge from the bottom border; and a particle that moves out of the right border will emerge from the left border and vice versa. This creates a seamless motion of particles.

We implemented three behaviours: (i) flocking behaviour according to the boid framework, (ii) swarming toward a specific target position, and (iii) returning to the original position i.e., forming an original 2D image. Behaviour (i) was obtained by the boid framework. This is an emergent behaviour of the swarm. Behaviour (ii) was obtained by (in each time frame) randomly steering 10% of the swarm population $\mathbf{P}^{10}$ toward the desired target T. The remaining 90% of the swarm population $\mathbf{P}^{90}$ was still controlled by the boid framework.

$$\forall_{i \in \mathbf{P}^{10}} \quad \mathbf{v}_i = k_2(T.loc() - p_i.loc())$$

$$\forall_{i \in \mathbf{P}^{90}} \quad \mathbf{v}_i(t+1) = \mathbf{v}_s(t) + \mathbf{v}_c(t) + \mathbf{v}_a(t) - \mathbf{v}_i(t)$$

Behaviour (iii) was obtained by (in each time frame) steering each particle i back to its original position o_i.

$$\forall_{i \in \mathbf{P}} \quad \mathbf{v}_i = k_3(o_i.loc() - p_i.loc())$$

where k_2 and k_3 were constants that regulated the speed of behaviours (ii) and (iii).

Fig. 3 Ten snap shots from two animation clips, for each clip, from topleft and in a clockwise direction: (i) original image, (ii, iii, iv) the swarm of cubes is simulated, and (v) the swarm integrate back to form the original image.

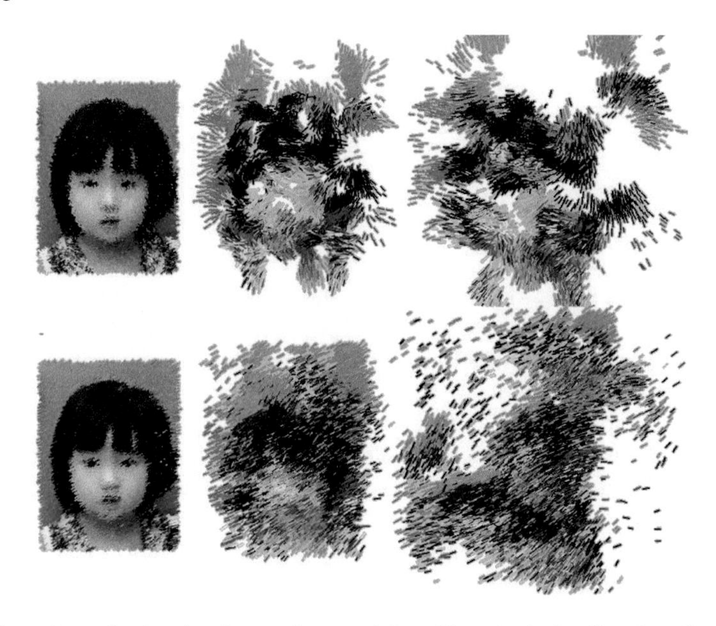

Fig. 4 Snap shots of animation frames, from topleft and in a clockwise direction: (i) original image, (ii, iii, iv, v) the swarm of cubes is simulated, and (vi) the swarm integrate back to form the original image.

4 Discussion

The purpose of this work is to generate a swarm movement using a particle system. A given image was segmented into many small segments and their pixels' information was used to initialise a swarm of particle; which would be programmed into any desired shape.

We successfully implemented a controller for three different behaviours of the particle swarm. These parameters were controlled by users using shortcut-keys. In this implementation, the default behaviour was the flocking behaviour according to the boid framework. If the user pressed key 'T', then the particle swarmed into two alternate active targets on the bottom left and the right area of the screen. If the user pressed key 'H', then each particle swarmed into its original position. If the user pressed key 'S', then the particles resumed their flocking behaviour.

We simulated two swarm appearances: swarm of mosaics and swarm of line strokes (see https://www.youtube.com/watch?v=a-V6h0VzTZ8). Figure 3 shows 10 snapshots (5 from each example) of a swarm of mosaics. The exercise in this figure was motivated by the idea to create a swarm movement of mosaics. An original image disintegrated into a swarm of cubes, traversing in a 3D space before integrating back to its original position. We arbritarily set the width of each cube at 7 pixels as we felt that this produced an appealing swarm movement.

Figure 4 shows 5 snapshots of a swarm of line strokes. This exercise was motivated by the idea to generate a swarm of line strokes that could recombine into drawings or paintings.

5 Conclusion and Future Work

We have shown two examples of swarm-based visual effect: a swarm of cubes and a swarm of lines. The swarming pattern is infinitely complex since each particle is randomly initialised with $< v_x, v_y, v_z >$. The swarming visual FXs are common ingredients in games, films, and advertisements. The footage can be further composited with other materials.

Many interesting areas can be extended from the current work, for examples, stylistic swarm movements, rich particle shapes, etc. In this work, we also explore the use of camera to create various perspective of the swarm. However, the camera positions were manually coded and it was a challenge to select appropriate dynamic camera posoitions. This can be further explored in the future work e.g., to analyse the swarm formation and automatically position the camera at appropriate positions at different times.

Acknowledgments We wish to thank anonymous reviewers for their comments, which help improve this paper. We would like to thank the GSR office for their financial support given to this research.

References

1. McClean, S.T.: Digital Storytelling: The Narrative Power of Visual Effects in Film. The MIT Press (2008)
2. Ho, C.S., Nguyen, Q.H., Ong, Y.S., Chen, X.S.: Autonomous multi-agents in flexible flock formation. In: Proceedings of the Third International Conference on Motion in Games (MIG 2010), pp. 375-385 (2010)
3. Phon-Amnuaisuk, S.: Exploring particle-based caricature generations. In : Proceedings of the International Conference on Informatics Engineering and Information Science (ICIES 2011), Malaysia, pp. 37–46 (2011)
4. Phon-Amnuaisuk, S.: Investigating a hybrid of Tone-Model and particle swarm optimization techniques in transcribing polyphonic guitar sound. Applied Soft Computing **29**, 211–220 (2015)
5. Reynolds, C.W.: Flocks, heards, and schools: A distributed behavioral model. Computer Graphics **21**(4), 25–34 (1987)
6. Maiorino, G.: The Portrait of Eccentricity: Arcimboldo and the Mannerist Grotesque. The Penn State University Press (1991)
7. Kennedy, J., Eberhart, R.C., Shi, Y.: Swarm Intelligence. Morgan Kaufmann (2001)
8. Hastings, E., Guha, R., Stanley, K.O.: NEAT particles: design, representation, and animation of particle system effects. In: Proceedings of the IEEE Symposium on Computational Intelligence and Games (CIG 2007), pp 154–160. IEEE, Piscataway (2007)

9. Kim, J., Pellacini, F.: Jigsaw image mosaics. In: Proceedings of the 29th Annual Conference on Computer Graphics and Interactive Techniques, SIGGRAPH 2002, pp. 657-664 (2002)
10. Battiato, S., Blasi, G.D., Farinella, G.M., Gallo, G.: A novel technique for opus vermiculatum mosaic rendering. In: The 14th International Conference in Central Europe on Computer Graphics, Visualization and Computer Vision (WSCG 2006), pp. 133-140 (2006)
11. Breder, C.M.: Equations descriptive of fish schools and other animal aggregations. Ecology **35**, 361–370 (1954)
12. Parrish, J.K., Viscido, S.V., Grünbaum, D.: Self-organized fish schools: An examination of emergent properties. The Biological Bulletin **202**, 296–305 (2002)
13. Hemelrijk, C.K., Hildenbrandt, H.: Some causes of the variable shape of flocks of birds. PLoS ONE **6**(8), e22479 (2011). doi:10.1371/journal.pone.0022479

Empirical Analysis of Mobile Augmented Reality Games for Engaging Users' Experience

Dendi Permadi and Ahmad Rafi

Abstract This paper presents the results of an empirical analysis of mobile augmented reality (AR) games focusing on elements of user engagement. The area covered in the analysis was based on the user engagement literature review in classical games as well as AR technology. The results showed that five major elements that affected the user engagement were social, perceived usability, challenge, satisfaction and clear goals. This finding is suggested as one of the key considerations prior in developing mobile AR game.

Keywords Augmented reality game · Mobile game · User engagement

1 Introduction

Current progress in mobile augmented reality (AR) games has been remarkable with potentially millions of users in the future. However there are still no killer AR apps appear yet [1]. Pioneer research works on mobile AR games have started from year of 2000 [2] but the first high quality content mobile AR game that considers able to reach commercial games level just successfully developed in 2009 [3]. Mobile AR game combined real and virtual experience, a different experience from other casual games. It allowed player to interact with the virtual game object that overlaying on the top of the real environment, stimulates imagination as if the players are really interact with the real environment. In mobile AR games, player's attention especially will focus in the real world rather than on the screen [4]. There are also several guidelines presented by [5] to design player's experience for AR games which include the experience, stick to the theme, do not stay digital, use the real environment, keep it simple, create shareable experience, use various social elements, show reality, turn weakness

D. Permadi · A. Rafi(✉)
Faculty of Creative Multimedia, Multimedia University, Cyberjaya, Malaysia
e-mail: {dendi,rafi}@mmu.edu.my

K. Lavangnananda et al. (eds.), *Intelligent and Evolutionary Systems*,
Proceedings in Adaptation, Learning and Optimization 5,
DOI: 10.1007/978-3-319-27000-5_28

into strengths, do not just convert, create meaningful content and choose tracking wisely.

The literature revealed that AR technology itself often affecting in the enhancement the value of engagement. For instance, the study showed that people quickly become engaged in their evaluated AR games [6]. They have discovered that the AR interaction technique enables unique gaming experience but requires simple design so as to prevent the user from paying too much attention to technology instead of the game itself. In fact, they suggested to focus on the game content and gameplay in order to sustain users' engagement levels with the virtual and real elements in the game. Another study developed an AR game for education to increase learning and engagement [7]. They found that the game should encourage students to have fun as they learn, and that the technology does much more to engage student rather than with an experiment outside. There are also studies that discovered three cognitive issues related to the effectiveness of mobile AR systems from identifying the property of engagement [8]. Firstly is the information presentation of the amount, representation, placement and view combination. Secondly is the physical interaction, such as navigation, direct manipulation and content manipulation. Lastly, shared experience in the context of social, bodily configuration, artifact manipulation and display space. User engagement has been considered as a key factor for understanding general user and task behavior in social networking tools, traditional education environment, work-oriented information retrieval, games and game-based learning [9]. One study has classified engagement into two categories [11]. First, a pragmatic quality represented usefulness and usability of the system. Second, hedonic quality suggested motivation, stimulation and challenge for the user. It stated that hedonic quality became the highlight among researchers in video game and game based learning environment. However, detailed studies to investigate engagement were still lack especially addressed for mobile AR games. Thus, one question that could be highlighted i.e. – to what extent user engagement in mobile AR games needs to be investigated. This is similar to a literature mentioned that the need to study user engagement for mobile AR games [12].

This research empirically analyzed elements that contributed to the user engagement in mobile-based AR games and relate the current situations of engagement ideas in existing mobile AR games. It examined the existing mobile AR games available in the digital market based on identified engagement criteria from the literature review.

2 Literature Review

The review process comprised the fundamental areas that were identified from the previous study related to the engagement on video game and AR domain. O'Brien and Toms [13] classified engagement scale as aesthetics, endurability, felt involvement, focused attention, novelty and perceived usability. These classifications have been investigated to fit video game environment. It was then further revised into four additional factors namely focused attention, perceived usability, aesthetics and satisfaction [14]. In addition to this, Flow Theory (or also known as optimal engagement) was used as the basis in analyzed the construct of focus attention value. It is a mental state of operation performed by people in their activity to fully immerse, focus, involve and enjoy [15]. This theory recommended to analyze game experience, as [17] suggested concept of positive psychology for game developer. Apart from this, [18] adapted the Flow Theory for game experiences by establishing Game Flow model. Eight core elements were constructed namely concentration, challenge, skills, control, clear goals, feedback, immersion, and social.

As for AR engagement, [19] presented the Positive Engagement Evaluation Model (PEEM) that designed to incorporate qualitative experience and holistic for interactive and mobile applications. They created an evaluation matrix [12] based on eight domains namely goals, attention, interaction, concentration, content, identity collaboration and emotional outcome or satisfaction. Other researchers recommended mixed fantasy triad consisting virtual content, real content and imagination [20]. They believed that imagination that blend with virtual and real content should be able trigger player emotional, thus increase the user's engagement. Good relation between virtual and real content can be achieved by designing a seamless merge of those virtual and physical worlds in augmented reality interface [7].

Based on the analysis from these literature reviews, this paper focused the eight elements of user engagement that has strong correlation with the augmented reality games namely clear goals, satisfactions, focused attention, mixed fantasy, perceived usability, challenge, interaction and social.

3 Game Selection

In October 2014, using the xyo.net app search engine, the researchers discovered over 220 mobile augmented reality games (MARG) available in Apple and Play stores. 76 MARG in apple store and 144 MARG in play store were identified and currently available in the market to be downloaded for smartphones usage. Xyo.net is an app that provides sales data for android and ios applications. According to [10], it is a reliable data source, and is used for analysis in their research studies [16]. It was observed that MARG games were less downloaded as compared to other type of games i.e. not reaching even one million global download. The closest to this number was Candy Flick, which had 938,000 global downloads.

In this context, the researchers selected 5 MARG games that rated more than 7 Xyo, a rating score that represented high-rated games, 5 MARG ranked between 6 to 7, representing mid-rated games and 5 MARG rated below 5 to represent low-rated games, which summed 15 MARG. The researchers also founded that there were no MARG games rated more than 8, thus MARG with more than 7 rating score are classified as the highest rated for MARG. This was also supported based on Xyo showing that MARG games that rated more that 7 were rated with 5 stars (highest) in official Android and Apple stores. Only games published in 2009 (Figure 1) were selected for the study due to the fact that the first successful MARG was commercialised in year 2009 [3]. The summary of the selection criteria is as follows:

(a) Ranked by Xyo Score.
(b) Published starting year 2009.
(c) Selected on the top most downloadable online game stores– i.e. Apple and Play stores.

Fifteen mobile AR games were selected based on these criteria comprised of eight marker-based games (53.3%), six gyro-based games (40%) and one edge-based tracking game (6.7%). Table 1 showed the game samples that the researchers have selected.

Table 1 Selected mobile AR games

No.	AR Game	Total of Global Download (Apple and Play store)	Rating Stars (Apple / Play store)	Xyo Score	Tracking Method
1.	Ghostbusters™ Paranormal Blast	233.9K	5	7.7	Gyro
2.	Reality Hoops	8.6K	5	7.5	Gyro
3.	Fairy Magic	18K	5	7.4	Gyro
4.	Bugs Mayhem	6.4K	5	7.4	Gyro
5.	AR Space Ship	19K	4	7.0	Marker
6.	Warp Runner	36.2K	4	6.9	User defined target marker
7.	AR Invaders	52K	4	6.8	Gyro
8.	Fly Hunter	71K	4	6.5	Gyro
9.	ARBasketball	102K	4	6.2	Marker
10.	Candy Flick	938K	4	6.1	Marker
11.	ARSoccer	78K	4	5.9	Edge
12.	Augmentron	63K	4	5.6	Marker
13.	Fight of the Castle AR	1K	4	5.2	Marker
14.	AR Defender 2	155K	4	5.1	Marker
15.	Akodomon	230K	4	4.8	Marker

Fig. 1 A few snapshots of MARG (Candy Flik, Reality Hoops and Warp Runner – left to right)

4 Results of Mobile AR Games Analysis

The results from this empirical analysis exhibited the current situation of mobile AR games in terms of elements that affecting engagement. Most importantly, these rendered the elements hat able to engage lasting player experience for the mobile AR games.

4.1 Clear Goals

Expectation and goals were the significant motivators in all human behavior which inherent in any of interaction between human-to-human or human to device [21] [22]. In the flow theory, goal required mental energy and appropriate skills. As mentioned by [18], the process towards goal is the main source of the reward experience in achieving optimal engagement. Goal is clearly one of the components that should be established in games. According to [18], overriding and intermediate goals will require players to continue playing the games to find out the answers. It is suggested by them to provide clear overriding goal early in game such through introductory cinematic that established the background story while Intermediate goals often described through a mission briefing. Overall, these goals needed to be well defined inside game as the players wanted to know the reason why they need to finish the task.

This analysis indicated that only three games (20%) had having clear overriding goals. Ghostbusters game for example informed the players to help the mayor of city to catch all ghosts, Fairy Magic game indicated to catch all fairies and Augmentron Game explained that the robot was given the mission to defend from the galactic fiend. All these however were not fulfilled, as highlighted by [18] to establish the overriding goal as cinematic overview (i.e. background story).

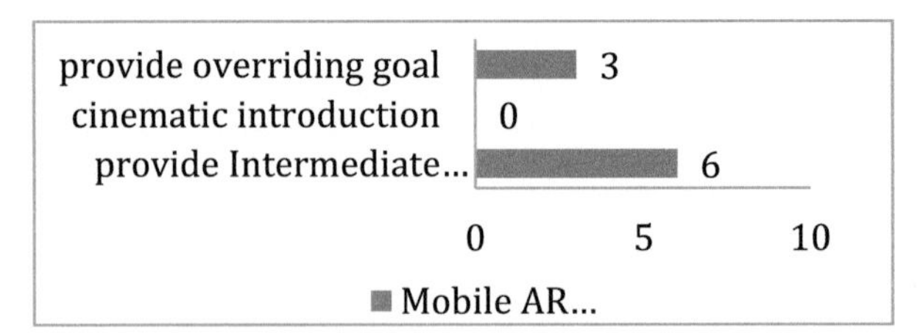

Fig. 2 Clear goal elements in selected mobile AR games

In the context of intermediate goal, only six out of fifteen games (40%) prepared this in the games. The game designers commonly used missions briefing to inform the players about the goal for every stage or task. In contrast, nine games did not provide any mission briefing thus the players need to figure it out by themselves on actions required in the game play. As a result, players were expected to explore all stages which only gained as high score points.

4.2 Satisfaction

Satisfaction is one of the conditions for user engagement that covered novelty, endurability and felt involvement. Undoubtedly, novelty aspect is important to attract and satisfy users. Study has showed that the behavior of mobile users to seek novel content [23]. In simple terms, novelty is triggered by curiosity for something unfamiliar or new. Novelty however, not mainly indicated as the AR interaction technique [24]. Novelty expressed the mystery and puzzlement provided to make people wanting to learn more [25]. This analysis shows that nine out of fifteen games (60%) suggested mystery that cannot be predicted.

Another important consideration is endurability. This basically described the concept of like hood to returns, worthwhile and future recommendation to other [13]. According to [23], mobile users have a special behavior to automatically checking the device to seek rewards as well as rewards that could not be obtained in real daily life situation [26]. This analysis showed that only six out fifteen games (40%) provided such rewards for the users.

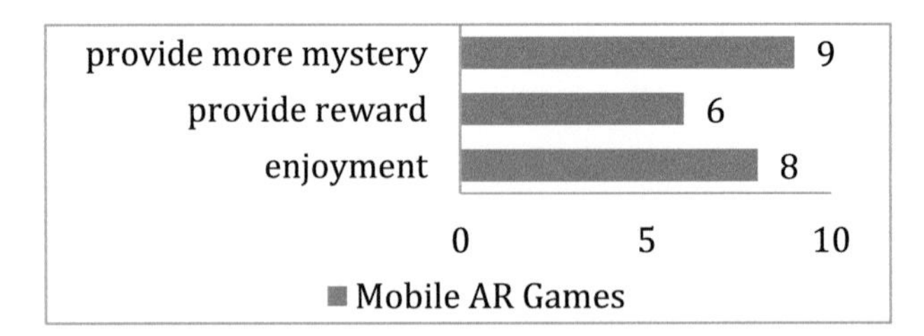

Fig. 3 Satisfaction elements in selected mobile AR games

Some developers extended this as user's involvement and enjoyment [27] while interacting with the game. This analysis indicated that eight games (53.3%) fulfilled this aspect. However, it was a challenge for AR shooting game with fast pace of movement where users were easily tired, discomfort and dizzy especially when this game involved 360 degrees maneuver as part of the game plays.

4.3 *Focused Attention*

Focused attention referred to a mental activity for concentrating on only one stimulus attention and ignoring all others [28]. [13] suggested this as the compulsory element to attract and keep user attention for the engagement in any technology usage. They further explained that continued attention is the one that shaped concentration by introducing an easy task with clear purposes and feedbacks to ease the players.

In this analysis, eleven out of fifteen games (73.3%) portrayed a lot of visual and audio elements with feedback to gage users while playing the games. Several developers in fact provided high quality virtual content in the form of unique animation, sound, speech and graphics primarily to stimulate the interest. Om Nom Candy Flick for example implied cuteness and animation of virtual objects to draw the user attention. Thirteen games (86.7%) provided clear audio and visual content to assist the players. However, none of these games provided background story and only five games (33.3%) somehow suggested variety of mission [18] even though these elements were deemed important. Nevertheless, all games (100%) offered feedback to guide the players.

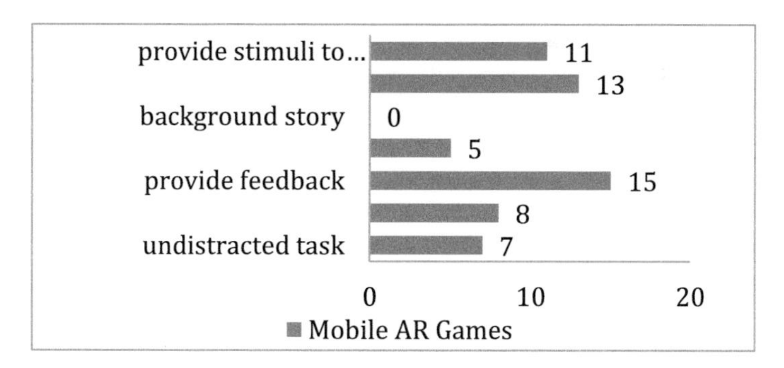

Fig. 4 Focused attention elements in selected mobile AR games

The need to increase player's workload is significant to keep player's concentration while still appropriate to player's perceptual, cognitive and memory limits [31]. This analysis showed that eight games were high in workload due to numerous tasks to perform and monitor. Among these were Reality Hoops, AR Defender and Warp Runner games. Apart from this, game needs to consider simple design, not distract user interface control and not to give overburdened game task [21]. This analysis showed seven games (46.7%) are run with undistracted task. Many games were distracted due to hard navigation, unclear task and tracking problem. The virtual object appeared very shaky thus easily got lost from tracking and a bit hard to adjust the view alignment. In the Candy Flick game for instance, while the control is easy, the marker tracking error movement and angle view resulted virtual objects vanished from the game.

4.4 *Mixed Fantasy*

Undoubtedly, mixed fantasy becomes a unique element only for AR games in games area. According to [20], mixed fantasy model creates immersive experience that stimulates emotional aspect of player imagination that lead to user engagement. In this case the challenge lies in the design of the mobile AR game so that the environment will be as seamless as possible between virtual and real objects. This is similar to the findings gained from [7] research on virtual object in real world setting. This research found that eleven games (73.3%) successfully developed the content with seamless merge. These games used visual effects (87%) such as shadow and flying animation to create virtual object blended with real environment. Several games introduced audio effects (87%) such as sound of flying mosquito, footstep and talking character to trigger the player's imagination as if the virtual objects are really appeared in the real world. Twelve out of fifteen games (80%) extended the games by incorporating the concept of meaningful content coined by [20] as memorable while playing video games. This approach was demonstrated in the Ghostbuster game that engaged the player's role as ghost hunter and flight robot coming from the marker as gate in Augmentron game.

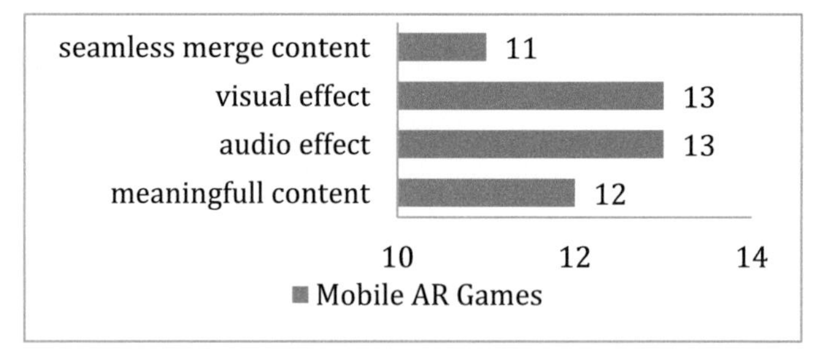

Fig. 5 Mixed fantasy elements in mobile AR games

4.5 *Perceived Usability*

This empirical analysis showed that marker and image target tracking were not the preference for mobile games development. Most of the games in particular with higher download scores excluded marker and image target tracking. Such decision is similar to [7] findings on the impracticality to print and bring the marker while playing the games. Additionally, [29] found that people were more comfortable doing simple activities on mobile device. Eight games (53.4%) suggested the players to prepare the setup prior to playing the game, particularly to print the marker and imager for the tracking purpose. In contrast, seven games (46.7%) excluded this. Alternatively, a simple and stable tracking [7] could be a solution for AR games primarily to cover sensor unreliability and connectivity loss [30]. Seven games (46.7%) used this approach with stable tracking, mostly established as the gyro method. The gyro-based games were much more stable as compared to marker-based.

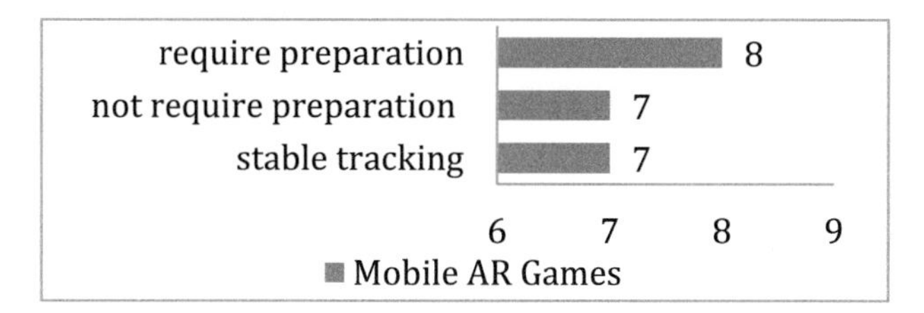

Fig. 6 Perceived usability elements in selected mobile ARgames

4.6 Challenge

The results showed that all games prepared with certain level of challenges to match the players' skill, as highlighted by [18]. According to them, challenges contribute to user engagement and avoid games boredom. As such the levels of difficulties have to increase with variety of choices. This study however found that only nine games (60%) managed to increase the level of challenges and six games (40%) provided with new challenge.

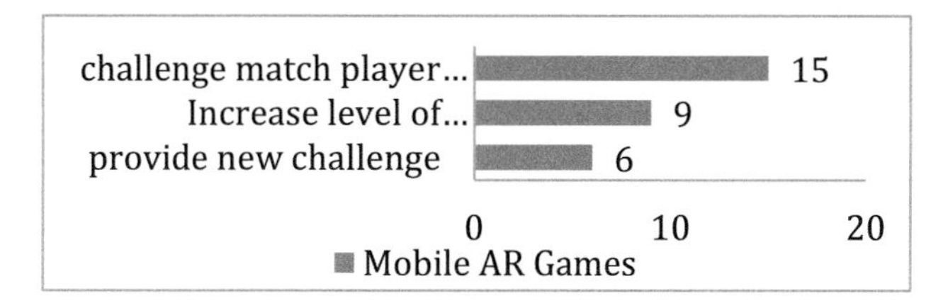

Fig. 7 Challenge elements in selected mobile AR games

4.7 Interaction

Design guidelines suggested to design simple and easy to better understand interaction for AR games [8]. Apart from this, the interaction for mobile AR games is different with other games as it involves with physical and social surroundings [5]. Good interaction will result a more enjoyable feeling thus contributes to the engagement [18] [30].

In this analysis twelve (80%) games demonstrated with clear and simple interaction without any form of interference. In contrast, the interface of Bug Mayhem game often became unclear due to marginal color differences of the virtual objects with the real environment. The importance of having good interaction was highlighted by [18], concerning that designer to introduce personalization, activity choices and information message for the players. This analysis found that six games (40%) provided personalization for the virtual object whereas twelve (80%) games implied activity choices and only three games (20%) design the information message with human feeling.

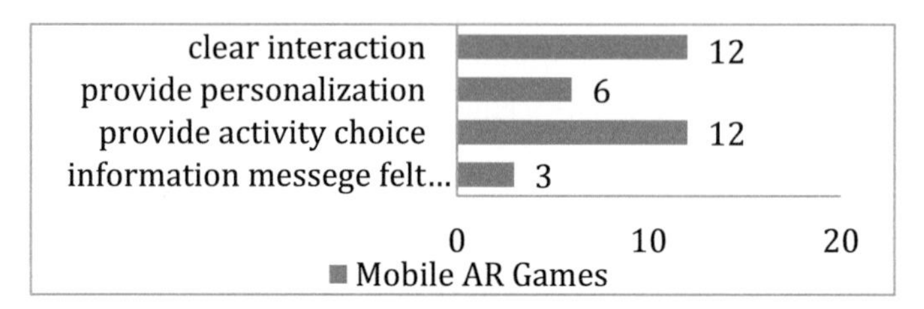

Fig. 8 Interaction elements in selected mobile AR games

4.8 *Social*

The result showed that majority of the mobile AR games disregarded social-based game play. Only one game (6.7%) created opportunity for competition and only three games out of fifteen (20%) provided collaboration in the form of multiplayer function. Providing social element such as multi-player function for competition and cooperation are necessary to the engagement especially for AR as defined by [19] which believed that this has the persuasion power. However this analysis found that only six (40%) games offered social network link, seven games (46.7%) introduced leaderboard, and none (0%) of the game provided social communities for the game discussion.

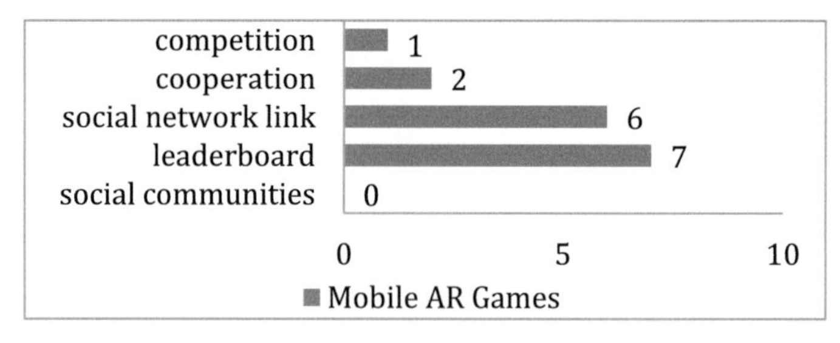

Fig. 9 Interaction elements in selected mobile AR games

5 Discussion

Generally, the selected mobile AR games were good in terms of mixed fantasy as most of the designers successfully developed the visual and audio effects to create seamless merge content. The only attention however is to create more meaningful content in order to complete this element. Similarly for the interaction element, most of games provided clear and simple interaction. But there are still necessities to enhance on the personalization and information message felt human. Focused attention elements have good and bad aspect. Several games successfully provided worthy and clear visual audio content to stimulate and ease the overall understanding. It also showed that most of the games provided feedback. However, quite a number of the games lacked of background story and variety of mission. In fact only half offered high workload and undistracted task.

Several major problematic elements were also identified. Firstly, the social element was indeed low in the mobile AR game applications. In fact only a few games proposed multi-player function such as competition and cooperation, social network link community and leaderboard. Secondly, in the context of perceived usability, the used of marker was impractical for the mobile AR game purpose. As suggested in the literature, that game needs to provide convenient and ease of use for the player to experience, thus printing and bringing markers would be a cumbersome. The marker usability problem such as unstable tracking also affected the focus of attention. Thirdly, most of the mobile AR games did not prepare users to have level up and provide new challenge due to the fact that the solutions were to attempt for easy and simple. Fourthly, in term of satisfaction, many games missed to provide reward for the players. Only a few provided puzzlement and enjoyment tasks. Lastly, most of the analyzed games were not having a clear main goal and intermediate goal. None of these games were created with cinematic introductions to establish a background story as suggested in the literature review.

6 Conclusion

This research has identified elements of user engagement for reviewing mobile AR games from related literatures. The empirical analysis classified elements that were able to give an impact on the user engagement. Five major issues that were ignored even though important to be considered in the mobile AR games – namely social, perceived usability, challenge, satisfaction and clear goals. Three other elements, which are mixed fantasy, interaction and focused attention, were well incorporated in the game play but still have some minor problems. These identified issues are proposed to game designers as awareness aspect for the mobile AR game development. The finding from the analysis is suggested as one of the key considerations prior in developing mobile AR games.

References

1. Zheng, R., Zhang, D., Yang, G.: Seam the real with the virtual: a review of augmented reality. In: Information Technology and Mechatronics Engineering Conference, no. 7, pp. 77–80 (2015). Atlantis Press
2. Thomas, B., Close, B., Donoghue, B.J., Squires, J., De Bondi, P., Morris, M., Piekarski, W.: ARQuake: an outdoor/indoor augmented reality first person application. In: Proceedings of the 4th International Symposium on Wearable Computers, pp. 139–146 (2000)
3. ARhrrr! Georgia Institute of Technology. http://www.augmentedenvironments.org/lab/research/handheld-ar/arhrrrr/
4. Koh, R.K.C., Duh, H.B.L., Gu, J.: An integrated design flow in user interface and interaction for enhancing mobile AR gaming experiences. In: Proceedings of IEEE International Symposium on Mixed and Augmented Reality-Arts, Media, and Humanities (ISMAR-AMH), pp. 47–52. IEEE Computer Society, Washington DC (2010)
5. Wetzel, R., McCall, R., Braun, A. Broll, W.: Guidelines for designing augmented reality games. In: Proceedings of the 2008 Conference on Future Play: Research, Play, Share, Toronto, Ontario, Canada (2008)
6. Broll, W., Lindt, I., Herbsr, I., Ohlenburg, A., Braun, A., Wetzel, R.: Toward next-gen mobile AR games. Journal of IEEE Computer Graphics and Application, 40–48 (2008)
7. Handheld Augmented Reality Project (HARP) & Alien Contact! Overview. http://isitesharvard.edu/fs/docs/icb.topic135310.files/AlienContactOverview012907.pdf
8. Li, N., Duh, H.B.L.: Cognitive issues in mobile augmented reality: an embodied perspective. In: Human Factors in Augmented Reality Environments, pp. 109–135. Springer, New York (2013)
9. Boyle, E.A., Connolly, T.M., Hainey, T.: The role of psychology in understanding the impact of computer games. Journal of Entertainment Computing **2**, 69–74 (2011)
10. Cheng, S., Meszaros, B.: The Influence of Online Product Reviews on the Downloading Decision for Mobile Apps. Master Thesis, Blekinge Institute of Technology (2015)
11. Hassenzahl, M., Diefenbach, S., Göritz, A.: Needs, affect, and interactive products – Facets of user experience. Interacting With Computers **22**(5), 353–362 (2010)
12. Enrique, L., Rutledge, P., Neal, M.: Assessing user experience in augmented reality applications using the positive engagement evaluation model. In: The 2012 EEE International Conference on e-Learning, e-Business, Enterprise Information Systems, and e-Government, Las Vegas (2012)
13. O'Brien, H.L., Toms, E.G.: The development and evaluation of a survey to measure user engagement in e-commerce environments. JASIST **61**(1), 50–69 (2010)
14. Wiebe, E.N., Lamb, A., Hardy, M., Sharek, D.: Measuring engagement in video game-based environments: Investigation of the User Engagement Scale. Computers in Human Behavior **32**, 123–132 (2012)
15. Csikszentmihalyi, M.: Flow: The psychology of optimal experience. Harper & Row, New York (1990)
16. De Pablos-Heredero, C., López-Berzosa, D., Sánchez-Gonzalez, G.: Open business models and platform mediated networks: an application in the mobile industry. In: Procedia Technology, vol. 5, pp. 122–132 (2012)

17. Pavlas, D.: A Model of Flow and Play in Game-based Learning: The Impact of Game Characteristics, Player Traits, and Player States. Ph.D. Dissertation, University of Central Florida (2010)
18. Sweetser, P., Wyeth, P.: GameFlow: A model for evaluating player enjoyment in games. Computers in Entertainment **3**, 1–24 (2005)
19. Rutledge, P., Neal, M.: Positive engagement evaluation model for interactive and mobile technologies. In: The 2012 EEE International Conference on e-Learning, e-Business, Enterprise Information Systems, and e-Government, Las Vegas, NV (2012)
20. Stapleton, C.B., Hughes, C.E., Moshell, J.M.: Mixed fantasy: exhibition of entertainment research for mixed reality. In: International Symposium on Mixed and Augmented Reality, pp. 354–355. IEEE Computer Society, Tokyo (2003)
21. Dweck, C.S.: Leggett. S.: A Social- Cognitive Approach to Motivation and Personality. Psychological Review **95**, 256–273 (1988)
22. Buss, D.M.: Human Social Motivation in Evolutionary Perspective: Grounding Terror Management Theory. Psychological Inquiry **8**, 22–26 (1997)
23. Oulasvirta, A., Rattenbury, T., Ma, L., Raita, E.: Habits make smartphone use more pervasive. Personaland Ubiquitous Computing **16**, 105–114 (2012)
24. Lamantia, J.: Inside out: Interaction design for augmented reality. Uxmatters. http://www.uxmatters.com/mt/archives/2009/08/inside-out-interaction-design-for-augmented-reality.php
25. Malone, T.: What makes things fun to learn? A study of intrinsically motivating computer games. Technical Report CIS-7, Palo Alto, CA: Xerox PARC (1980)
26. Boyle, E.A., Connolly, T.M., Hainey, T., Boyle, J.M.: Engagement in digital entertainment games: A systematic review. Computers in Human Behavior **28**, 771–780 (2012)
27. O'Brien, H.L.: Toms. E.G.: What is user engagement? A conceptual framework for defining user engagement with technology. Journal of the American Society for Information Science and Technology **59**(6), 938–955 (2008)
28. Matlin, M.W.: Cognition, 3rd edn. Harcourt Brace, Orlando (1994)
29. Nylander, S., Lundquist, T., Brännström, A., Karlson, B.: It's just easier with the phone–a diary study of Internet access from cell phones. Pervasive Computing, 354–371. Springer, Heidelberg (2009)
30. Dixon, D., Kiani, S.L., Ikram, A.: Experiences with AR plots: Design issues and recommendations for augmented reality based mobile games. Communications in Mobile Computing **2**(1), 1–6 (2010)
31. Lazzaro, N., Keeker, K.: What's my method? a game show on games. In: Extended Abstracts of the 2004 Conference on Human Factors in Computing Systems, pp. 1093–1094. ACM Press, New York (2004)

Automated Differential Evolution for Solving Dynamic Economic Dispatch Problems

Saber Elsayed, Md Forhad Zaman and Ruhul Sarker

Abstract The objective of a dynamic economic dispatch problem is to determine the optimal power generation from a number of generating units by minimizing the fuel cost. The problem is considered a high-dimensional complex constrained optimization problem. Over the last few decades, many differential evolution variants have been proposed to solve this problem. However, such variants were highly dependent on the search operators, control parameters and constraint handling techniques used. Therefore, to tackle with this shortcoming, in this paper, a new differential evolution framework is introduced. In it, the appropriate selection of differential evolution operators is linked to the proper combination of control parameters (scaling factor and crossover rate), while the population size is adaptively updated. To add to this, a heuristic repair approach is introduced to help obtaining feasible solutions from infeasible ones, and hence enhancing the convergence rate of the proposed algorithm. The algorithm is tested on three different dynamic dispatch problems with 12 and 24 hours planning horizons. The results demonstrate the superiority of the proposed algorithm to the state-of-the-art algorithms.

Keywords Dynamic economic dispatch problem · Differential evolution

1 Introduction

The optimal utilization of fossil fuel in power generation has become an important research topic [1]. Since the operating costs of different generating units significantly vary, it is a challenging problem to schedule the right mix of generation from a number of units to serve a particular load demand at a minimum cost, which is known as

S. Elsayed(✉) · Md.F. Zaman · R. Sarker
School of Engineering and Information Technology,
University of New South Wales, Canberra, Australia
e-mail: {s.elsayed,r.sarker}@adfa.edu.au, md.zaman@student.adfa.edu.au

K. Lavangnananda et al. (eds.), *Intelligent and Evolutionary Systems*,
Proceedings in Adaptation, Learning and Optimization 5,
DOI: 10.1007/978-3-319-27000-5_29

a power economic dispatch problem. Its objective is to allocate the total generation required among the available thermal generating units, assuming that the thermal unit commitment has been previously determined, which is also named as a static static economic dispatch [2]. However, the demand may change from one hour to another. As a consequence, the solution may become infeasible due to the new ramp limits. This scheduling problem is known as a dynamic economic dispatch (DED) problem. This problem has dynamic (e.g., ramp limits), inequality (e.g., generation capacity) and equality (e.g., load balance) constraints. In reality, large steam generators have a multi-fuel option and some fluctuations appear on the cost function while the steam is admitted through the valve (the valve-point effect (VPE)) [3]. This make the objective function becomes quadratic, non-smooth, non-convex and multi-modal [4]. All of these factors make a DED problem difficult to solve.

Over decades, conventional methods, evolutionary algorithms (EAs) and combinations between them based approaches have been used to solve DED problems. Among EAs, genetic algorithms (GA) [5] and differential evolution (DE) [6] are popular in solving such problems. Like any other EAs, the choice of DE's control parameters (scaling factor (F), crossover rate (Cr) and population size (PS)) and search operators plays a crucial role in its success. A trial-and-error approach is a possible way to define the control parameters and search operators. However, such an approach is known tedious. As a matter of fact, one combination of control parameters and/or search operators may work well for a set of problems, but may not perform the same for another range of problems.

As a consequence, different research studies have been introduced to adapt DE's control parameters and/or search operators. However, using such algorithms to solve DED problems has not been fully covered yet. To add to this, the existing approaches, which are based on adapting DE operators and parameters do not take into consideration the relationship between the control parameters and the success of any operator, and vice-versa. In other words, in current research studies, the automatic selection of the best DE operators is done without considering the control parameters assigned to those operators to succeed.

For instance, Elsayed et al. [7] proposed a general framework that divided the population into four sub-populations. Each sub-population used one combination of search operators. During the evolutionary process, the sub-population sizes were adaptively varied based on the success of each operator, which was calculated based on changes in the fitness values, constraint violations and the feasibility rate. The algorithm performed well on a set of constrained problems. Sarker et al. [8] proposed a DE algorithm that used a mechanism to dynamically select the best performing combinations of parameters Cr and F for a problem during the course of a single run. The results demonstrated that the proposed algorithm was superior to other state-of-the-art algorithms. Note that the algorithm used a single DE mutation operator. Zamuda and Brest [9] introduced an algorithm that employed two mutation strategies DE and the population size was adaptively reduced during the evolutionary process. The algorithm was tested on 22 real-world applications, and showed better performance than two other algorithms. In [10], with a probability (q), one set of control parameters was selected out of 12 available sets, and during the evolutionary process, q was updated based on the success rate in the previous steps. The authors used a single DE mutation, two values of F and three values of Cr.

Another motivation behind this work is the fact that DE consumes a considerable computational effort to reach the feasible space of a DED problem. Therefore, proposing a mechanism that is able to convert infeasible solutions to feasible ones is helpful.

Motivated by the above-mentioned facts, in this paper, a new DE framework is introduced. In it, three sets (F_{set}, Cr_{set} and SO_{set}) are considered, which represent the scaling factor, crossover rate and search operators, respectively. Then, each individual in the population is assigned a random combination of (F, Cr and SO). The success rate of each combination is accumulated over generations. Then, the number of combinations is linearly reduced along with the population size. To add to this, a new heuristic repair method is introduced to convert an infeasible solution to a feasible one. The performance of the proposed algorithm is tested on three 5-units and 10-units DED problems with a 24 scheduling cycle and a 10-units DED problem with a 12 hours scheduling horizon. The algorithm shows its superiority to the state-of-the-art algorithms.

The rest of this paper is organized as follows. Section 2 describes the mathematical model of DED problems considered in this paper. A brief overview of DE is then given in Section 3. The proposed algorithm and the heuristic repair method are illustrated in Section 4, while the experimental results and conclusions are discussed in Sections 5 and 6, respectively.

2 DED Problem Mathematical Model

The mathematical model of the DED problem is as follows[2]:

$$\text{Minimize} \quad FC = \sum_{t=1}^{T} \sum_{i=1}^{N_T} FC_i(P_{GT_{i,t}}) \tag{1}$$

where

$$FC_i(P_{GT_{i,t}}) = a_i + b_i P_{GT_{i,t}} + c_i P_{GT_{i,t}}^2 + \left| d_i \sin\left\{ e_i (P_{GT_i}^{min} - P_{GT_{i,t}}) \right\} \right| \; i \in N_T, t \in T \tag{2}$$

Subject to:

$$\sum_{i=1}^{N_T} P_{GT_{i,t}} = P_{D_t} + P_{loss_t} \quad \forall t \in T \tag{3}$$

$$P_{loss_t} = \sum_{i=1}^{N_T} \sum_{j=1}^{N_T} P_{GT_{i,t}} B_{ij} P_{GT_{j,t}} \quad \forall t \in T \tag{4}$$

$$P_{GT_i}^{min} \le P_{GT_{i,t}} \le P_{GT_i}^{max} \quad \forall i \in N_{T,}, \ t \in T \tag{5}$$

$$P_{GT_{i,t}} - P_{GT_{i,t-1}} \le UR_i \quad \forall i \in N_T, \ t \in T \tag{6}$$

$$P_{GT_{i,t-1}} - P_{GT_{i,t}} \le DR_i \quad \forall i \in N_T, \ t \in T \tag{7}$$

$$\sum_{i=1}^{N_T} P_{GT_i}^{max} - \left(P_{D_t} + P_{loss_t} + SR_t\right) \ge 0 \quad \forall t \in T \tag{8}$$

$$\sum_{i=1}^{N_T} \min\left(P_{GT_i}^{max} - P_{GT_{i,t}}, UR_i\right) - SR_t \ge 0 \quad \forall t \in T \tag{9}$$

$$\sum_{i=1}^{N_T} \min\left(P_{GT_i}^{max} - P_{GT_{i,t}}, UR_i/6\right) - SR_t^m \ge 0 \quad \forall t \in T \tag{10}$$

The objective function, as presented in equation (1), is to minimize the sum of all fuel costs (FC) for the thermal power plants under consideration (N_T) during the operational cycle (T), where a_i, b_i, c_i, d_i and e_i are the cost coefficients. Equation (3) refers to the power balance constraint in each cycle. Using the transmission loss coefficients B, the power loss (P_{loss}) of each period is expressed in equation (4). Equation (5) is the capacity constraints, where $P_{GT_i}^{min}$ and $P_{GT_i}^{max}$ are the minimum and maximum output power of the ith unit, respectively. Equations (6) and (7) represent the upper and lower ramp rate limits, respectively, where UR and DR are the upward and downward transition limits, respectively. Equations (8) and (9) are used to satisfy the one-hour reserve requirements to meet any uncertainty, while equation (10) describes 10-minutes reserve requirements, in which the ramp is arithmetically considered as $UR/6$. Here, SR and SR_m are the spinning reserves for 1 hour and 10 minutes, respectively.

3 Differential Evolution

DE was originally introduced by Storn and Price [11] for solving continuous domain problems. It has three main operators (mutation, crossover and selection), and three control parameters (F, Cr and PS) [12]. A brief description is discussed below.

- **Mutation:** The simplest mutation form is DE/rand/1 [13], in which a mutant vector is generated by multiplying F by the difference between two random vectors and the result is added to a third random vector, as shown in equation 11.

$$\vec{v}_z = \vec{x}_{r_1} + F(\vec{x}_{r_2} - \vec{x}_{r_3}) \tag{11}$$

where r_1, r_2, r_3 are different random integer numbers $\in [1, PS]$ and $r_1 \neq r_2 \neq r_3 \neq z \, \forall z = 1, 2, ..., PS$. As the mutation operator plays a vital role in the success of any DE variant, many mutation operators have been proposed in the literature. For more details, readers are referred to [12].

- **Crossover:** There are two well-known crossover schemes, binomial and exponential. The former is known to better than the later [7]. In it, a trial vector $\overrightarrow{u}$ is generated as follows:

$$u_{z,j} = \begin{cases} v_{z,j} & if(rand \leq Cr \text{ or } j = j_{rand}) \\ x_{z,j} & \text{otherwise} \end{cases} \tag{12}$$

where D is the problem dimension, $j_{rand} \in 1, 2, ..., D$ a randomly selected index, which ensures that $\overrightarrow{u_z}$ gets at least one component from $\overrightarrow{v_z}$.

- **Selection:** A tournament between $\overrightarrow{u}_z$ and $\overrightarrow{x}_z$, $\forall z = 1, 2, ... PS$, takes place, and the winner (based on the objective value and/or constraint violation) survives to the next generation.

4 Automated Differential Evolution Algorithm

Here, the proposed automated DE (ADE) framework is described as well as the heuristic repair method.

4.1 ADE

It has been proven that the relative performance of a DE operator and a set of parameters may work well on a specific problem, and may perform badly on another [7]. This motivated researchers to introduce ensemble of DE operators and parameters, and used some adaptive mechanisms to put emphasis on the best performing operators. However, such studies did not take into consideration the relationship between the success of any search operator and the control parameters assigned to it, or vice-versa. In other words, after selecting the best operator, it might not perform well as expected. On reason for this is the change in the control parameters values. Therefore, here, a new framework is proposed which keeps track of the best combinations of operators and control parameters. The proposed algorithm is presented in Algorithm 1.

To begin with, three sets are defined as: F_{set}, Cr_{set} and SO_{set}, where F_{set} and Cr_{set} contain nf and ncr discrete values, each discrete value represents a range of continuous values, respectively. For example, in a case of $F = 8$, and $Cr = 9$, the values are $0.8 \leq F < 0.9$ and $0.9 \leq Cr < 1$, respectively. $SO_{set} = \{SO_1, SO_2, ..., SO_{nso}\}$ is a set of different DE variants. This means that the total number of combinations (NoC) is equal to $(nf \times ncr \times nso)$.

First, PS random individuals are generated within the variables bounds. Each individual in the population $(\overrightarrow{x_z})$ is then assigned a combination that has three values $(F_z$, Cr_z and $SO_z)$. To make it clear, each combination is assigned to at least one individual. In case of NoC is less than PS, the remaining $PS - NoC$ individuals are

Algorithm 1. General framework of ADE

1: $PS \leftarrow PS_{max}$; define PS_{min}; $F_{set} \leftarrow F_1, F_2, ..., F_{nf}$; $Cr_{set} \leftarrow Cr_1, Cr_2, ..., Cr_{ncr}$; $SO_{set} \leftarrow SO_1, SO_2, ..., SO_{nso}$; $cfe \leftarrow 0$; $iter \leftarrow 0$;
2: Generate an initial random population $((\overrightarrow{x_z} \forall z \in [1 - PS])$;
3: Calculate the fitness value and constraint violation of $(\overrightarrow{x_z})$;
4: Update all infeasible solutions using the proposed heuristic (Section 4.3);
5: Sort the whole population.
6: **while** $cfe < cfe_{max}$ **do**
7: Each individual is assigned a random combination of parameter segments F, Cr and SO;
8: Convert discrete segments of F and Cr to continuous values.
9: **for** $z = 1 : PS$ **do**
10: Generate a new individual $(\overrightarrow{u_z})$ using its assigned combination;
11: Calculate the constraints violation $\Theta(\overrightarrow{u_z})$;
12: **if** $\Theta(\overrightarrow{u_z}) > 0$ // the individual is infeasible **then**
13: repair $\overrightarrow{u_z}$ (Section 4.3) and update $\Theta(\overrightarrow{u_z})$;
14: **end if**
15: Calculate the fitness value $(fit(\overrightarrow{u_z}))$ and update cfe;
16: **if** $\overrightarrow{u_z}$ is better than $\overrightarrow{x_z}$ **then**
17: $\overrightarrow{u_z}$ survives to the next generation; $com_{y,suc} \leftarrow com_{y,suc} + 1$; $SO_{p,suc} \leftarrow SO_{p,suc} + 1$;
18: **end if**
19: Update and sort the new population.
20: **end for**
21: Calculate the rank of each combination based on Equation 13.
22: Reduce PS and the number of combinations, if required.
23: $iter \leftarrow iter + 1$; and go to step 9;
24: **end while**

assigned random combinations. If any solution is infeasible, the solution is repaired using the proposed method (as will be described in Section 4.3). Then, for each $\overrightarrow{x_z}$, a new offspring $(\overrightarrow{u_z})$ is generated by using its assigned combination of operators and parameters. If $\overrightarrow{u_z}$ is infeasible, it is converted to a feasible one. Then, a tournament takes place, such that if $\overrightarrow{u_z}$ is better than $\overrightarrow{x_z}$, it will survive to the next generation and the success of the corresponding combination $(com_{y,suc})$ is increased by 1, where $y = 1, 2, ..., NoC$. At the end of each generation, the ranking of any combination (R_y) is calculated using Equation (13), where N_y is the number of individuals updated by a combination y. Note that the initial value of every R_y is 0.

$$R_y = \frac{com_{y,suc}}{N_y} \tag{13}$$

At the same time, a linear reduction of PS takes place, i.e., PS is set at a large value at the start of the evolutionary process and then linearly reduced (by removing the worst individuals), such that

$$PS_{iter} = \text{round}(((\frac{PS_{min} - PS_{max}}{FFE_{max}}) \times cfe) + PS_{max}), \tag{14}$$

where, PS_{max} and PS_{min} are the maximum and minimum values of PS, respectively, and FFE_{max} is the maximum fitness evaluations.

At the same time, all combinations are sorted based on their ranks, and the worst $(PS_{iter-1} - PS_{iter})$ combinations are removed. The process continues until an overall stopping criterion is met.

4.2 The Constraint Handling Mechanism

In reference to [14], in the selection process, three conditions exist: (1) between two feasible candidates, the fittest one (according to fitness function) is selected; (2) a feasible point is always better than an infeasible one; and (3) between two infeasible solutions, the one with a smaller sum of constraint violations (Θ) is chosen, where Θ of an individual ($\overrightarrow{x_z}$) is calculated such that:

$$\Theta_z = \sum_{k=1}^{K} max(0, g_k(\overrightarrow{x_z})) + \sum_{e=1}^{E} max(0, |h_e(\overrightarrow{x_z})| - \epsilon_e) \tag{15}$$

where $g_k(\overrightarrow{x_z})$ is the kth inequality constraint, $h_e(\overrightarrow{x_z})$ is the eth equality constraint, while ϵ_e is initialized with a large value and then reduced to $1.0\text{E} - 06$.

The equality constraints are relaxed such that

$$\varepsilon(iter) = \begin{cases} \varepsilon_0\left(1 - \frac{iter}{N_{G_c}}\right), & \text{if } 0 < iter < N_{G_c}; \\ 1e - 6, & \text{otherwise} \end{cases} \tag{16}$$

where ε_0 is the sum of constraints violation (Θ) at the initial generation, while $iter$ and N_{G_c} are the current generation and the level at which ε is set at $1.0\text{E} - 06$, respectively. Here, $N_{G_c} = iter_{max}/2$.

4.3 A New Heuristic Repair Method

As previously mentioned, the thermal DED is a highly constrained optimization problem that involves a number of equality and inequality constraints. The solutions generated by EAs may not satisfy all constraints. To add to this, a feasible solution which is obtained in one generation, may become infeasible in the following generation due to the dynamic nature of the ramp constraint. In order to overcome this deficiency, a heuristic repair method is proposed to convert an infeasible solution into a good-quality feasible one. The pseudo-code of the heuristic is shown in Algorithm 2.

In its process, the T-hour load cycle is divided into T sub-problems, with the electricity production is allocated to meet the load demand in each hour. The allocation starts from different random hours instead of the first hour of the operational cycle. If the allocation starts from the first hour, it may become infeasible at a later stage, due to the ramp constraint and any significant changes in demand (i.e., peak demand

Algorithm 2. Pseudo code of heuristic for DED constraints

1: Transform an infeasible individual $\overrightarrow{x}$ into a matrix P in $T \times N_T$ size.
2: Randomly select an hour $t \in T$, and its generation $P_t \in P$, and start the forward process. Save $t_{start} = t$.
3: Set, capacity limits, $P_{t,i}^{max} = P_{GT_i}^{max}$ and $P_{t,i}^{min} = P_{GT_i}^{min}$ $\forall i$.
4: **for** $t = t_{start} : 1 : T$ **do**
5: Satisfy the generation limits, such that

$$P_{t,i} = \begin{cases} P_{t,i}^{max}, & \text{if } P_{t,i} > P_{t,i}^{max}; \\ P_{t,i}^{min}, & \text{if } P_{t,i} < P_{t,i}^{max}; \\ P_{t,i}, & \text{otherwise;} \end{cases} \quad \forall i$$

6: **while** $\left| \sum_{i=1}^{N_T} P_{t,i} - (P_{D_t} + P_{loss_t} \right| \leq \varepsilon$ **do**
7: Satisfy demand constraints by repairing a random unit ($n_d \in N_T$), as:

$$P_{t,n_d} = \max\left[P_{t,n_d}^{min}, \min\left\{\left(P_{D_t} - \sum_{i=1, i \neq n_d}^{N_T} P_{t,i}\right), P_{t,n_d}^{max}\right\}\right]$$

8: **end while**
9: Update the capacity limits, such that

$$P_{t+1,i}^{max} = \min\left[P_i^{max}, \left(P_{t,i} + UR_i\right)\right], \quad P_{t+1,i}^{min} = \max\left[P_i^{min}, \left(P_{t,i} - DR_i\right)\right] \forall i$$

10: **end for**
11: Set, $t = t_{start-1}$, and start the backward process,
12: **for** $t = t_{start-1} : -1 : 1$ **do**
13: Satisfy the demand constraint using the steps 6, 7 and 8.
14: Update capacity limits, as:

$$P_{t+1,i}^{max} = \min\left[P_i^{max}, \left(P_{t,i} - UR_i\right)\right], \quad P_{t+1,i}^{min} = \max\left[P_i^{min}, \left(P_{t,i} + DR_i\right)\right] \forall i$$

15: **end for**
16: Reconstruct the individual $\overrightarrow{x}$ from the updated P matrix.
17: Return feasible $\overrightarrow{x}$.

period), as the power generation limit in any hour depends on the power generation of its immediate past hour.

5 Experimental Results

In this section, the computational results of ADE are discussed based on solving three DED test problems, and compared with the state-of-the-art algorithms (most of these algorithms used mechanisms to repair infeasible solutions). Note that the results of the state-of-the-art algorithms considered in the comparison are taken from their corresponding published papers. These problems involve up to 10-thermal units for a 24-hours planning horizon with a one-hour long time period. Based on the availability of data, these problems can be solved with and without the consideration of the power-loss constraints. The problems are briefly described below.

- Case 1: A 5- unit thermal system with P_{loss} for a 24-hours planning horizon [15];
- Case 2: A 10-unit thermal system without P_{loss} for a 12-hours planning horizon [16];
- Case 3: A 10-unit thermal system without P_{loss} for a 24-hours planning horizon [4];

30 runs are conducted for each test problem, where the stopping criterion is to run for up to 10, 000D, where D = 120, 120 and 240, for case 1, case 2 and case 3, respectively. The algorithm stops if one of the following criteria is met: (1) reaching the maximum number of fitness evaluations; or (2) the best solution did not change during the last 100 generations.

5.1 Parameters Setting

- $SO_{set} = \{\text{DE}_1, \text{DE}_2\}$, in which

1. DE_1: DE/φ-best/1/bin [8]

$$u_{z,j} = \begin{cases} x_{\phi,j} + F_z.(x_{r_1,j} - x_{r_2,j}) \; if(rand \leq Cr_z \text{ or } j = j_{rand}) \\ x_{z,j} \qquad \text{otherwise} \end{cases} \tag{17}$$

2. DE_2: DE/current-to-ϕbest with archive/1/bin [17]

$$u_{z,j} = \begin{cases} x_{z,j} + F_z.(x_{\phi,j} - x_{z,j} + x_{r_1,j} - \widetilde{x}_{r_3,j}) \; if(rand \leq Cr_z \text{ or } j = j_{rand}) \\ x_{z,j} \qquad \text{otherwise} \end{cases} \tag{18}$$

where $\varphi = 0.5$ as suggested in [8], $\phi = 0.1$ [17], $r_1 \neq r_2 \neq r_3 \neq z$ are random integer numbers, $\widetilde{x}_{r_2,j}$ is randomly chosen from the union $PS \cup AR$, i.e. the union of PS and the archive AR. The archive is initially empty; then, the unsuccessful parent vectors are added to the archive. To add to this, once the size of the archive size ($arch_{size}$) exceeds a threshold, randomly selected elements are deleted to leave a space for the newly inserted elements [17]. The reason for using DE_1 is to make a balance between diversity and intensification as described in [8], while DE_2 has a high convergence rate.

- F_{set} =$\{F_3 \in [0.3 - 0.4[, F_4 \in [0.4 - 0.5[, F_5 \in [0.5 - 0.6[, F_6 \in [0.6 - 0.7[, F_7 \in [0.7 - 0.8[, F_8 \in [0.8 - 0.9[, F_9 \in [0.9, 1[\}$.
- Cr_{set} =$\{Cr_2 \in [0.2{-}0.3[, Cr_3 \in [0.3{-}0.4[, Cr_4 \in [0.4{-}0.5[, Cr_5 \in [0.5{-}0.6[, Cr_6 \in [0.6 - 0.7[, Cr_7 \in [0.7 - 0.8[, Cr_8 \in [0.8 - 0.9[, Cr_9 \in [0.9, 1[\}$, hence $NoC_{total} = 7 \times 8 \times 2 = 112$ combinations.

- PS_{max} = 100, 200 and 300, for case 1, case 2 and case 3, respectively, while $PS_{min} = 30$.
- $arch_{size}$ 1.4PS.

5.2 Case-1

Here, a 5-unit thermal system for a 24-hours planning horizon is considered. The power loss is considered in this case, while the loss coefficients (B) and generators data can be found in [18]. The results obtained by the proposed ADE algorithm and those from literature, genetic algorithm (GA) [19], particle swarm optimization (PSO) [5], Hybrid PSO (H-PSO) [15], artificial bee colony (ABC) [5] and artificial immune system (AIS) [19] are listed in Table 1. From the results, it is clear that ADE outperforms all of the state-of-the-art algorithms.

5.3 Case-2

In this case, the new England system, which has 10 generating units, 39 buses and 46 transmission lines over a period of 12 hours scheduling horizon is considered. The detailed data of this system can be found in [16]. Due to the unavailability of data and comparison purposes, the test problem without P_{loss} is solved using ADE and compared with other algorithms in the literature, such as quadratic programming (QP) [20], augmented Lagrange hopfield network (ALHN) [20], a hybrid evolutionary programing and sequential quadratic programing based algorithm (EP-SQP) [21], PSO and different variants of DE [16]. The results are shown in Table 2, which reveal the superiority of ADE to the state-of-the-art-algorithms.

5.4 Case-3

In this case, another 10-unit test system for a 24 hours scheduling horizon is considered. In this system, P_{loss} is neglected and can be found in [22]. Similarly, the results of the proposed algorithm are compared with those found in literature, such

Table 1 A summary of the results obtained by different algorithms for a 5-unit 24 hours system with P_{loss}, where N/A means the corresponding result is not available

Method	Production cost ($)			
	Minimum	Average	Maximum	STD
GA	44862.00	44922.00	45894.00	N/A
PSO	44253.00	45657.00	46403.00	N/A
ABC	44046.00	44065.00	44219.00	N/A
AIS	44385.00	44759.00	45554.00	N/A
H-PSO	43223.00	43732.00	44252.00	274.95
ADE	**42523.60**	**42576.11**	**42673.9**	**38.15**

Table 2 A summary of the results obtained by different algorithms for a 10-unit 12 hours system without P_{loss}, where N/A means the corresponding result is not available

Method	Production cost ($)			
	Minimum	Average	Maximum	STD
QP	2,185,413	N/A	N/A	N/A
ALHN	2,185,413	N/A	N/A	N/A
EP - SQP	2,196,439	N/A	N/A	N/A
Standard PSO	2,186,264	2,186,757	2,187,148	N/A
CSDE/rand/1	2,185,406	2,185,416	2,185,428	N/A
CSDE/target-best/1	2,185,403	2,185,411	2,185,421	N/A
CSDE/rand/2	2,185,405	2,185,412	2,185,428	N/A -
CSDE/best/2	2,185,403	2,185,414	2,185,425	N/A
DE/rand/1	2,199,770	2,203,816	2,208,763	N/A
DE/best/1	2,186,229	2,186,519	2,186,951	N/A
DE/target-best/2	2,186,334	2,186,661	2,186,979	N/A
DE/rand/2	2,200,936	2,202,651	2,204,665	N/A
DE/best/2	2,189,920	2,190,625	2,191,287	N/A
CSDE/best/1	2,185,400	2,185,408	2,185,421	N/A
ADE	**2,185,335.1**	**2,185,335.5**	**2,185,340**	**0.32**

Table 3 A summary of the results obtained by different algorithms for a 10-unit 24 hours system without P_{loss}, , where N/A means the corresponding result is not available

Method	Production cost ($)			
	Minimum	Average	Maximum	STD
GA	1033481	1038014	1042606	N/A
DE	1036756	1040586	1452558	3225.8
MDE	1031612	1033630	N/A	N/A
PSO	1027679	1031716	1034340	N/A
CE	1022702	1024024	N/A	N/A
ECE	1022272	1023335	N/A	N/A
CSDE	1023432	1026475	1027634	N/A
ABC	1021576	1022686	1024316	N/A
CDE	1019123	1020870	1023115	1310.7
ICPSO	1019072	1020027	N/A	N/A
ICA	1018468	1019291	1021796	N/A
H-PSO	1018159	1019850	1021813	826.94
ADE	**1017160**	**1017582**	**1018200**	**261.90**

as GA [5], DE [23], modified DE (MDE) [24], chaotic DE (CDE) [23], chaotic sequence based DE (CSDE) [25],, cross-entropy (CE) [26], enhanced CE (ECE) [26], PSO [5], ABC [5], improved chaotic PSO (ICPSO) [27], imperialist competitive algorithm (ICA) [26], hybrid bare-bones PSO [15]. The detailed results, presented in Table 3, show that the algorithm is better than all other algorithms considered.

6 Conclusions

In this paper, a new DE framework was introduced, which adaptively configured the best combination of DE operators and control parameters for solving DED problems. In it, three sets (F_{set}, Cr_{set} and SO_{set}) were initiated. Then, each individual in the population was assigned a random combination of (F, Cr and SO). The success rate of each combination was accumulated over generations. Then, the number of combinations was linearly reduced along with the population size. To help reaching the feasible space quickly, and hence increasing the convergence rate, a heuristic repair method was used to deal with infeasible solutions.

The algorithm was tested on three DED problems: (1) a 5-unit thermal system with P_{loss} for a 24-hours planning horizon; (2) a 10-unit thermal system without P_{loss} for a 12-hours planning horizon; and (3) a 10-unit thermal system without P_{loss} for a 24-hours planning horizon. The results obtained were compared with the state-of-the-art algorithms, which showed the superiority of the proposed algorithm.

Acknowledgments This work was supported by an Australian Research Council Discovery Project (Grant# DP150102583) awarded to A/Prof Ruhul Sarker.

References

1. Wood, W.G.: Spinning reserve constrained static and dynamic economic dispatch. IEEE Transactions on Power Apparatus and Systems **PAS–101**(2), 381–388 (1982)
2. Zaman, M., Elsayed, S., Ray, T., Sarker, R.: Evolutionary algorithms for dynamic economic dispatch problems. IEEE Transactions on Power Systems **PP**(99), 1–10 (2015)
3. Niknam, T., Azizipanah-Abarghooee, R., Zare, M., Bahmani-Firouzi, B.: Reserve constrained dynamic environmental/economic dispatch: A new multiobjective self-adaptive learning bat algorithm. IEEE Systems Journal **7**(4), 763–776 (2013)
4. Victoire, T.A., Jeyakumar, A.E.: Reserve constrained dynamic dispatch of units with valve-point effects. IEEE Transactions on Power Systems **20**(3), 1273–1282 (2005)
5. Hemamalini, S., Simon, S.P.: Dynamic economic dispatch using artificial bee colony algorithm for units with valve-point effect. European Transactions on Electrical Power **21**(1), 70–81 (2011)
6. Zhang, H., Yue, D., Xie, X., Hu, S., Weng, S.: Multi-elite guide hybrid differential evolution with simulated annealing technique for dynamic economic emission dispatch. Applied Soft Computing **34**, 312–323 (2015)
7. Elsayed, S.M., Sarker, R.A., Essam, D.L.: Multi-operator based evolutionary algorithms for solving constrained optimization problems. Computers & Operations Research **38**(12), 1877–1896 (2011)
8. Sarker, R., Elsayed, S., Ray, T.: Differential evolution with dynamic parameters selection for optimization problems. IEEE Transactions on Evolutionary Computation **18**(5), 689–707 (2014)
9. Zamuda, A., Brest, J.: Population reduction differential evolution with multiple mutation strategies in real world industry challenges. In: Rutkowski, L., Korytkowski, M., Scherer, R., Tadeusiewicz, R., Zadeh, L., Zurada, J. (eds.) Swarm and Evolutionary Computation, pp. 154–161. Springer (2012)

10. Tvrdík, J., Polakova, R.: Competitive differential evolution for constrained problems. In: IEEE Congress on Evolutionary Computation (CEC), pp. 1–8. IEEE (2010)
11. Storn, R., Price, K.: Differential evolution-a simple and efficient adaptive scheme for global optimization over continuous spaces, vol. 3. ICSI Berkeley (1995)
12. Das, S., Suganthan, P.N.: Differential evolution: A survey of the state-of-the-art. IEEE Transactions on Evolutionary Computation **15**(1), 4–31 (2011)
13. Storn, R., Price, K.: Differential evolution-a simple and efficient heuristic for global optimization over continuous spaces. Journal of Global Optimization **11**(4), 341–359 (1997)
14. Deb, K.: An efficient constraint handling method for genetic algorithms. Computer Methods in Applied Mechanics and Engineering **186**(2), 311–338 (2000)
15. Zhang, Y., Gong, D.W., Geng, N., Sun, X.Y.: Hybrid bare-bones PSO for dynamic economic dispatch with valve-point effects. Applied Soft Computing **18**, 248–260 (2014)
16. Niu, Q., Li, K., Irwin, G.: Differential evolution combined with clonal selection for dynamic economic dispatch. Journal of Experimental & Theoretical Artificial Intelligence **27**(3), 325–350 (2015)
17. Zhang, J., Sanderson, A.C.: Jade: adaptive differential evolution with optional external archive. IEEE Transactions on Evolutionary Computation **13**(5), 945–958 (2009)
18. Panigrahi, C.K., Chattopadhyay, P.K., Chakrabarti, R.N., Basu, M.: Simulated annealing technique for dynamic economic dispatch. Electric Power Components and Systems **34**(5), 577–586 (2006)
19. Hemamalini, S., Simon, S.P.: Dynamic economic dispatch using artificial immune system for units with valve-point effect. International Journal of Electrical Power & Energy Systems **33**(4), 868–874 (2011)
20. Dieu, V.N., Ongsakul, W.: Economic dispatch with emission and transmission constraints by augmented lagrange hopfield network. Transaction in Power System Optimization (GJOT) **1** (2010)
21. Babu, G.S., Das, D.B., Patvardhan, C.: Dynamic economic dispatch solution using an enhanced real-quantum evolutionary algorithm. In: Joint International Conference on Power System Technology and IEEE Power India Conference. POWERCON 2008, pp. 1–6. IEEE (2008)
22. Attaviriyanupap, P., Kita, H., Tanaka, E., Hasegawa, J.: A hybrid EP and SQP for dynamic economic dispatch with nonsmooth fuel cost function. IEEE Power Engineering Review **22**(4), 77 (2002)
23. Liao, G.C.: A novel evolutionary algorithm for dynamic economic dispatch with energy saving and emission reduction in power system integrated wind power. Energy **36**(2), 1018–1029 (2011)
24. Yuan, X., Wang, L., Yuan, Y., Zhang, Y., Cao, B., Yang, B.: A modified differential evolution approach for dynamic economic dispatch with valve-point effects. Energy Conversion and Management **49**(12), 3447–3453 (2008)
25. He, D., Dong, G., Wang, F., Mao, Z.: Optimization of dynamic economic dispatch with valve-point effect using chaotic sequence based differential evolution algorithms. Energy Conversion and Management **52**(2), 1026–1032 (2011)
26. Immanuel Selvakumar, A.: Enhanced cross-entropy method for dynamic economic dispatch with valve-point effects. International Journal of Electrical Power & Energy Systems **33**(3), 783–790 (2011)
27. Wang, Y., Zhou, J., Qin, H., Lu, Y.: Improved chaotic particle swarm optimization algorithm for dynamic economic dispatch problem with valve-point effects. Energy Conversion and Management **51**(12), 2893–2900 (2010)

Part VI
Smart Workspace and Image Processing

An Agent-Based Model of Smart Supply Chain Networks

Tomohito Okada, Akira Namatame and Hiroshi Sato

Abstract A global industrial enterprise is a complex network of different distributed production plants producing, inventory, and distributing products. Agent-based model provides the approach to prove complex network problems of independent actors. A global economy and increase in both demand fluctuation and pressures for cost decreasing while satisfying customer services have put a premium on smart supply chain management. It is important to make risk-benefit analysis of supply chain design alternatives before making a final decision. Simulation gives us an effective approach to comparative analysis and evaluation of the alternatives. In this paper, we describe an agent-based simulation tool for designing smart supply chain networks as well logistic networks. Using an agent-based approach, supply chain models are composed from supply chain agents. The agent-based simulation tool can be very useful for predicting the effects of local and system-level activities on multi-plant performance and improving the tactical and strategic decision-making at the enterprise level. Specifically, this model can reveal the optimal method to ship the inventory on some situations which are demand fluctuation and network disruption. The demand fluctuation effects the inventory management. The network disruption restricts the logistics. This model evaluates supply chain management from the viewpoints of the amount of inventory, the way of shipping and cost.

Keywords Agent based model · Multi-Agents · Supply chain management · Logistics · Multi-echelon · Manufacturing · Risk management · Demand fluctuation · Network disruption · Supply chain cost

---

T. Okada(✉) · A. Namatame · H. Sato
Department of Computer Science, National Defense Academy of Japan, Yokosuka, Japan
e-mail: {em53040,nama,hsato}@nda.ac.jp

K. Lavangnananda et al. (eds.), *Intelligent and Evolutionary Systems*,
Proceedings in Adaptation, Learning and Optimization 5,
DOI: 10.1007/978-3-319-27000-5_30

1 Introduction

Supply chain and logistics networks become bigger and more complex as the result of globalization and new initiatives. This tendency makes it hard to manage the supply chain and to meet the demand for market. The collaboration of many independent contractors and suppliers is necessary to meet the demand for market. The company having the supply chain should have the alternative strategic solution for their supply chain to continue their company. One of these solutions is to make not only one-plant manufacturing facilities but also multi-plant enterprise. This change gives the company many advantages which are the low cost raw materials, the flexibility of change the product, and the changing the network of product flow [1]. On the other hand, the supply chain network involves many actors and it is operated collectively. The logistic network, physical network of transportation, also forms an interconnected complex system where it affects the behavior of the supply chain network.

Many analytical methods for modeling and optimizing different scenarios in a multi-plant enterprise are proposed [2]. They discuss mathematical formulations for operation management in multi-plant industrial networks. These works propose the methods for solving the combined production and distribution scheduling problem in multi-plant environments by using mathematical programming approaches.

2 An Architecture of Agent-Based Supply Chain

Lee and Billington [3] provided an insightful survey of common pitfalls in supply chain management practices. Some studies provide that the relationships of market and supplier depend on factors of quality, delivery time, flexibility in contract, as opposed to the factor of cost. From the point of analytical, there are so many researches of inventory problems in multi-echelon supply chain. Svoronos and Zipkin [4] study a multi-echelon system which having multiple tiers in the supply chain. The multi-echelon system is assumed that the company manages the supply chain with centralized control.

Towill et al. [5] applied a simulation technology to the evaluation of the effect of different supply chain strategies under the situation of demand expansion. Swaminathan et al. [6] present a modeling and simulation framework for developing decision support tools for supply chain management. They develop a framework that has two basic elements: object modeling of supply chain flows and agent modeling of supply chain entities.

Just-In-Time (JST) philosophy changes the typical style in supply chain management. The supply chain is effected by this philosophy becomes globalization, use of third party, and reducing the lead-time. These trends reduce the cost in supply chain and give the company competitiveness against other company in the same market. The merits of the reducing lead time are the reducing the inventory and defective, the removing the waste, and the problems to be clear [7].

The disruption is the important elements in supply chain management. The disruption in supply chain gives significant cost to the company. The company that experiences the disruption in supply chain will face significant declines in sales growth, stock returns, and more [8]. The disruption in supply chain takes place the network problem. The player and relationship between players can be expressed to the node and link. Considering a supply chain as a network problem, a disruption means the shutdown of node or the cut of link.

Supply chains are defined as a collection of business centers through which products pass at various stages of completion from the provision of raw materials to final sales. A supply chain can make the products for markets and delivery it to the markets. The individual company in supply chain can only grasp the limited visibility situation of supply chain. This is difficult to make the demand estimation in each player. The players in supply chain depend on the information which is obtained by own. Their information may be different from the information by headquarter management. The amount of order is larger than the headquarter one. The more amplified the amount of order. These problems occur under the dynamically supply chain. As a result of these problems, each company makes incorrect demand estimations and the amount of inventory is larger than the one by correct strategy. This is the well-known Bullwhip effect [6].

One of the solutions proposed to deal with the bullwhip effect is to have information sharing across the companies in the supply chain. There are unique characteristics required for information systems that support supply chain management. First, they should be able to support distributed collaboration among companies. Second, a single company cannot manage multi-players in supply chain directly, but there are need to coordinate each company by autonomy. Toyota motor adopts the automation of each player for a part of removal of waste [9]. Third, It is required the high intelligence for strategy, planning, and flexibility adaptation. For these reasons, agent modeling is suitable to support the supply chain management.

Many kinds of supply chains exist in the real world. Most share some common elements. For example, within each supply chain, material flows from a raw material state to an end-user [10].

Multi-agent technology has many beneficial features for autonomous, collaborative, and intelligent systems in distributed environments, which makes it one of the best candidates for complex supply chain management. Agent-based modeling (ABM) is a suitable approach to analyze the system influenced by autonomous agents. The system behavior effected from the behavior of the player in the system and their interactions. The reason of this is that the agent-based model is made of decentralized agent and the network which is made by them. Generally speaking, agents have their own decision making. They can make decision under any situation without centralized management. They are able to change their decision in any environment.

A multi-plant enterprise is modeled as modular, decentralized, changeable agent networks. The network of like this enterprise has so many kinds of agent and so many amount of agent. In addition, all agents set their aim in their supply chain by oneself.

The agent-based model can be used to analyze different stages of the supply chain in order to define what could happen under different scenarios, in example if aid were not enough to supply the "demand", and therefore understand possible side effects or delayed consequences such as bullwhip effect, distorted information from one end of a supply chain to the other can lead to remarkable inefficiencies. In the case of the aid supply chain, inventory moves up in the chain and it fluctuates more with donations recollected and in the distribution centers creating a distortion on the demand information. The model is built to represent the flow of emergency goods, and how it is affected by the information feedbacks, that explain the presence of bullwhips effects on the chain at different periods of time depending on the initial stocks of good storage on the different parts of the chain and on exogenous variables.

The elements of the ABM system are involved with production and transportation of products. The structural elements follow as:

***Factory Agents*:** Factory agent makes the inventory for the demand in supply chain network. When the downstream agent orders the inventory to the factory agent, the factory agent ships it to the downstream agent. The factory agent calculates the amount of product by the order data from downstream agent. If the amount of inventory in the factory is under the point of product, the factory agent makes the inventory.

Retailer Agents: A retailer agent consumes the inventory by customer. This agent has the parameter of demand which is unpredictability. This agent calculates the amount of the safety stock, the amount of order point, and the amount of order. When the inventory in this agent is under the order point, this agent orders the inventory to upper stream agent.

Distribution Center Agent: A distribution center agent is role of buffer of stock in supply chain. This agent reduces the inventory of supply chain and makes a flow of inventory effectively. When this agent receives the order of inventory from downstream agent, this agent ships it to the downstream agent. If the inventory in this agent is under the order point, this agent orders the inventory to upper upstream agent.

Inventory Control: These elements control inventory at a particular production element by considering inventory levels at that entity in the supply chain. The basic control strategy is (s,S) policy in which ordering is done when the inventory levels goes below *s* [and orders are placed so that inventory is brought up to *S*.

Links: The link means the transportation route. The links effect the management at each agent, and the calculation of cost. The reason is that the link has the parameter lead time. The lead time from ordering agents to the clients has two features. The one is the almost route to delivery under 1 day. This is the transportation by the land route. This transportation is used to delivery from factory to distribution center or from distribution center to retailer. The cost by the land route is higher than others. This transportation has been not developing yet by reason of this. Another is that the lead time is over one day. This transportation is mainly sea route. The sipping distance by sea route is so longer than the one of land route. This transportation can ship the so many amount of inventory and can decrease the shipping cost by the packed many inventory into one. The cost of transportation par one inventory is cheaper than the cost by land route. The sea routes develop for long ago by the reason that low cost, massive transportation, and simple to maintain the equipment (the land routes need to maintain the all load. But, the sea routes need to keep up the ports and the ships.).

A Multi-enterprise in ASEAN

Asia in the current age is always growing up as the result of globalization, outsourcing and new strategy. Global company builds the factory in these places and opens up a new market in Asia. Developing areas and development one exist at the same time.

In ASEAN areas, the Mekong River in land area is developed. The reasons of this, these areas are the central of South-East Asia of shipping route. The sea route is developed and traditional route. The transportation of ships are able to delivery many inventory, to cost is less cheap than land route, and to maintain only ship. But in these days, ASEAN countries become rich by economic development. They make the land route to ship the inventory. This purpose is to short the lead time. This trend makes the network complex.

Disruption in Supply Chain

In South-East Asia, there is so large damage resulting from floodwaters. Despite there are many damage by floods, there is very little anti-disaster operations that deal with floods[11]. We should consider the network disruption by the disaster. Disasters bring the network disruption to supply chain network. For example, disaster makes the shutdown of factory, distribution center, and market. The streets stop the out of transportation function. In supply chain network, these mean that the shutdown of the nodes and the cutting of the links.

3 Model and Simulation

This simulation is to reveal the way of election the client when the demand fluctuation occurs at retailer agent. This simulation runs 3 times. We use the average data of 3 simulations. Each simulation has 1500 steps (this simulation define that one day is four steps. So, 1500steps means about one year.). The model using for simulation is consisted of two structures. First, the agent model structure is made by four kinds of agents which are factory agent (FC), distribution center agent (DC), and retailer agent (RT). The factory agent makes products when the amount of their inventory is under the order-point parameter. Ordered by distribution center agent, the factory agent ships the inventory. The cost of production is $10,000. The distribution center agent manages the inventory flow. This agent stocks, orders and ships the inventory to related agent. This agent ship the inventory to retailer ordered it. When their amount of inventory is under their order-point parameter, the distribution center orders the shortage inventory. In the retailer agent, the inventory is consumed by customer. The profit of 1 product is $20,000. When the amount of inventory is under the order-point parameter, the retailer agent orders the inventory. The delivery cost is $25 per one inventory by one step between each agent. The store cost is $1 per one inventory by one step in all agents.

The location of agents in this model is from the real location. In the following we describe the agent location in Figure 1 and Table 1. The data of FC location are based on TOYOTA motors. The locations of DC are the international port in Southeast Asia and the new land route junction. The cities of RT are the large city in Southeast Asia[12][13].

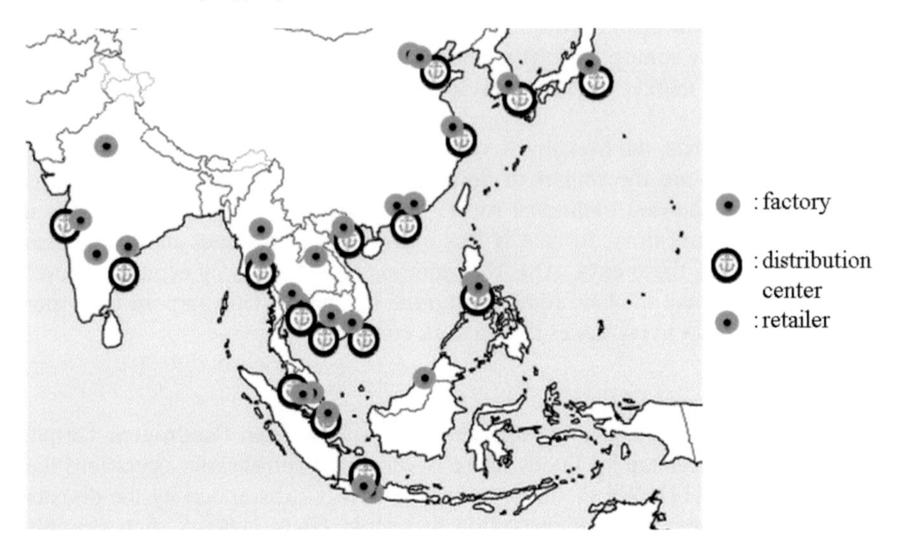

Fig. 1 The relation of the agents

Table 1 The city of agents

ID	FC	DC(departure)	DC(arrival)	RT
0	Kirloskar	Nhava sheva	Nhava sheva	New deli
1	Thilawa SEZ	Chen nai	Chen nai	Munbai
2	Bangkok	Thilawa SEZ	Thilawa SEZ	Chennai
3	KL	Leam chabang	Leam chabang	Nepido
4	Hanoi	sianukviles	sianukviles	Yangon
5	Jakarta	Ho chi minh	Ho chi minh	Bangkok
6	Hong kong	Hanoi	Hanoi	Vienchan
7	Beijing	KL	KL	Punonpen
8	Maynila	Singapore	Singapore	Ho chi minh
9	-	Jakarta	Jakarta	Hanoi
10	-	Beijing	Beijing	KL
11	-	Shanghai	Shanghai	Singapore
12	-	Hong kong	Hong kong	Jakarta
13	-	Maynila MITC	Maynila MITC	Burnai
14	-	Busan	Busan	Maynila
15	-	Tokyo	Tokyo	Beijing
16	-	land route Bangkok	land route Hanoi	Shanghai
17	-	-	-	Hong kong
18	-	-	-	Busan
19	-	-	-	Tokyo

The network in this model is constructed that the inventory flow is from FC to RT via DC.

Each agent makes transportation network with the link between each others. The link is drawn by the decision making of the agent. The agents make their decision by the calculation the order from the states of around them. The agents can see the parameter of next agents which are upper stream and down stream. They always calculate the safety stock (ss), the order point (op), and the amount of order (odr) to maintain their amount of stock[14][15]. These formulas of calculation follow as:

ss = service level × (the average of consume × lead time)
op = ss + (the average of consume × lead time)
odr = op - the amount of stock +(the average of consume × lead time)

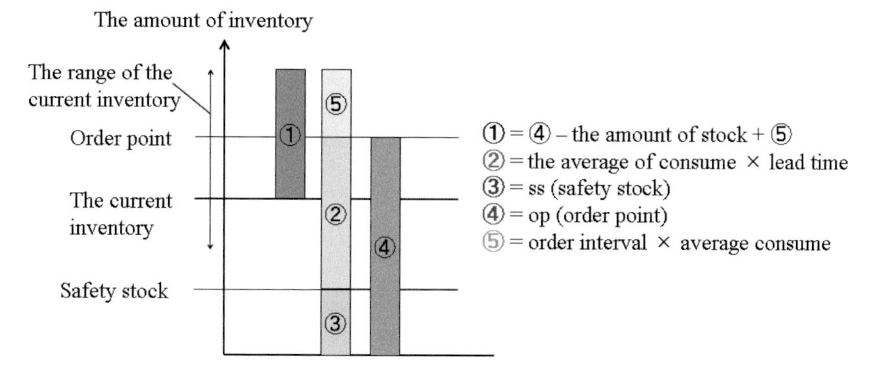

Fig. 2 The calculation of safety stock, order point and the amount of order

About the demand fluctuation in retailer, we use the two types of the demand value at the retailer in this model. These are based on the constant (cnst) and the log-normal distribution (lgnml). The reason of using the constant value is that comparing the way to elect the client under demand fluctuation with the only electing client simulation. We use the result of constant demand simulation as the standard elect client model. The constant demand in this model means that the same value through the all step. The value of consume by one step is 1. This means that the customer always buy one inventory at each of the retailer by steps.

When we simulate the model that the demand is based on the log-normal distribution, this value is given by

$$D_{t+1} = \mu + \rho_t D_t$$
$$\mu = 5$$
$$\rho_t = N(0, var), \rho_t \in (-1, 1), var \in (0.25, 0.5, 1.0)$$

The value of demand at the t step is Dt, The value of μ means the basement value of consume at the t step (This value is 5 in this simulation.), and the value of ρt represents the value from -1 to 1 following normal distribution (The average = 0, The variation is variable number among 0.25, 0.5, and 1.0 in this simulation.). Fig.1 shows the output of this formula on simulation. This means the value of demand fluctuation in this supply chain model (Simulation runs for 500 steps, the average of demand is 1.225).

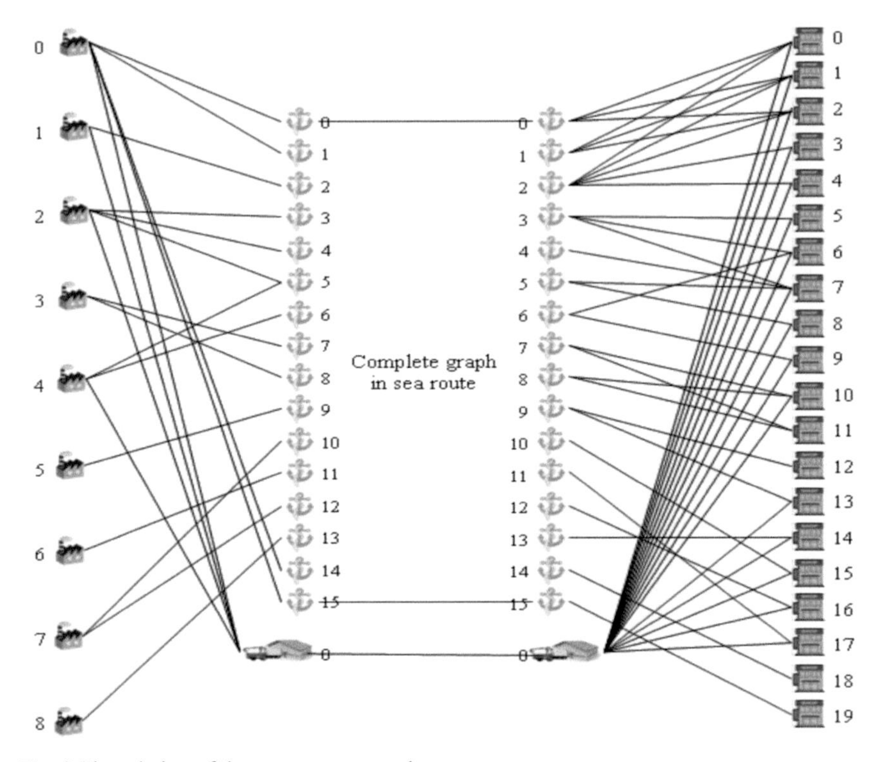

Fig. 3 The relation of the transport network

Another structure is the rule of how to decide making to manage supply chain network. In this model, this structure influences the choosing client for each agent. The clients which are elected by agents to order inventory have two features. First, the amount of inventory at the client change through time. Second, they are not the same lead time route in supply chain network. The agents have the three ways to elect their client. It is (1) random, (2) the same country, (3) the client which has the most inventory in all client candidate (shortest lead time), and (4) the client which has the shortest lead time in all client candidate (max stock). The agents which can deal with clients have this function. Figure 3 illustrates the shape of the network that the agents create. The numbers in the figure are the ID number of each agent.

Land Route: The junctions of land route are in Bangkok and Hanoi. The agents in continent are able to use the land route. In this model, the election of land route is one of the delivery methods. There are no agents to depend on the land route only in Southeast Asia. These land routes are built by a reference to [16] and [17].

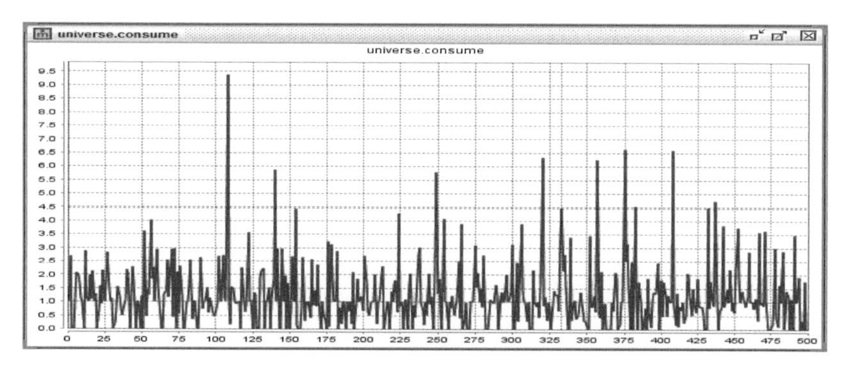

Fig. 4 The state of the demand Changes in market

Sea Route: The sea routes are 240 routes in this model (the number of the departure ports and the arrival ports are each 16 ports. This means that 16*15=240.). The all agent can use the distribution agent (port) to delivery the inventory. These values of transport dates by ship are based on the route via the Singapore port. The Singapore port is the most famous hub port in South-East Asia. So many ships go to their destination via Singapore port. The term of transport dates from arrival port to Singapore port is based on Jetro data. The term of transport dates from Singapore port to Arrival port is based on MAERSKLINE.

When the agent elect the client by the way of (1) random, (2) the same country, (3) the client which has the shortest lead time in all client candidate (shortest lead time), and (4) the client which has the shortest lead time in all client candidate (max stock) with the demand based on the function of Figure 4, the results of these simulation show table 2 and Figure 5.

Table 2 The average of stock and the probably of out of stock (OoS) at a retailer by simulation

Demand is constant		Random	Country	Lead time	Max stock
Only sea route	stock	3.0058	3.1922	2.7277	1.2775
	OoS (%)	14.86	0.09	43.92	66.68
Sea & land route	stock	3.3591	3.1708	2.2792	1.5001
	OoS (%)	13.76	1.39	51.62	64.20

Demand is based on log-norm		Random	Country	Lead time	Max stock
Only sea route	stock	4.3820	5.0035	3.4794	2.7260
	OoS (%)	15.07	1.68	42.96	48.86
Sea & land route	stock	4.8352	4.8876	3.0963	2.4181
	OoS (%)	3.42	3.10	44.73	51.88

In this model, the inventory calculates the profit and cost required at each location. When the transportations finish their role to delivery the inventory, the account calculates their profit and cost. The way to calculate is that accumulated profit = accumulated sales – accumulated cost. The Figure 5 shows the accumulated sales in each simulation by steps. We can see the difference of the increase rate in accumulated profit from figure 5.

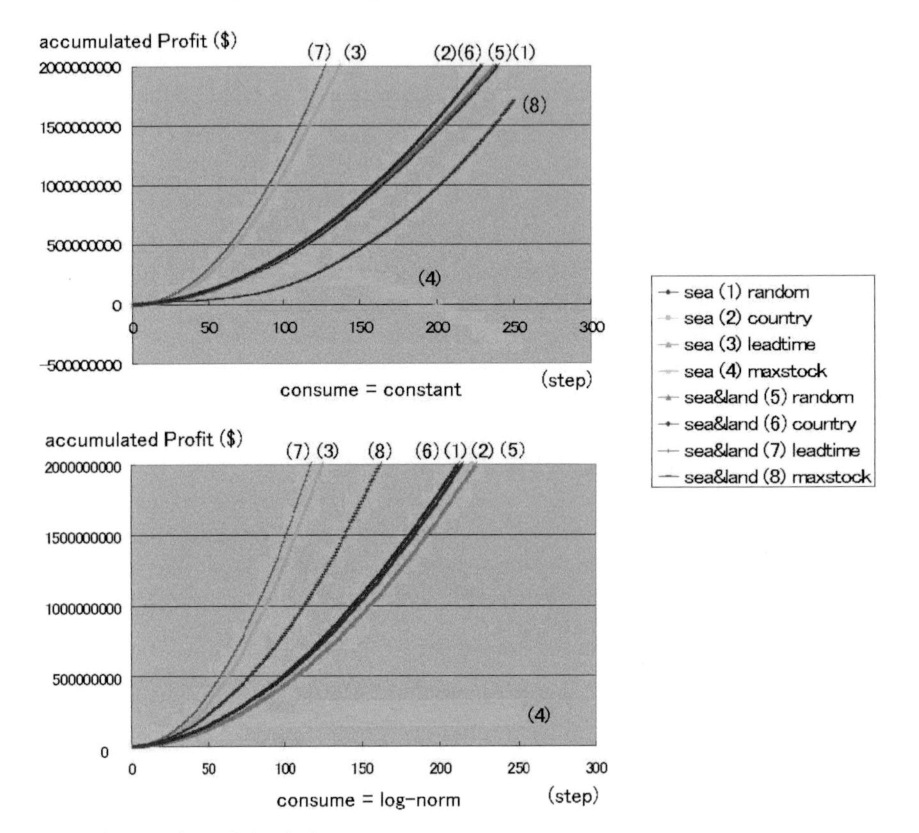

Fig. 5 The cost data of simulations

About the amount of stock, the ratio of the stock under the constant and log-norm distribution are no difference so much (Random:Country:leadtime:Max-stock≒4:4:3:2). When we choose the way to elect the max stock client, the amount of stock in all is the least in any other simulation. The reason is that two or three agents are working while changing and others are suspended. It is possible to reduce the amount of stock at the suspended agent. The way to reduce the amount of stock is to short the lead time of client.

About the out of stock, choosing the client with the way to elect the client randomly or the client in the same country, we can avoid out of stock better than other ways. The reason is the flexibility of choosing the client. The flexibility by random does not influence the previous decision making. The choosing new client which is not related previous client is the important element to make the network away from the previous influence most.

About the route, the results in 6 out of 8 simulations are that the amount of stock by sea route only is less than the one by sea & land route. Especially, the results under the demand based on log-norm distribution show that all network by sea&land route are less amount of stock than the one by sea route only.

About the cost, the result of “sea(1) and sea&land(5)”, “sea(2) and sea&land-(6)”, and “sea(3) and sea&land(7)” are resemble. “sea(4) and sea&land(8)” show the different lines. Choosing the client with the way to elect the client which has the shortest lead time in all client candidate (max stock), the profits under the constant demand are less than any other way. Especially, situation(4) is the least profit in all. The result of profit using the way to choose the max stock agent is that the value of profit under the log-norm distribution consume is more than the one under the constant consume. The reason is the difference in the average consumption. The average consumption under the log-norm distribution is more than the one under the constant. This value changes the result of the profit between these simulations. The way to get the biggest profit among all simulations is the choosing the client which has the shortest lead time in all client candidate (shortest lead time). The reason is that the short lead time decreases the cost of delivery and storage. When we choose the client which has the most stock in any other clients, the lead time is not the shortest in all.

4 Conclusion

These simulation results mean that the network made by the concept of short lead time is to reduce the amount of stock and cost, that we can avoid the out of stock at retailer by out of influence of previous client which is shortage stock (by random), and that we should make sure that a land route can also be used. We have some attention. The lead time by land route in real world is very short (The lead time by sea route is almost over 2 times of the one by land route). But the cost is very high. The problem which merit is better for the supply chain management depends on a making decision by manager.

5 Extension and Future Works

The proposed model in this paper is the application to help the decision making in the multi-plant supply chain problems. Especially, this model focuses the inventory flow from factory to retailer. If the model expands the region of flow, the future model is able to simulate the flow from row material to product in the customer hands. The future model which is made from this model can assist to make the supply chain management. The future model will be able to solute the optimal delivery route and location under various situations and the demand from the enterprise manager. When the future model get some function and expanded the data, this can correspond to demand fluctuation, disruption, and so many and different compromise plan.

This research reveals the potential of using the agent-based modeling for supply chain management. Decision making in supply chain network is effected of various types of agent, the amount of agent, and the relationship of each agents. The agent- based modeling approach is able to help the decision making to manage the supply chain under the dynamically network and environment. Considering unpredictability change and our request under the any situation, agent-based modeling approach reveals the optimal decision.

Acknowledgements The research is partially supported by the research fund, KAKENHI, 25330277.

References

1. Behdani, B., Zofia, L., Arief, A., Rajagopalan, S.: Agent-based modeling to support operations management in a multi-plant enterprise. In: International Conference on Networking, Sensing and Control (ICNSC 2009) pp. 323–328 (2010)
2. Fermando, D.M., Gonzalo, G., Antonio, E., Puigjaner, L.: An agent-based approach for supply chain retrofitting under uncertainly (2007)
3. Lee, H.L., Billington, C.: Managing supply chain inventory: pitfalls and opportunities. Slone Management Review **33**(3), 65–73 (1992)
4. Svoronos, A., Zipkin, P.: Evaluation of one-for-one replenishment policies for multiechelon inventory systems. Management Science **37**(1), 68–83 (1991)
5. Towill, D.R., Naim, M.M., Wikner, J.: Industrial dynamics simulation models in the design of supply chains. International Journal of Physical Distribution & Logistics Management **22**(5), 3–13 (1992)
6. Swaminathan, J.M., Smith, S.F., Sadeh, N.M.: Modeling supply chain dynamics. Decision Sciences **29**(3), 607–632 (1998)
7. Takuya, K., Ryo, H., Satohiko, M.: Let's make the environment to increase benefit with decreasing lead time (2011)
8. Snyder, L.V., Shen, Z.J.M.: Managing disruptions to supply chains. The Bridge (National Academy of Engineering) (2006)
9. Tomoyuki, M.: The problems and merit of JIT and Toyota manufacture. Ryukoku Business Review No.13 (2012)
10. John, A.M., David, H.M., James, A.R., Dwight, E.C.: Guidelines for collaborative supply chain system design and operation. Information Systems Frontiers **3**(4), 427–453 (2001)
11. Sodhi, M.S., Tang, C.S.: Buttressing supply chains against floods in Asia for humanitarian relief and economic recovery. Production and operations management **23**(6), 938–950 (2013)
12. JETRO.: New information about logistics in ASEAN and Mecon area (2013)
13. JETRO.: Thailand-Japan Cooperation and Prospect for Efficient Logistics Network in ASEAN (2008)
14. Kengo, K.: Supply Chain Model under emergency. Business review in university of Hyogo march 2012 No.1-2 (2012)
15. Mikio, K.: Logistics industry, Asakura Publish (2001)
16. Yasuhiko, O. (Ministry of Land, Infrastructure and Transport): Research of the logistics in ASEAN. Research Number.115, (2014)
17. Ryuichi, S. (Ministry of Land, Infrastructure and Transport): Building the dynamic simulation model of international marine container flow in East Asia (2009)

Low Cost Parking Space Management System

Azhan Ahmad and Somnuk Phon-Amnuaisuk

Abstract Managing parking lots usually involve tasks that should provide important information such as parked car counts, and available parking spaces and their locations. This can be used to direct drivers in real-time towards empty spaces which will minimise the time spent looking for one and thus reduce traffic congestions. Using an image-based integrated parking system is an effective way to automatically track a parking lot without exhausting time and manual resources. In this paper, we present a low-cost vision-based parking system to manage a closed area parking lot by using cameras that takes real-time footage of the parking lot. The footage is processed using HSV-based histogram technique and the resulting models are compared against pre-trained models. These models define either a Parked or an Empty class. The parking spaces within the processed footage are then categorised using this two classes based on their matching probability.

Keywords Car parking · Image processing · Probabilistic reasoning

1 Introduction

Multilevel parking lots found in offices, shopping centers or malls have a high chance of experiencing traffic congestion without proper management. This is normally due to cars being allowed in without considering the number of available parking spaces. Another contributing factor is that cars kept driving around looking for spaces in an already filled area. One obvious solution is to provide a clear count of available parking spaces which can control the influx of admitted cars, as well as minimising the amount of time a car spend driving around by directing them towards the locations of empty parking spaces. However, implementing this manually would exhaust time and increase labour resources which may detract parking administrators from adopting it. To overcome this problem, using an intelligent parking management system might be feasible in terms of cost. Intelligent parking systems normally provide integrated features of tracking incoming and outgoing cars, as well as identifying vacancy of parking spaces.

A. Ahmad(✉) · S. Phon-Amnuaisuk
Media Informatics Special Interest Group, School of Computing and Information Technology, Institut Teknologi Brunei, Gadong, Brunei
e-mail: {azhan.ahmad,somnuk.phonamnuaisuk}@itb.edu.bn

K. Lavangnananda et al. (eds.), *Intelligent and Evolutionary Systems*,
Proceedings in Adaptation, Learning and Optimization 5,
DOI: 10.1007/978-3-319-27000-5_31

In regards to parking space detection, most intelligent parking systems used various technologies such as sensor-based or image-based. Sensor-based systems use various types of sensors to detect parking spaces accurately, but are usually expensive and time-consuming to install. A single sensor can only monitor one single parking space [1]. Systems that uses image-based on the other hand can be quite effective, where a single camera can cover and monitor a wider area as opposed to a single sensor. By using image-based systems, the cost can be further brought down significantly and still provide a comparable accuracy as compared to sensor-based.

In this paper, we present a low-cost image-based parking system using an image processing technique. Models of Parked and Empty classes are learned from training data and employed to classify new input based on probability matching. This approach makes use of installed cameras to take real-time footages of every parking division in the parking lot. We then convert the footages using HSV model, after which we compare the resulting model against the trained models. The matching probability produced by the comparisons will determine the status of a parking space within the footage as either being *Parked* or *Empty*.

The paper is organised into the following sections. Section 2 gives an overview of related works. Section 3 discusses our proposed concept and gives the details of the techniques used. Section 4 provides samples of the output of our proposed system. Finally, the conclusion and further research are presented in section 5.

2 Related Work

Managing big parking facilities can be a challenge when it comes to providing smooth operations for drivers and administrators convenience. One of the usual problems found in these facilities involve traffic congestions, which are most likely caused when cars admitted into the facility exceeds the number of parking spaces available. Another contributing factor lies in the time spent by drivers driving slowly and stopping regularly looking for parking spaces [2]. The use of intelligent parking management system can definitely alleviate these issues by providing important data that can help manage these facilities efficiently. These important data include the number of empty parking spaces and their locations, as well as the number of cars are enough to control the admission of cars and reduce the amount of time spent by drivers looking for a free parking space by directing them to the empty spaces. Various researches have looked into this matter, using different approaches such as image-based system or using multi-sensor devices.

In [1], ultrasonic sensors were installed on the ceiling of every individual parking space of an indoor parking facility to detect the presence of vehicles. This system however would be costly to install especially in a huge multi-level parking facility. And since it uses the echo-location to determine if an object is present, the system does not seem to be able to distinguish vehicles or non-vehicular objects. In another system proposed in [3], RFID is used to allocate cars towards empty parking spaces. This requires vehicles to possess RFID tags on their cars, and might be more suitable for private parking facility such as those found in apartments or offices. Temporary RFID tag however can be given through parking tickets issued when entering the parking facility and make it a feasible approach for public parking facilities.

In terms of image-based parking detection system, it has the advantage for being low-cost, and able to perform the two required tasks of parking space detection and car tracking. Various image-recognition techniques have been explored. In this paper however, we focus on the parking space identification feature using image recognition techniques.

In one such implementation of using image-based parking space detection [4], recorded images of a parking area are processed by making use of mean square value and variance of the difference image, variance of the ratio of background and foreground, linear dependency of the background image and test image, and the marginal density of the image. Another existing system provides identifying vacant parking space in an outdoor parking facility [5] using multiple cameras. In this system, video footage of parking spaces captured by cameras installed on the facility is processed by using an edge based scheme and a color-based model. This system also makes use of two geometrical models (ellipses and grids) to define a parking space. These weighting of parking space is used to overcome the problem when a parking space is occluded from neighbouring parked cars.

One major concern regarding image-based parking detection system is the different lighting conditions. In [6], median filtering and Sobel edge detection are used to process shadows that might cause false detection of parking spaces. A more complex system was developed in [7], where it uses a 3D scene model to detect parking spaces. The direction of sunlight at a given time is determined, and a vehicle and shadow models are then simulated to create an intensity model. The parking area is then processed using classifiers generated from the intensity model to identify the state of parking spaces. In [2], the system creates an adaptive background to overcome lighting changes in an outdoor parking facility. The background of a parking space is dynamically created at regular intervals based on the color of the road near each space. The created background image is then used to extract the foreground image to determine whether it is vacant or not by determining the pixel differences. The system also employs a shadow detection mechanism which removes RGB color pixels in the foreground which are caused by shadows. A more robust implementation [8] used three processing techniques: edge pixel counting, object counting and foreground/background in- formation. Results from each of these techniques are then integrated into a final result using a majority voting rule. The system is shown to work well in both indoors and outdoors parking areas.

While most system uses multiple features in their approach to identify empty parking spaces, other literatures used only one set of features. System proposed by [9] makes use of edge detection to process camera footage of parking spaces and identify empty parking spaces.

3 Low Cost Parking Space Management System

3.1 Overview of the System

As illustrated in Figure 1, our proposed system used a single camera to monitor an area of 4-5 parking spaces. Position of the camera is fixed after installation and calibration, which allow the bounding region on every parking space to be easily set up and monitored. Results from the parking space detection subsystem

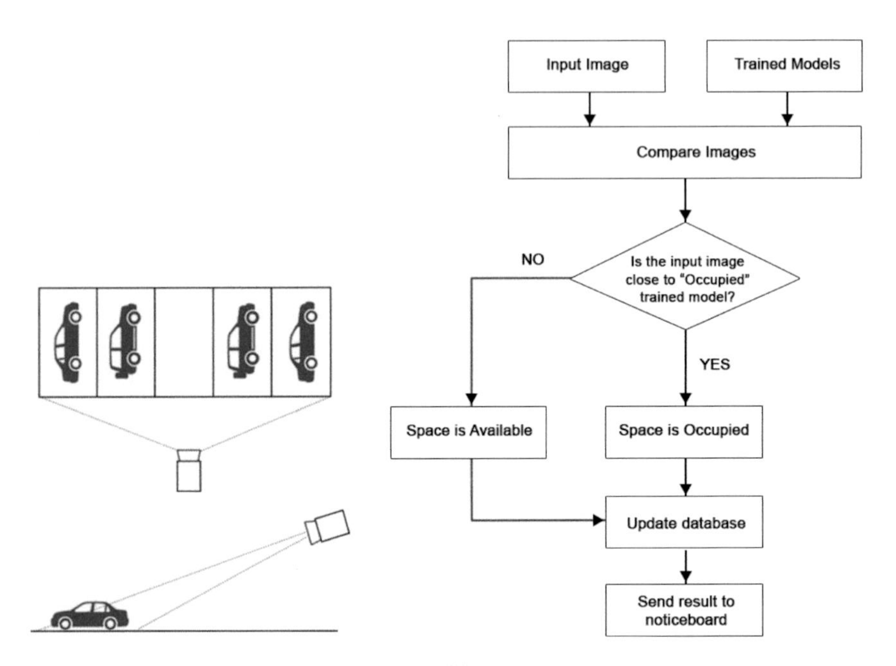

Fig. 1 The overall concept of the low cost parking management system

will be relayed to a display board to indicate the number of available parking spaces. Figure 1 (right pane) shows the overall flowchart of the system.

3.2 Parking Space Detection Subsystem

It seems reasonable to approach the detection of an empty parking space by designing the system according to the target operating condition. Although a system that handles various operating conditions can be implemented with a higher cost e.g., open parking space, indoor parking space, etc., it does not seem to be a fruitful approach.

We propose a system that works on indoor parking spaces using image processing technique. The parking space areas are well located and low cost cameras are fitted such that a row of 4 to 5 parking spaces can be captured within a single view. This means 1000 parking spaces only require up to 200 low cost cameras.

The detection, based on an image processing technique, is fast and inexpensive. We expect to update the information of empty parking spaces to the display board twice a minute. Figure 2 highlights the main process in the parking space detection system.

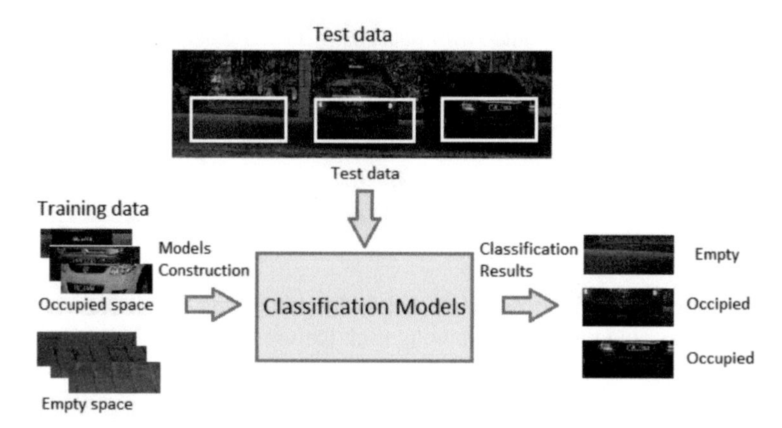

Fig. 2 The overall concept of the parking space detection subsystem

Knowledge Representation. Let us define a rectangular area of each parking space on a 2D image as a region of interest (ROI). Each ROI is centered at point $p = (x, y)$ and let us define the region using a feature vector $R = \{R(u)\}_{u=1..m}$ where u is the bin index number. In this implementation, the vector $\boldsymbol{R}$ is constructed from the hue value of pixels in ROI. The probability of feature $R(u)$ in a model can be expressed as:

$$R(u)_p = C_h \sum_{b(x_i)=u} k(\|\frac{x_i - p}{h}\|^2) \tag{1}$$

where C_h is the normalisation factor such that $\sum_u R(u) = 1$, $\{x_i\}_{i=1..n}$ denotes target pixel locations; $b(x_i)$ denotes the bin index of pixel at x_i; and $k(.)$ is a Gaussian kernel (similar to [10]).

Model Construction Let a model **M** be a tuple (μ, σ^2) describing the mean and variance of the features vector **R**. In our domain, two types of models are constructed; one for an empty parking space $\mathbf{M_e}$ and the other for an occupied parking space $\mathbf{M_o}$. Let N be the number of available training examples of empty parking spaces (as well as the occupied parking space). We compute the mean and variance as follows:

$$\mu_e = \frac{1}{N}\sum_{i=1}^{N} \mathbf{R}_e^i; \text{ and } \mu_o = \frac{1}{N}\sum_{i=1}^{N} \mathbf{R}_o^i; \tag{2}$$

$$(\sigma_e)^2 = \frac{1}{N}\sum_{i=1}^{N}(\mathbf{R}_e^i - \mu_e)^2; \text{ and } (\sigma_o)^2 = \frac{1}{N}\sum_{i=1}^{N}(\mathbf{R}_o^i - \mu_o)^2; \tag{3}$$

Classification using MDC and NDBC We implement two classifiers *Minimum Distance Classifier (MDC)* and *Normal Density Bayes Classifier (NDBC)*. MDC classifies a new observed region **X** having the feature vector **R** by measuring the distance between the feature **R** and the features μ_e and μ_o of the models M_e and M_o. The observed space, **X**, is classified as empty parking spaces if

$$\|\mathbf{R} - \mu_e\| < \|\mathbf{R} - \mu_o\| \tag{4}$$

MDC classifies a new observation solely by comparing the distance but does not take into account the variance among each feature u. A NDBC classifies a new observation **X** as empty spaces if

$$P(\mathbf{M}_e|\mathbf{X}) > P(\mathbf{M}_o|\mathbf{X}) \tag{5}$$

$$where \quad P(\mathbf{M_e}|\mathbf{X}) \approx \frac{1}{\prod_{u=1}^{m}\sigma(u)_e} e^{[-\frac{1}{2}\sum_{u=1}^{m}(\frac{X(u)-\mu_e(u)}{\sigma_e(u)})^2]} \tag{6}$$

4 Experimental Results and Discussion

4.1 Experimental Design

In this preliminary study, the dataset consisted of 100 images, 50 were positive examples (empty parking spaces) and 50 were negative examples (occupied parking spaces). The images were prepared by photographing the actual parking site on both a sunny day and a cloudy day. Due to the small data size, 90% of the training samples were randomly selected from the dataset and the remaining 10% were used for testing. Here we selected 45% from each class for the training and 5% from each class for the testing. This process was repeated 20 times, so there were a total of 100 classification results for each class. The accuracy reported is averaged over 10 repetitions of 20 runs (200 runs in total).

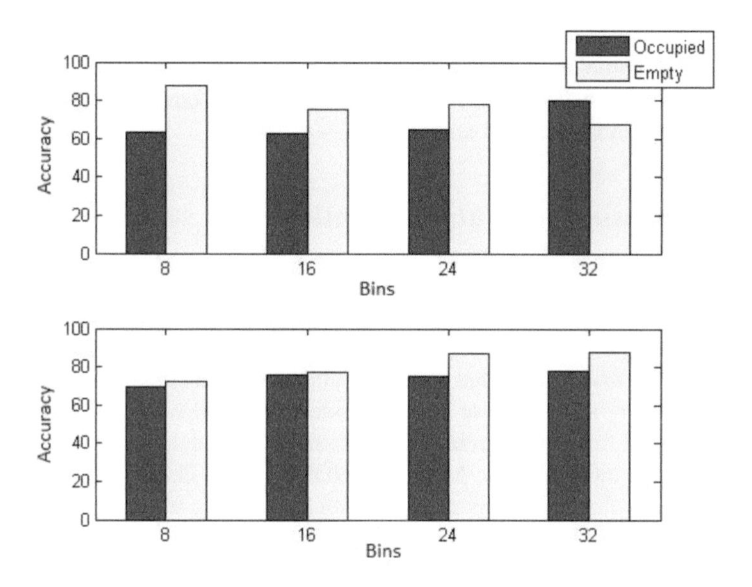

Fig. 3 Bar plots of classification results from different number of bins. Top pane: results from minimum distance classifiers (MDC). Bottom pane: results from normal density Bayes classifiers (NDBC).

4.2 Analysis and Discussion

The feature employed in this experiment was a hue value since it was robust to changes in brightness. Four different bin numbers were employed: 8, 16, 24 and 32 bins. Figure 3 and Table 1 summarise the experimental results. The classification results are affected by the number of bins. In general, more bins seem to improve the average accuracy of the system. However, the rate of increment may not be worth the computational cost.

Table 1 Classification results from different number of bins, the values reported here are averaged over 10 repetitions.

Model	Bins	Occupied	std	Empty	std	Average accuracy
MDC	8	63.2	4.9	87.9	2.7	75.5
	16	62.5	3.7	75.3	3.4	68.9
	24	64.8	4.6	78.1	4.8	71.5
	32	79.8	5.1	67.3	5.9	73.5
NBC	8	69.7	4.2	72.4	3.9	71.1
	16	75.5	2.8	77.5	2.8	76.5
	24	75.4	3.7	87.2	2.7	81.3
	32	77.8	5.4	87.6	2.5	82.7

The NDBC classifiers show a better accuracy as compared to MDC (82.7% and 75.5%). This could be from the fact that the MDC did not incoporate variance information in its process. The NDBC on the other hand employed this knowledge in its process, see Equations 4 and 6.

5 Conclusion and Future Directions

We have developed a low-cost parking space management system that is specifically tailored for indoor parking areas. This system allows a single low-cost camera to monitor one parking strip which can contain around 4-5 parking spaces. Trained models are first created by using sample images consisting of vacant and occupied parking space. These trained models are then compared against input images captured from the camera. Two classification models are used to determine the state of the input images; MDC and NDMC. Both classification models are computationally fast and inexpensive, and we have shown to provide a satisfactory average accuracy of 75.5% and 82.7% respectively.

However, one drawback of this system can be attributed to the limited captured area of a single camera. While there are other systems available that allows the use of a single camera to monitor wider parking areas, it has to accommodate the issues of occlusions such as between cars or other overlapping objects. Another area of concern is regarding different illumination settings. While the system does not adapt to changing in lightings, it can however be set up to work under specific lightings during installation.

In future work, other sub-systems can be incorporated such as vehicle identification through license-plate recognition. This can be applied for various applications in a more complex intelligent management system, such as theft avoidance or car tracking in huge multi-storey parking facilities. Parking spaces can be allocated with id numbers via the system, and parked cars can easily be associated to these id numbers. This can provide navigation assistance by directing vehicles towards available parking spaces.

References

1. Kianpisheh, A., Mustaffa, N., Limtrairut, P., Keikhosrokiani, P.: Smart parking system (SPS) architecture using ultrasonic detector. International Journal of Software Engineering and Its Applications **6**(3), 55–58 (2012)
2. Lin, S.F., Chen, Y.Y., Liu, S.C.: A vision-based parking lot management system. In: Proceedings of the IEEE International Conference on Systems, Man, and Cybernetics (SMC 2006), vol. 4, pp. 2897–2902, October 2006
3. Wei, L.X., Wu, Q.H., Yang, M., Ding, W., Li, B., Gao, R.: Design and implementation of smart parking management system vased on RFID and internet. In: Proceedings of the International Conference on Control Engineering and Communication Technology (ICCECT), pp. 17–20, December 2012

4. Deng, W., Luo, X.Q., Jiang, L., Luo, Y.W.: Research on video-based monitoring algorithm of parking space. In: Proceedings of the Third International Conference on Multimedia Information Networking and Security (MINES), pp. 261–264 (2011)
5. Chen, L.C., Hsieh, J.W., Lai, W.R., Wu, C.X., Chen, S.Y.: Vision-based vehicle surveillance and parking lot management using multiple cameras. In: Proceedings of the Sixth International Conference on Intelligent Information Hiding and Multimedia Signal Processing (IIH-MSP), pp. 631–634, October 2010
6. Bong, D.B.L., Ting, K.C., Lai, K.C.: Integrated approach in the design of car park occupancy information system (COINS). IAENG International Journal of Computer Science **35**(1) (2008)
7. Huang, C.C., Wang, S.J.: A hierarchical Bayesian generation framework for vacant parking space detection. IEEE Trans. Circuits and System for Video Technology **20**(12), 1770–1785 (2012)
8. Liu, J.Z., Mohandes, M., Deriche, M.: A multi-classifier image based vacant parking detection system. In: Proceedings of the IEEE 20th International Conference on Electronics, Circuits and Systems (ICECS), pp. 933–996, December 2013
9. Banerjee, S., Choudekar, P., Maju, M.K.: Real time car parking system using image processing. In: Proceedings of the 3rd International Conference on Electronics Computer Technology (ICECT), vol. 2 (2011)
10. Fukunaga, K., Hostetler, L.D.: The estimation of the gradient of a density function with applications in pattern recognition. IEEE Transactions on Information Theory **21**(1), 32–40 (1975)
11. Phon-Amnuaisuk, S.: Classify event-related motor potentials of cued motor actions. In: Proceedings of International Conference on Neural Information Processing (ICONIP 2008), pp. 232–239 (2008)

Campus Access Control and Management System

Mei Jun Voon, Sy Mey Yeo and Nyuk Hiong Voon

Abstract Advanced centralised access control (CAC) systems are widely deployed on campuses as a means to provide security and track movements. There is concern that some campuses are not using such protective systems, hence this paper attempts to resolve this weaknesses in such institutions' by developing a simpler CAC system using the Radio Frequency (RF) and contactless smart card technologies. The scope of the developed system is not only limited to access control but also to utilise the gathered data to automate and potentially to support other processes of the institution, such as lecture scheduling and attendance tracking.

Keywords Access control · Radio Frequency · Contactless smart card · Multifunctional

1 Introduction

Access control strategies have always been a necessity in university campuses. There has always been a need to limit access to sensitive areas and protect resources and assets in various scenarios and locations, including classrooms and special laboratories. Not only that, unfortunate incidents such as thefts, vandalism of campus property, cars being broken into [1], and even fatal assault case [2] had happened in local universities. This paper introduces a multifunctional centralised access control and management system using Radio Frequency (RF) and contactless smart card technologies. Campus security is the prime candidate for such technologies as the nature of campuses are easy access by a relatively high number of people in a place. The proposed system will benefit universities with no access control in place or are deploying standalone system in various strategic

M.J. Voon · S.M. Yeo(✉) · N.H. Voon
School of Computing and Informatics, Institut Teknologi Brunei,
Jalan Tungku Link, Gadong, Brunei Darussalam
e-mail: Symey.yeo@itb.edu.bn

K. Lavangnananda et al. (eds.), *Intelligent and Evolutionary Systems*,
Proceedings in Adaptation, Learning and Optimization 5,
DOI: 10.1007/978-3-319-27000-5_32

areas on their grounds. Standalone systems have limitations in their features and one main weakness is their ineffective ability to trace back individuals who gained access into the rooms, thus making investigations on campus grounds difficult. A background study to explore the various technologies such as barcodes, QR codes, Bluetooth, biometrics, RFID, contactless smart cards, and NFC, was carried out by reviewing relevant academic journals, papers and online technology articles. Research showed that the best possible technology to use for the implementation is the contactless smart card. The focus of this paper is to highlight the design and development of a Centralised Access Control (CAC) system for campus environment that is also capable of fully utilising its gathered data for other useful tasks such as lecture attendance automation, room booking and scheduling, and real time occupancy validation.

The rest of the paper is organised as follows: Section 2 provides the background research of relevant leading technologies in today's CAC systems and their deployment in universities. Section 3 describes the research methodology. Section 4 gives an overview of the implementation of the prototype system. Section 5 identifies potentials of the developed system and Section 6 provides the conclusion.

2 Background Research and Related Work

Ever since the concept of electronic access control system first appeared, it has been tested and implemented with various different technologies. Different approaches come with their own set of advantages and disadvantages. The following discusses prevailing technologies:

- Barcodes and Quick Response (QR) codes
  Both are technologies that require direct line-of-sight which could cause some delays especially when users have difficulty positioning the codes properly for scanning. Although they are low-cost access control solution, they are low-security technologies as the codes can be duplicated very easily.

- Bluetooth
  Several commercial products such as Kevo Smart Lock and EC Key which turn Bluetooth enabled devices into a key are already on the market. In 2014, HID Global announced the completion of a mobile access control pilot featuring Bluetooth Smart technology at Vanderbilt University [3]. The main concern of this approach is the battery consumption of the user credential. In order for users to travel fast and conveniently through the entry points, their Bluetooth are encouraged to be turned on and remain on discoverable mode at all times. This drains the battery of the mobile devices and backup plans should be considered in cases of mobile devices failure or devices running out of battery.

- Biometrics
  Biometrics provides highest form of security because it eliminates specific credential devices, thus providing access control that cannot be transferred unlike

keys or cards. However, with the high security it provides, it comes with a high cost for implementation as well. Deploying biometrics for campus security in areas with large amount of users and high traffic might not be wise as its nature of authentication might cause bottlenecks at entry points.

- Radio Frequency Identification (RFID) Tags
  An increasingly prevalent technology since 1970s as the technology becomes more affordable. One of its major advantages is the fact that it does not require line of sight and is capable of high read range, therefore making it one of the best candidates for object identification and tracking. However, using RFID tags that operates at high frequency and therefore having high read range will not be able to prevent tailgating. So ideally, tags operating at low frequency (LF) would be a better choice for access control systems.
- Contactless smart card
  Contactless smart cards employ radio frequency between card and reader which requires no physical insertion of the card as reading is done by passing it along the exterior of the reader. These cards conform to the ISO14443 standard, with variations of type A, B, and C. Equipped with the memory storage and ability to encrypt make these cards an ideal option for applications that require certain level of security. Santander's smart card is an example of using contactless smart cards for secure applications in educational institutions at a large scale [4].
- Near Field Communication (NFC)
  NFC is an emerging technology that has enabled smartphones to be used as user credential. A pilot program involving the deployment of NFC in access control system was done at Arizona State University in 2011 by HID Global [5]. But the lack of standardisation among cell phone carriers, handset manufacturers and security manufacturers is the biggest obstacle to the adaptation of the technology [6].

A brief comparison of the discussed technologies is shown on Table 1.

Table 1 Comparisons between technologies

Technology	Cost	Security	Read Range	Power consumption
Barcodes and QR codes	Inexpensive	Very low	Line of sight required	None
Bluetooth	Lower than NFC	Medium	approx. 10 m (Class 2)	High
Biometrics	Very High	Very high	Contact	None
RFID tags	Low	Low	Variable, up to 100 m	Depends on type of tag
Contactless smart card	Low	High	< 10 cm	None
NFC	Higher than Bluetooth	High	10 cm or less	Low

To date, CAC systems are already in use by many campuses around the world. These systems usually employ multifunctional contactless smart cards as user credential. Aside from serving as official ID and access cards, they provide access to other on-campus services and facilities, and are also used as electronic wallets. ONEcards from University of Alberta [7] and TigerCard from Princeton University [8] are existing examples of such applications.

There has been several research projects [9], [10] which are focused on student attendance automation, and these authors have opted for using RFID technology for the implementation. Using a technology that is only capable of providing low security and high read range locks the proposed system out from the potentials of integrating various services that might require very secure transactions in the future, such as access controls and electronic cash applications. This project furthers the scope of these works by integrating not only smart attendance automation, but also lecture and room scheduling services to a CAC system, using a different approach by emphasizing on new use of the system's gathered data.

Ononiwu and Nwaji (2012) [11] had done similar research work. Their project was done with a low frequency RFID reader with a hardware motor unit to simulate an automatic door. A time attendance management system was also developed using visual basic.Net. The management system, however, was limited to only three major functions – showing attendance, adding and deleting users. The system did not provide for higher level of access control management; the door unit grant access to any registered users in the database and was also not designed for managing multiple doors. It did not have the ability to relate attendance to lessons as well.

3 Research Methodology

The research stage of the project involves reviewing relevant academic journals, papers, technology articles as well as commercial products, to study the various technologies used to develop modern digital access control systems in order to come up with alternate design options for the proposed system. Next, a relevant university is identified to conduct a survey to gauge the interests of multifunctional student smart card and access control system on campus, as well as the possibility of integrating NFC as user credential.

The short survey, consisting of twelve questions, is designed to solicit responses from university students, as they make up the largest portion of the system's end users. It is distributed randomly and a total of eighty-five university students responded; ranging from freshmen to seniors, across all departments of ITB. The rationale for picking ITB as the study bed is due to the fact that ITB fits the profile for universities which can benefit from such a system and also due to its proximity to the author.

An analysis is then conducted upon the gathered survey responses. The findings suggested high interests in the deployment of an access control system (75.3%) and use of multifunctional student smart card (84.7%), especially in electronic cash on campus (82.3%). More than half of the students (62.4%) also think that not being able to check their own attendance online is an inconvenience. The data also revealed that 60% of the students do not own an NFC device. These findings ultimately influenced the design rationale of the final product, which is based on the contactless smart card technology.

4 Implementation

An Internet Protocol (IP) reader is installed at every entry points on the campus and all of them are connected to a TCP/IP network via LAN cables. An Ethernet switch is used to connect all of them to a host computer. By scanning a MIFARE contactless smart card at the IP reader, the host computer will decide whether to grant or deny the user's request accordingly. Aside from handling authentication processes, the host computer is also responsible for running an attendance tracker (i.e. a scheduled PHP script) daily at midnight to mark attendances for all lessons that are conducted during the day time. The web-based management system runs on an online web server so all users can gain access through the Internet.

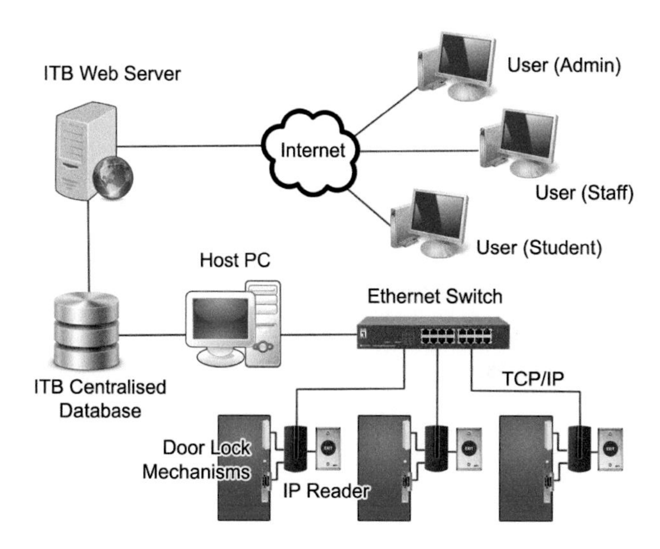

Fig. 1 Architecture Design of System for Intended Deployment

4.1 Access Control System

The access control consists of RFID reader which operates at 13.56 MHz and smart cards that conform to the ISO 14443A standard. The access control program is written in Java language and employs JDBC as middleware to facilitate the communication between the program and MySQL database when it is needed for authentication processes. The authentication processes verify that the presented card is authorised to the system, the user has sufficient privilege to enter the restricted area and the access request is made within the allowed hour of access.

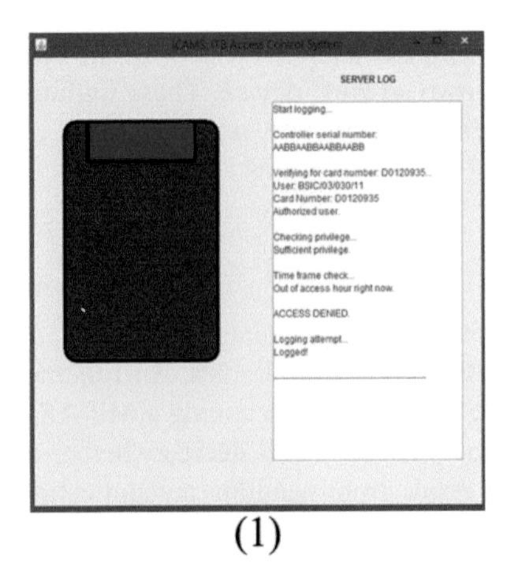

(1)

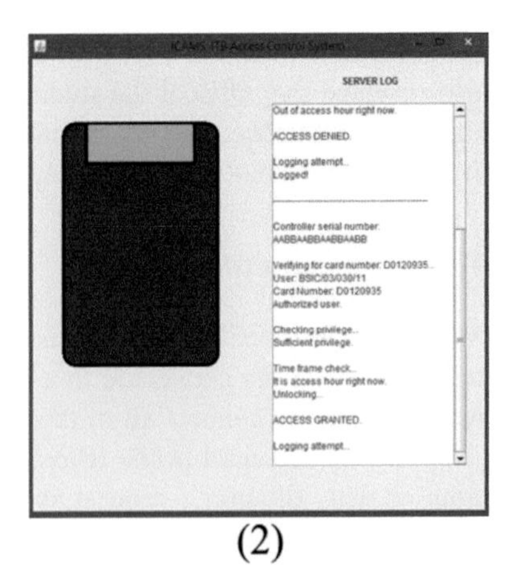

(2)

Fig. 2 Screenshot of the access control program: screen (1) shows a request denied due to request being requested outside of access hour and screen (2) shows an access request being granted because it met all the access criteria.

4.2 Features of the Management System

There are three categories of account types: - administrator, staff and student. The following describes all features available to the administrator account holders:

- Card management feature
  This feature allows staff to assign and deactivate access cards from students and staff members, as well as registering new blank cards into the system. To quicken the otherwise time-consuming card management procedures, the functions available under this feature are capable of smart and automatic card assignment or deactivation by batch. This help support high reusability of the student cards.

- Access management feature
  The purpose of this feature is to allow administrator to adjust reader's IP and access hour settings, user privileges by group or individual as well as tracking access logs.

- Semester management feature
  Administrators may provide details of the current semester (e.g. starting dates of semester breaks) as well as assigning staff members to handle lectures for modules of different courses. It is extremely important to provide these essential information accurately as these data will be used to assist auto lecture scheduling and lecture attendance tracking.

Other features include:

- Schedule Management feature (for staff accounts)
  Pre-schedules, reschedules or cancels lectures efficiently. Pre-schedule is capable of smart auto lesson scheduling for staff according to provided criteria (e.g. lecture day and number of lectures) by making use of the important semester dates provided by the administrators to avoid scheduling lessons on public holidays and semester break. Lesson schedules viewable by involved students through their personal accounts.

- Real time occupancy validation (for staff accounts)
  Allow staff to validate the occupancy of a specific location in real time, in cases of last minute cancellations of booking schedules which are not reflected in the system. Staff can also see the current and upcoming booking details of that room.

- Attendance feature (for both staff and student accounts)
  For staff, an overview of class attendance for lectures that have been conducted up-to-date is presented by each module they are teaching. The report can also be exported in Excel spreadsheet format. For student, they may view their accumulated lecture attendance for all modules they are registered in.

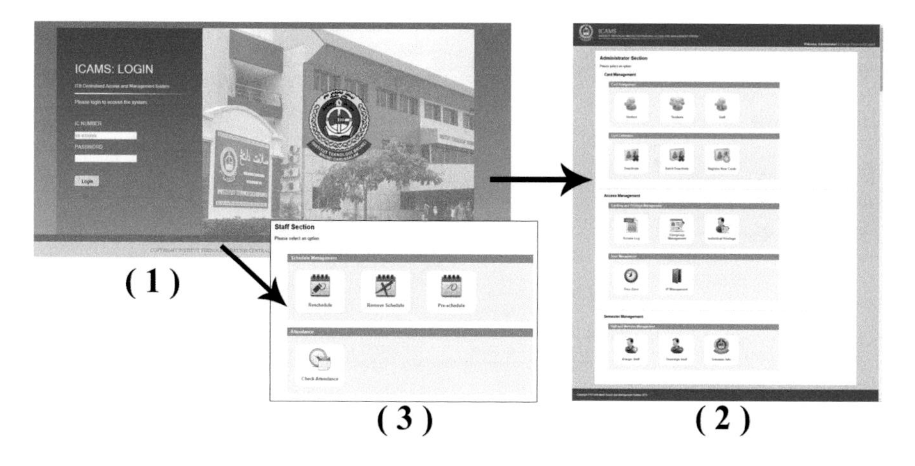

Fig. 3 User is requested to login at the index page (1) before they can access to the home page (2) of the system. Features that are available to the user changes accordingly to the user's account type. The administrator accounts are allowed to adjust various system settings and configurations, and perform administrative management. The staff accounts (3) are essential for schedule management. Attendance tracking for whole class and individual student are accessible by both staff and student accounts respectively.

5 Results and Analysis

Rather than implementing a specific system for attendance registration as suggested by many previous research projects [12],[13], the developed system marks attendance by making use of the access logs by running the attendance tracker script the end of the day. Since the database contains all lesson schedules arranged by staff members as well as students' registered modules, the attendance tracker identifies a list of lessons that were conducted during the day before identifying the lists of students that are involved for each lesson. Then for each student, the attendance trackers tracks if they have created any access logs at the lesson venue between one hour before and after the lesson starts to mark their presence for the class. Employing an attendance tracker at the end of the day to automate attendance marking is efficient and also makes good use of the server during its most idle time because access traffic on campus would be low to almost none during midnight.

The highlight of the developed CAC system is in its infrastructure design to maximise the use of its gathered data, hence enabling its scope to reach beyond just campus security, including not only smart lecture attendance, but also room booking, validation of room occupancy, facility management, and other data possibilities; Since this is analysing data already gathered, there should be no significant operational costs increment.

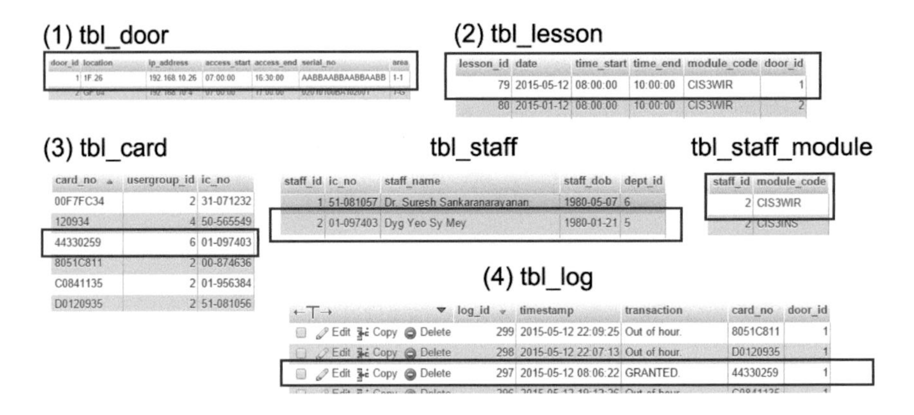

Fig. 4 Data simulation of the validation of occupancy feature: User checks for real-time occupancy of lecture room 1F.25 at 8 AM. The system checks for the door ID of that room is '1' (1) and identify if there is a lesson scheduled at that time. Schedule stored in database reveals that CISWIR module is scheduled at 8 AM that day (2). System then checks for the details of the lecturer who booked the room (3), and found out that CISWIR is taught by staff with ID of '2', who goes by the name of 'Dyg Yeo Sy Mey', holding the ID card with number '44330259'. The system then proceeds to validate the immediate occupancy of the room by checking if the teaching staff member is already in the room by going through the access logs (4). In this example, the staff member has made an access at 8.06 AM to the room, and therefore the system deduces that the room is already occupied. If no access log was found, the system would deduce that the room is not occupied so it is free to be used by the requesting user.

Fig. 4 illustrates the validation of room occupancy by analysing live data. Since all details of lessons (i.e. venue and time) are necessary to be kept in database to support smart lecture attendance, these data can also be used to illustrate occupancy timeslots of lecture rooms. This offers a good overview at one glance to staff members, making room booking a more convenient process. The availability of rooms, especially laboratories equipped with special facilities, can also be confirmed in real time by the use of access logs. The system validate the occupancy by going through the access logs of a specific room to determine if the teaching staff who booked the room for the timeslot beforehand is already present at the venue. This feature would be useful in the event of any last minute lesson cancellations which are not reflected in the system.

The access logs can also be utilised to support facility management. Maintenance workers or teams (e.g. air conditioning technicians, I.T. technicians) can be given another account to distinguish them from other users of the CAC system. The system can then track when was the last visits these maintenance workers came to perform maintenance work around the campus, and remind the administrators when it is time for the next maintenance work to be scheduled.

The potential of this system can be expanded beyond the examples given above. The introduction of multifunctional student smart card also opens up opportunities for integrating other services on campus such as integration with the library management system, canteen food voucher management system, campus parking management services and e-payment management services.

The proposed system is validated by a small group of 6 students and 2 staff members. The prototype consisted of only one USB RF reader connected to a laptop which was running the Java program to simulate the door locking mechanism and log for data collection. The reader was set with the serial number that was assigned to a specific door on campus. Students were distributed with a contactless smart card and asked to interact with the system. The staff members were also asked to perform various tasks on the management system with their own accounts. Feedbacks from the testers were used as inputs to refine mainly the user interface of the system. The testing session validated that the system was performing as expected, and was able to handle the good amount of data that was collected in the database properly.

6 Conclusions

Implementing a CAC system can bring in numerous benefits to a university, including making the campus a smarter and more secure space. This paper has discussed on how to maximise the utilisation of the data gathered by a CAC system to integrate several useful features and services quite efficiently in terms of its stored data. The fact that the management system is web-based makes it very flexible as additional functions can be incorporated into the system with ease. Selecting smart cards as user credentials is not only a security strategy to reduce tailgating, but also to allow for future applications that require secure transactions. A multifunctional smart card would surely relief the burden of students and staff having to carry several different cards at the same time. The system, however, is still quite simple and therefore prone to trickeries such as proxy attendance and tailgating issues. The system could be bolstered with facial recognition and biometrics to address such weaknesses.

Acknowledgements This paper is developed by an undergraduate (first author) from a final year project in the Bachelor of Computing, School of Computing & Informatics, Institut Teknologi Brunei.

References

1. Ismail, H.: Handbag reported stolen from UBD student's car. Borneo Bulletin, March 8, 2012. http://www.brusearch.com/news/107443 (accessed: July 10, 2015)
2. Hamit, R.: Students want UBD to strengthen security. The Brunei Times, December 23, 2014. http://bt.com.bn/frontpage-news-national/2014/12/23/students-want-ubd-strengthen-security (accessed: July 10, 2015)
3. Security Today: HID Global, Vanderbilt University pilot uses mobile access control featuring Bluetooth smart technology (2014). http://security-today.com/articles/2014/10/06/hid-global-vanderbilt-university-pilot-uses-mobile-access-control-featuring.aspx (accessed: July 11, 2015)
4. Santander UK: Santander smart card in the UK. Satntander UK (n.d.). http://www.santander.co.uk/uk/santander-universities/smartcard (accessed: July 11, 2015)
5. HID Global: HID Global completes NFC mobile access control pilot at Arizona State University (2012). http://www.hidglobal.com/press-releases/hid-global-completes-nfc-mobile-access-control-pilot-arizona-state-university (accessed: July 11, 2015)
6. Gray, R.H.: Is NFC the future of access control? (2011). http://www.campussafetymagazine.com/article/nfc-the-next-step-in-access-control/P2 (accessed: July 11, 2015)
7. University of Alberta: ONEcard (n.d.). http://onecard.ualberta.ca/ (accessed: July 11, 2015)
8. Princeton University: Welcome to Tigercard! (n.d.). http://www.princeton.edu/tigercard/ (accessed: July 11, 2015)
9. Saparkhojayev, N., Guvercin, S.: Attendance Control System based on RFID-technology. International Journal of Computer Science Issues **9**(3), 227–230 (2012)
10. Patel, R., Patel, N., Gajjar, M.: Online Student's Attendance Monitoring System Classroom Using Radio Frequency Identification Technology: A Proposed System Framework. International Journal of Emerging Technology and Advanced Engineering **2**(2), 61–66 (2012)
11. Chiagozie, O.G., Nwaji, O.G.: Radio Frequency Identification (RFID) Based Attendance System with Automatic Door Unit. Academic Research International, 163–182 (2012)
12. Arulogun, O.T., Olatunbosun, A., Fakolujo, O.A., Olaniyi, O.M.: RFID-Based Students Attendance Management System. International Journal of Scientific & Engineering Research **4**(2) (2013)
13. Man, M., Law, Y.K.: TITO: utilizing MYKAD touch n go features for student attendance system. In: Proceeding of 1st International Malaysian Educational Technology Convention 2007, pp. 114–120 (2007)

Lip-Reading: Toward Phoneme Recognition Through Lip Kinematics

Ak Muhammad Rahimi Pg Hj Zahari

Abstract Heuristic parameters such as width and height are usually obtained in audio-visual speech recognition. However, the presence of noise has an impact on such system. In the paper, we present a mathematical study investigating whether descriptive parameters derived from lip shapes can improve the performance of the system through the use of a mathematical model. The video database used consists of five separate pronunciations of the numbers ranging from 0 to 9. Three categories of data have been successfully classified; the polynomial coefficient (curving of the lips), width and height (both inner and outer) and also the raw data (coordinates). The results showed that the best classifier is the curving of the bottom lip contour with an accuracy of 90.91% and the weakest classifier is from points on the right upper lip contour with accuracy of 12.24%.

Keywords Noise · Mathematical model · Polynomial coefficients · Classifier

1 Introduction

In the presence of audible noise, the performance of speech recognition systems becomes degraded. One method of reducing the effects of noise in such systems is to make use of visual information that can be obtained from the speaker and in particular the movements of the lips. Lip reading is therefore seen as a supporting process to speech recognition where its application in stand-alone process ranges from the use of mobiles phones in health application to video surveillance use for security. However detecting the lips has become more and more challenging, because of large difference between people in shapes, existence of facial hair, head movement and lighting. To be able to make use of the visual information, features derived from the lip should be extracted and therefore lip models needs to be built.

Ak.M.R.Pg.Hj. Zahari(✉)
Institut Teknologi Brunei, Jalan Tungku Link, Gadong, Brunei Darussalam
e-mail: rahimi.zahari@itb.edu.bn

K. Lavangnananda et al. (eds.), *Intelligent and Evolutionary Systems*,
Proceedings in Adaptation, Learning and Optimization 5,
DOI: 10.1007/978-3-319-27000-5_33

In this paper, we proposed a mathematical model of the lips that will be developed by obtaining measurements to estimate both the static parameters that are peculiar to individual speakers and the dynamic changes in these parameters that occur when specified words are uttered. Once a database of parameters has been developed, the models will be used to identify words spoken by further speakers and the results compared with existing approaches in the literature. A publicly available corpus of speakers will be used as well as a new high-definition corpus that is currently being established at Loughborough University. Although mathematical models of lips have been developed by previous researchers (and these may be re-used in this study), there appears to be no previous use of such models for audio-visual speech recognition.

The text is organized as follows: In Sect. 2, previous related works are studied as well as the introduction of the proposed method with classification techniques and in Sect. 3, discussion of the results. Lastly, conclusion of this work will be addressed with further suggestion in Sect. 4.

2 Lip Kinematics

In noisy environments, humans are able to reduce speech recognitions errors by using the speaker's lip movements and indeed many people with hearing difficulties rely on lip reading to provide majority of the speech information they receive. Most of the recent methods for extracting lip contours are based on image segmentation and color-based information of the lip region. However, a lip template can be used to describe lip contour and several curves and special points are employed to approximate actual lip shape in order to obtain geometric feature of the lip shape. Various lip models may be found in the literature. In [1], the lip is made up of two contours, outer and inner. Both of them are described by curves using nonlinear least square methods. In [2], combination of two semi-ellipses was proposed with the employment of 16-point geometric deformable model and initialization of evolving curves. In [3], the introduction of parameterized key points has proven to be the most important aspects for lip movement recognition. These points will help to calculate the height, width and area. Lastly in [4], multiple points' representation of the lip was introduced and the classifications of the words were implemented using Euclidean Distance.

2.1 The Proposed Method

The proposed method consists of three different phases with the use of a selected video database. The schematic depicted in Fig.1. Mainly, first phase deals with the extraction of parameter; second phase concerns with classification techniques and the last phase discusses and evaluates the classification results.

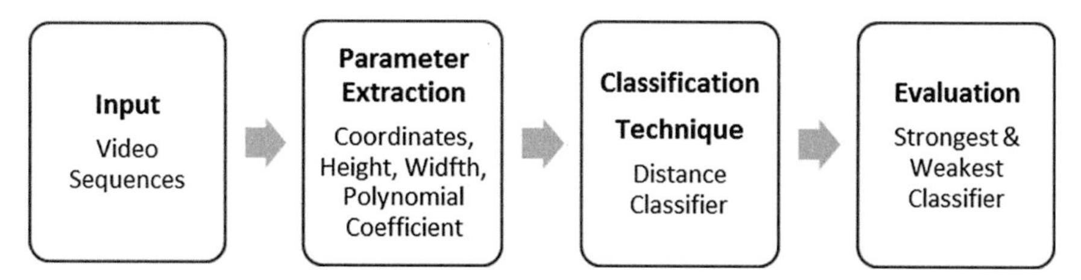

Fig. 1 Schematic Diagram of the proposed method

2.2 *Video Images Database*

In this paper, the video sequences are obtained from [5]. This has been made using a high definition camcorder. Selection of video was based on the success detection rate of the mouth region. Moreover, it consists of pronunciation of numbers ranging from '0' to '9' and each has 5 different sets labelled as 'a, b, c, d and e'. These sets will be used as comparison means.

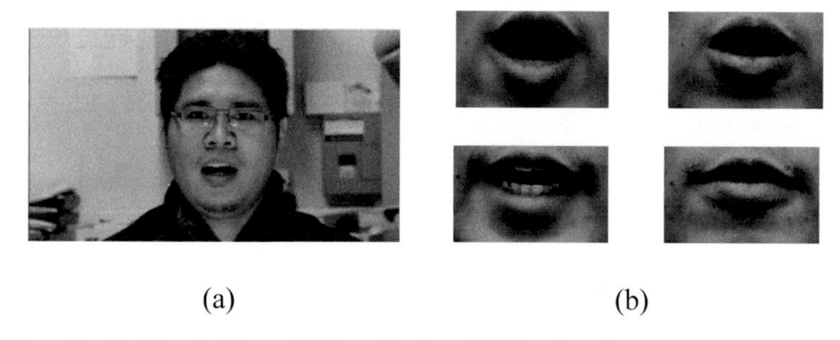

(a) (b)

Fig. 2 Sample of video database (a) Face Region (b) Mouth region

2.3 *Parameter Extraction*

In extracting the parameter, we proposed the idea of manually getting the outline. This implies that the user needs to choose the right point on the image. The implementation of the automatic method has shown its weaknesses in getting the right contour. For the method, specific functions have been made which corresponds to the name containing built-in MATLAB functions which perform image processing tasks. Coordinates from every frame and video will have to be selected and these videos will have a slightly different amount of frames. Manual Identification of the coordinates via user selection takes around 15-30 minutes for each video as each frame needs to be processed, and this also includes the processing time for the arrangement of the result into a database. Through the study of the mathematical model in the literature, we implemented a 21 point model as seen in Fig. 3 and the coordinates are in Table 1.

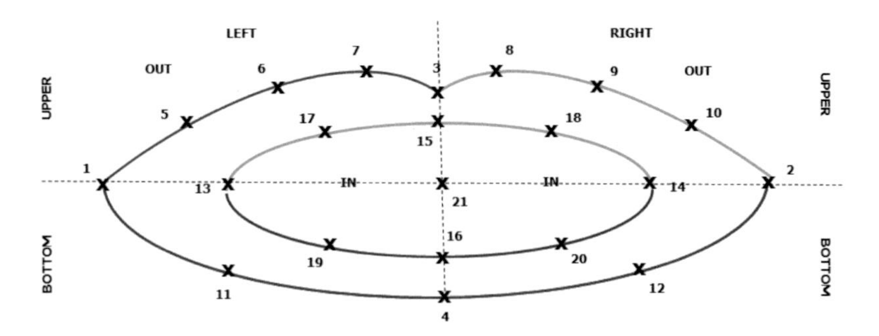

Fig. 3 21-point geometric proposed model

Table 1 Proposed coordinates

1	Outleft	8	UpperOutRight	15	UpperInMid
2	OutRIght	9	UpperOutRight1	16	BottomInMid
3	UpperOutMid	10	UpperOutRight2	17	UpperInLeft
4	BottomOutMid	11	BottomOutLeft	18	UpperInRight
5	UpperOutLeft	12	BottomOutRight	19	BottomInLeft
6	UpperOutLeft1	13	InLeft	20	BottomInRight
7	UpperOutLeft2	14	InRight	21	Mid

These coordinates were used to identify five more additional parameters; Outer and Inner height, Outer and Inner width and lastly, polynomial coefficient. Both height and width are calculated using Euclidean Distance based on Eq.1. Suppose we have two coordinates (x1, y1) and (x2, y2), the distance, D is as follows:

$$D = \sqrt{(x_1 - x_2)^2 + (y_1 - y_2)^2} \quad (1)$$

In order to get the polynomial coefficient, we implemented a 'least squares' method [6]. Given the coordinates, this method will minimize the squared error between the set of measured data and the curve. We then use the result to compute a 2 degree polynomial, i.e. quadratic polynomial. The order of polynomial relates to the number of turning points that can be accommodated. In this case, we eventually come across a turning point.

2.4 Classification Techniques

The obtained data will have some sort of patterns between each other. It is very important to develop proper methodologies to organize them. The idea behind this classification is to assign a class to an unknown or unknown pattern based on previously acquired knowledge about the objects and the classes to which they belong. However , designing such a pattern recognition system is usually an interactive process that involves the selection and computation of features from the objects that needs to be classified and the numerical data, for instance collections of feature vectors often necessary needs to undergo pre-processing before they can be inputs to any classifier. We have chosen a simple method known as minimum distance for the classification [7]. Assumptions were made to which each time when the test data set is applied to the training data it will give minimum two results. Prior knowledge suggests each of the data used has two correct sets. The process will remove any presence of duplication. As each video has five different sets, it can be divided as three training sets and two tests sets.

The classes that have been considered are divided into three categories:-

- Polynomial coefficient
- Width and Height
- Raw Data (or, Coordinates)

In this paper, the only preprocessing technique being applied is normalization method based on Eq. 2 which will result in the number of frames ranging from 0 to 1. This allows easier comparison of results between each of the videos. Suppose that we have a certain number of frames in a video, i. The variable which contains these values can be represented as f. Thus,

$$NormalizedValue(f_i) = \frac{f_i - f_{\min}}{f_{\max} - f_{\min}} \quad (2)$$

Where

$f_{\max}$ - Maximum number of frames

$f_{\min}$ - Minimum number of frames

Other than normalization, the obtained data is assumed to be correct, and this can be checked in the classification process. The proposed classification is illustrated in Fig. 4.

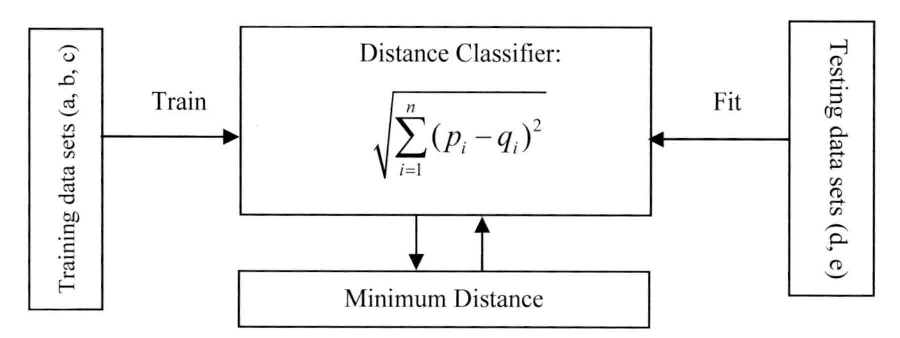

Fig. 4 Chosen classification method

2.5 *Evaluation*

To evaluate the results of the classification, we studied only the proportion of the total number of predictions that were correct through the use of confusion matrix [8]. The motivation behind this is to check which parameter has a significant change in the pronunciation of numbers and thus can be used to improve the accuracy of detection. The results will undergo two phases shown in Table 2: Phase I is aimed at whether it can be classified correctly to its respective data, basically to classify data correctly and Phase II suggests without the existence of a specific data set will it be able to classify that data from the rest of the data or successfully classify incorrect data. The classifier also can be categorized into two; the strongest classifier and the weakest classifier. The strongest classifier can be found by searching the maximum result of accuracy in the Phase I and also a minimum accuracy in Phase II. The reason behind the maximum suggests that the classifier manage to classify the data correctly whereas the minimum in Phase II shows that the classifier can classify unwanted data. However the weakest classifier will have the opposite characteristics; a minimum in Phase I and a maximum in Phase II.

Table 2 Proposed Coordinates

Phase	Training Sets (a, b, c)	Phase	Training Sets (a, b, c)	Test Sets (d, e)
I	0	II	1 till 9	0-9
	1		0,2-9	
	2		0-1,3-9	
	3		0-2,4-9	
	4		0-3,5-9	
	5		0-4,6-9	
	6		0-5,7-9	
	7		0-6,8-9	
	8		0-7,9	
	9		0 till 8	

3 Results

3.1 *Behavior of the Lip*

In dealing with such model, basic definition and terminology are quite useful. One of them is the coordinate system which for an image in MATLAB is different from a normal graph. The y-axis is reversed and the axes started from 1 instead of 0. Moreover, units used are in unit pixel. Video '1a' was chosen as an example to illustrate the movement of the lip uttering the number '1'. The video consists of 13 frames altogether. The analysis consists of three parts, the changes in the width, changes in height and as well as the changes in the polynomial coefficients.

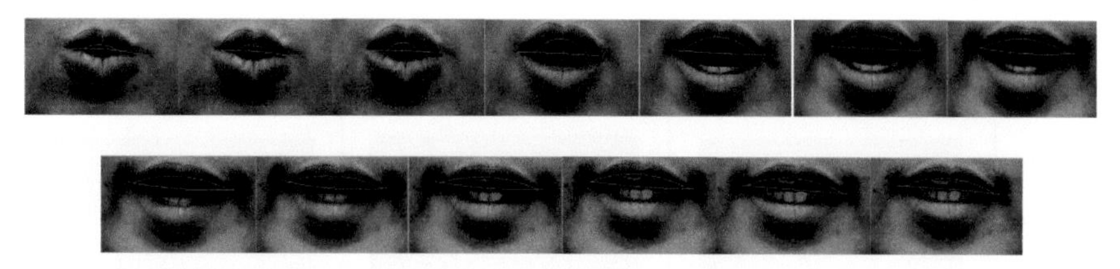

Fig. 5 Pronunciation of number '1' (top left to bottom right)

3.2 Changes in Parameter

The behavior of the width and height are seen in Fig. 6. Results suggest that both of the width decrease to a minimum and then slowly increase to a steady state. This implies that the mouth is in the state of protrude. In other words, the lip extends forward which gives pressure in pronouncing 'wuu'. And later the increase is when the lip stretches out, uttering 'unn'.

On the other hand, the height behaves oppositely where the height increases to a maximum; this is the effect after the mouth utters 'wuu'. It then returns to its original state.

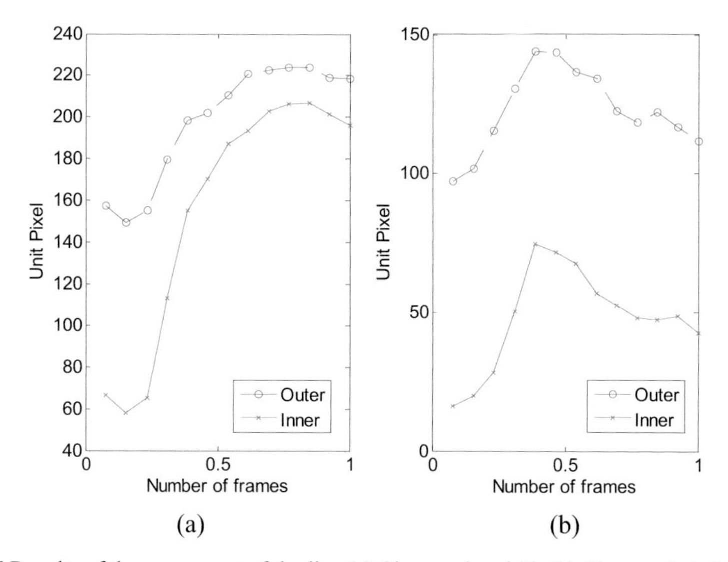

Fig. 6 Results of the movement of the lips (a) Changes in width (b) Changes in height

Aside from the two parameters, the effect of polynomial coefficients was investigated where only the y-intercept has been analyzed. This is due to visible large change in values. Curve representations and observation are illustrated in Table 3.

Table 3 Changes in the polynomial coefficient based on the respective curve

Video Sequence	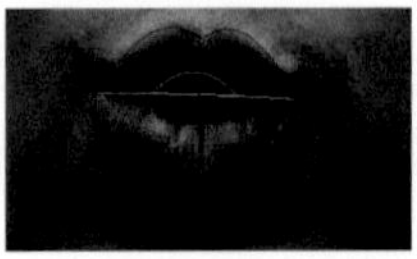		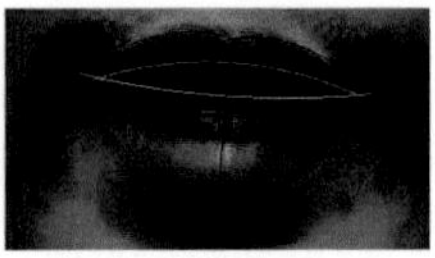
Curve 1 (solid line)\ Curve 2 (dahsed line)	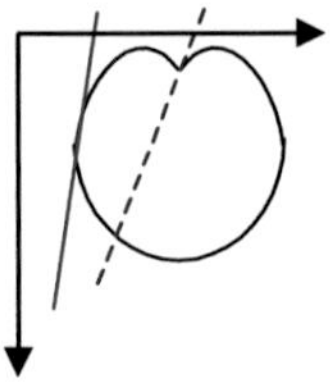	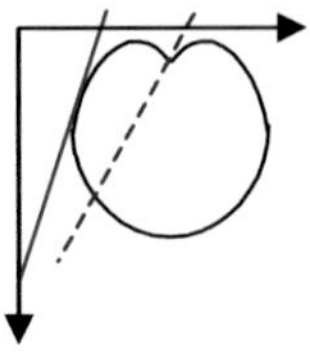	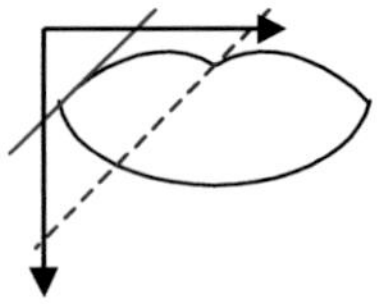
Curve 3 (solid line) Curve 5 (dahsed line)	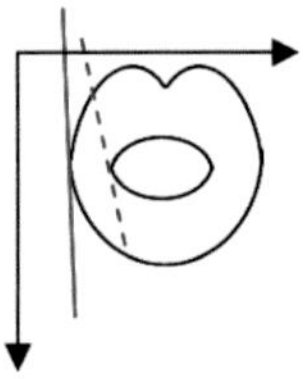	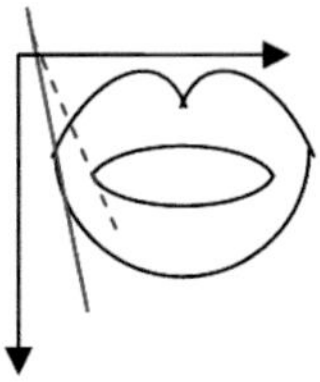	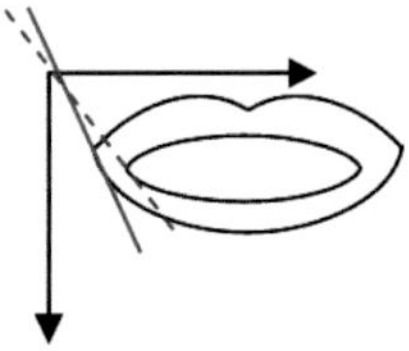
Curve 4 (solid line)	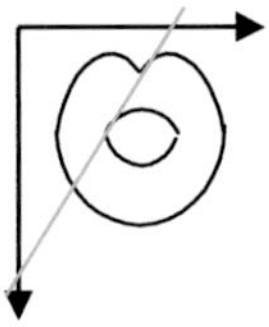	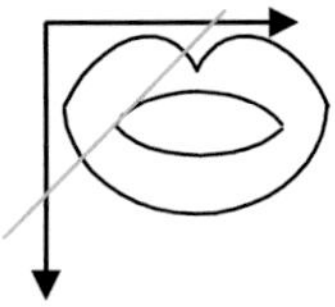	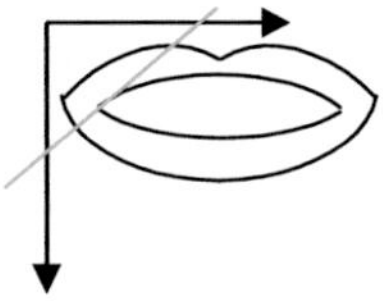

3.3 *Discussion*

The summary of the classification results based on only the strongest and weakest classifier from each of the three categories is shown in Table 4.

Curve 3, representing the bottom outer lip contour was found to be strongest classifier because of the highest percentage of accuracy compared to the other classifier within all of categories. This implies that each of the data sets has its own unique behavior for this specific coefficient, illustrated in Fig. 7. An example of the training dataset, i.e. 1b is plotted together with both test sets 8d and 1d. Although it seems that they behave in a similar way, the Euclidean distance between 1b and 1d has proven to be smaller than of 8d. Thus it has been successfully classify the data. Other datasets are not shown due to obscure result.

Table 4 Summary of the classification results

	Classifier	Accuracy (%)	
		Phase I	Phase II
Strongest	Curve 3	90.91	1.10
↓	Inner Height	83.33	1.64
	YBottomInMid	72.73	3.61
	XInLeft	72.00	2.84
	Outer Height	67.86	5.00
	Curve 2	64.29	4.49
	YUpperOutRight	20.00	9.52
Weakest	XUpperOutRight	12.24	10.27

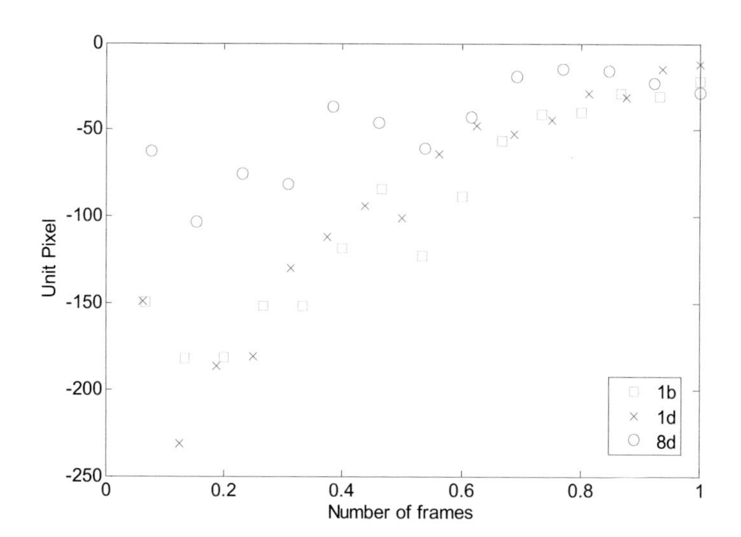

Fig. 7 Comparing of data resulting from the pronunciation of '1' and '8'

Constant movements of the lips up and down have major effect on the width and height. Results suggest that coordinate InLeft has a high accuracy, especially in the x-direction. This statement is supported by the coordinate which define the width (marked as 'o' in Fig. 8a). InRight, OutLeft and OutRight coordinates have high accuracy as well. The movement in y-direction however, relates to the impact is caused by the changed in height. The Y-Coordinate for UpperInMid and UpperOutMid (marked as 'o' in Fig. 8b) have lower accuracy compared to them as when the mouth is moving, only the bottom part of the lips played a major role. UpperOutRight coordinates, which are connected to the curving of the upper lip is the weakest classifier within the three categories and one factor might be because of the orientation of the image.

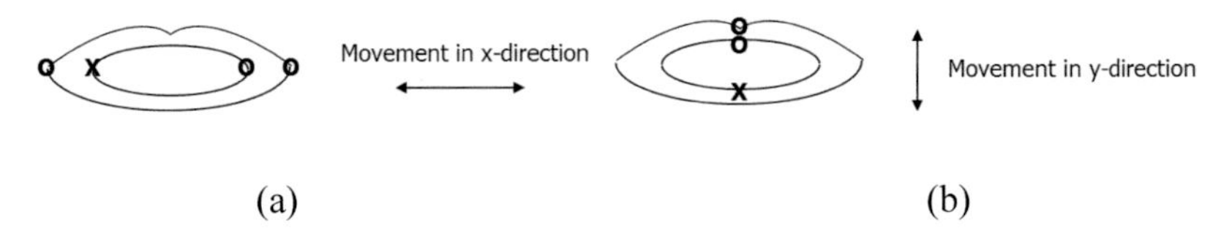

(a) (b)

Fig. 8 Illustrations of the lip (a) coordinate InLeft (b) coordinate BottomInMid

4 Conclusion

This paper presented a mathematical study of lip-reading parameters. We successfully implemented a method for extracting such parameter allowing them to be analyzed and classified depending on the training and test data sets used. The exact coordinates are located and tracked depending on the user selection. The system has been tested on the database that contains pronunciations of '0' to '9'. We have also discussed the result of the classification and the result has shown that the bottom outer lip contour is the strongest classifier.

Simple functions as well as the understanding of the different variations of behaviours of the lips are discussed throughout the paper which can aid in the distinguishing different between uttering of words. As future work, we intend to explore a higher degree of polynomial for the curve fitting, better classification methods as well as implementation of 3-Dimensional model.

References

1. Liu, H.: Study on lipreading recognition based on computer vision. In: Proceedings of the 2nd International Conference on Information Engineering and Computer Science (2010)
2. Liu, X., Cheung, Y.: A robust lip tracking algorithm using localized color active contours and deformable models. In: Proceedings of the IEEE International Conference on Acoustics, Speech and Signal Processing (ICASSP), pp. 1197–1200 (2011)
3. ur Rehman Butt, W., Lombardi, L.: A survey of automatic lip reading approaches. In: Proceednigs of the Eighth International Conference on Digital Information Management (ICDIM 2013), pp. 299–302 (2013)
4. Yargic, A., Dogan, M.: A lip reading application on MS Kinect camera. In: IEEE International Symposium on Innovations in Intelligent Systems and Applications, IEEE INISTA, pp. 1–5 (2013)
5. Ibrahim, M.Z.: A novel lip geometry approach for audio-visual speech recognition (2014)
6. Chi, E.C., Scott, D.W.: Robust Parametric Classification and Variable Selection by a Minimum Distance Criterion. Journal of Computational and Graphical Statistics **23**, 111–128 (2014)
7. Essenwanger, O.: Curve Fitting. Wiley StatsRef: Statistics Reference Online (2014)
8. Bowden, R., Cox, S., Harvey, R., Lan, Y., Ong, E.J., Theobald, B.J.: Recent developments in automated lip-reading. In: Proc. SPIE 8901, Optics and Photonics for Counterterrorism, Crime Fighting and Defence IX; and Optical Materials and Biomaterials in Security and Defence Systems Technology X (2013)

Object Matching Using Speeded Up Robust Features

Nishchal Kumar Verma, Ankit Goyal, A. Harsha Vardhan, Rahul Kumar Sevakula and Al Salour

Abstract Autonomous object counting system can help industries to keep track of their inventory in real time and adjust their production rate suitably. In this paper we have proposed a robust algorithm which is capable of detecting all the instances of a particular object in a scene image and report their count. The algorithm starts by intelligently selecting Speeded Up Robust Feature (SURF) points on the basis stability and proximity in the prototype image, i.e. the image of the object to be counted. SURF points on the scene image are detected and matched to the ones on the prototype image. The notion of Feature Grid Vector (FGV) and Feature Grid Cluster (FGC) is introduced to group SURF points lying on a particular instance of the prototype. A learning model based on Support Vector Machine has been developed to separate out the true instances of the prototype from the false alarms. Both the training and inference occur almost in real time for all practical purposes. The algorithm is robust to illumination variations in the scene image and is capable of detecting instances of the prototype having different distance and orientation w.r.t. the camera. The complete algorithm has been embodied into a desktop application, which uses a camera feed to report the real time count of the prototype in the scene image.

Keywords Object counting · SURF · Feature Grid Vector · Feature Grid Cluster · SVM

1 Introduction

Object counting is a common task performed in many industries, research laboratories, and other organizations. The information obtained from counting the number

N.K. Verma · A. Goyal · A.H. Vardhan · R.K. Sevakula(✉)
Department of Electrical Engineering, Indian Institute of Technology Kanpur, Kanpur, India
e-mail: srahulk@iitk.ac.in

A. Salour
The Boeing Company, St. Louis, MO, USA

K. Lavangnananda et al. (eds.), *Intelligent and Evolutionary Systems*,
Proceedings in Adaptation, Learning and Optimization 5,
DOI: 10.1007/978-3-319-27000-5_34

of objects is used to make quantitative and qualitative analysis. Counting objects manually is a tiresome and time consuming task, especially when the number of objects is large. Automating the process of counting could increase the efficiency of the overall process and also saves time and resources. Such automated object counting have a wide range of applications for e.g. in industrial manufacturing process for production, quality control and inventory management, and in medical research facilities for cell counting in digital microscopic images.

Lempitsky et al. [1] proposed a method for counting number of cells in a microscopic image or number of people in a surveillance video frame using supervised learning approach. In their method, density of objects i.e. number of objects per pixel was estimated by transforming density estimation into an optimization problem. Object count was finally obtained by integrating estimated density over the whole image. Barbedo et al. [2] performed object counting in nebular chambers by analyzing edges of chamber lines and objects and then determining the regions of interest. All edges were found using Sobel method. Guo et al. [3] proposed an automatic cell counting method using histogram dual-threshold to separate the background, followed by blob analysis and K-mean clustering to detect the cells. Fabic et al. [4] performed object counting for the estimation of fish population. For this, they perform Canny edge detection on underwater video frames to obtain fish contours and then employ blob counting to obtain the fish count. Mazei et al. [5] presented a method to analyze surveillance video and find the number of objects that enter or exit the given area. They perform background subtraction on the video frames and apply binary thresholding. The obtained binary image is used to determine blobs which represent objects and counting them gives the object count. In [6] Rabaud et al. tried to estimate the number of moving object in a scene by tracking feature trajectories.Pornpanomachai et al. in [7] proposed a system based on image processing to find the number of moving objects in scene. The system requires a background without the moving objects and the scene with moving objects.

In our previous work [8], counting was done matching the fuzzy color histogram of the prototype image and scene image to obtain the region of interest. Mean square error (MSE) values between the pixels of scene image and that of the prototype image are calculated within the region of interest. Local minimas among these MSE values were selected and two filters namely suppress filter and dip filter were applied on them to reject the false alarms. Performing these steps gives the location and the count of instances of the prototype present in the scene image. This method gave very accurate results for situations when alignment and scale of the objects present in the scale image were similar to that of the prototype image. The same was not true when objects with differential scales and alignment. The method proposed in this paper overcomes these drawbacks. Also it computationally cheaper and gives results in faction of a second.

Scale Invariant Feature Transform (SIFT) [9] and Speeded Up Robust Features (SURF) [10] are local feature descriptors extracted at interest points in an image. These descriptors can be used to detect extent of matching between two images or for the detection of a particular object in an image in which many other objects are present. For a particular object or pattern in an image, the features extracted corresponding to them should be similar even though they are extracted in different scale, illumination and noise in order to perform accurate recognition. SURF and SIFT are well known feature descriptors which gives features that are invariant to scale, rotation and illumination variation. These feature descriptors are used to perform tasks like image retrieval, recognition, mosaic etc. SIFT is more appealing as feature descriptor, but extraction of SIFT points in an image is compuatationally expensive. SURF on the other hand gives descriptors which are almost as robust as SIFT descriptors alongwith being computatuionally cheaper than SIFT [10]. For this reason SURF is preferred to SIFT. There are some really good works making use of SURF features [11, 12]. Juan et al. [12] proposed an algorithm to match scene using SURF. Huijuan et al. [11] proposed a fast image matching algorithm using improved SURF. There have been several other works which have used SURF for object classification and image stitching.

In this paper SURF is used to detect and count the instances of the prototype in the scene image. SURF points are extracted in both the prototype image and the scene image. Most relevant SURF points are selected from those detected in the prototype image and the rest discarded. These selected SURF points are matched to the ones detected in the scene image and then size filter is applied to remove the impossible matches. The concept of Feature Grid Vectors (FGV) and Feature Grid Cluster (FGC), which are introduced in preceding sections, is used in order the club the SURF points lying on a particular instance of the prototype. A Support Vector Machine (SVM) based model is used to classify whether a particular group of SURF points represents a true instance of the prototype or not. In this work we have assumed that the objects in the scene image are kept horizontally which is a reasonable assumption for objects in an inventory. Under this assumption the algorithm is robust to variation in distance and orientation of the object to be detected w.r.t to the camera. It is found that very few training examples are required for the algorithm to give good performance and thus the training is computationally cheap. Also the inference occurs in real time for all practical purposes.

This paper further proceeds in the following manner. Section 2 gives a brief introduction to SURF and SVM. In section 3 the step wise implementation of the proposed method is described. Section 4 has the description of the experimental setup and the results. Section 5 concludes the paper.

2 SURF and SVM

This section gives a brief overwiew of SURF and SVM.

2.1 Speeded Up Robust Feature (SURF)

There are two stages in obtaining a SURF descriptor, first detecting SURF point and then extracting the descriptor at the SURF point [10]. The detection of SURF point makes use of scale space theory [13]. For detection of SURF point, Fast-Hessian

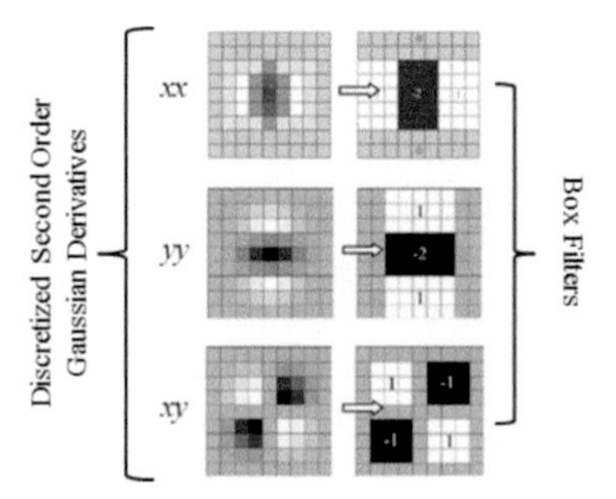

Fig. 1 Approximation for second order Gaussian derivatives by box filters

matrix is used. The determinant of Hessian matrix is used for deciding whether a point can be chosen as an interest point or not. In an image I, the Hessian matrix at point X and scale of σ, is defined by Eqn. 1. $L_{xx}(X, \sigma)$ is the convolution of Gaussian second order derivative with the image at point with coordinates (x, y).

$$H(X, \sigma) = \begin{bmatrix} L_{xx}(X, \sigma) & L_{xy}(X, \sigma) \\ L_{yx}(X, \sigma) & L_{yy}(X, \sigma) \end{bmatrix} \tag{1}$$

Gaussian second order derivative used in $L_{xx}(X, \sigma)$ is given as follows

$$\left.\begin{aligned} &\frac{\partial^2}{\partial x^2} g(\sigma) \\ \text{where, } &g(\sigma) = \frac{1}{2\pi\sigma^2} e^{\frac{-(x^2+y^2)}{2\sigma^2}} \end{aligned}\right\} \tag{2}$$

Similarly second order Gaussian derivatives for $L_{yy}(X, \sigma)$ and $L_{xy}(X, \sigma)$ are also found as

$$\frac{\partial^2}{\partial y^2} g(\sigma) \quad \text{and} \quad \frac{\partial^2}{\partial x \partial y} g(\sigma). \tag{3}$$

The Gaussian second order derivative needs to be discretized before performing convolution with the image. An approximation of this discretized Gaussian partial derivative with a box filter was proposed by Herbert [10]. This approximation provides us with almost similar results as in the previous case while making it less computationally complex. D_{xx}, D_{yy} and D_{xy} represent the convolution of box filters with the image. These approximated second order Gaussian derivative calculations is made fast by using integral images [10]. A pictorial representation of the discretized Gaussian partial derivative filters and their corresponding box filters approximation is given in Fig. 1. The determinant of approximated Hessian matrix is given by Eqn. 4.

$$det(H_{approx}) = D_{xx} D_{yy} - (0.9 D_{xy})^2 \tag{4}$$

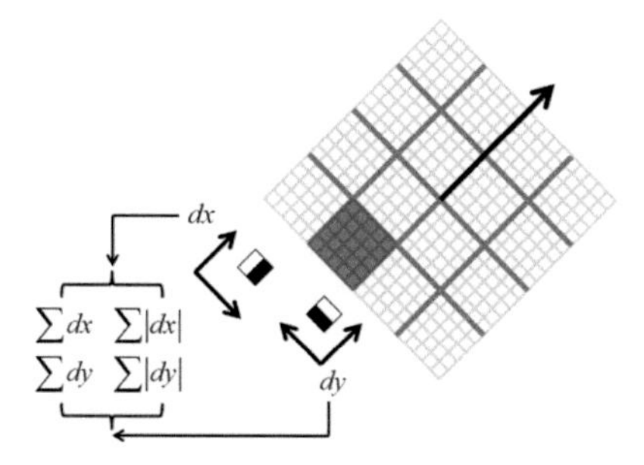

Fig. 2 Obtaining feature descriptor at an interest point

The scale space of the image is analyzed by changing the size of box filter. Generally Box filter begins with a default size of 9×9 which corresponds to Gaussian derivative with $\sigma = 1.2$. The filters size is later up scaled to sizes of 15×15, 21×21, 27×27 etc. The approximated determinant of Hessian matrix is calculated at each scale and the non-maximum suppression in $3\times3\times3$ neighborhood is applied to find the maxima. The SURF points location and scale, s is obtained with the maxima values [14].

Orientation for the obtained SURF point is assigned using Haar-wavelet response. In the neighborhood of SURF point i.e. within a radius $6s$, Haar-wavelet response is calculated in both x and y directions. Using these responses, a dominant orientation direction is determined. In the direction of dominant orientation, a square of size $20s$ centered at the SURF point is constructed. This is divided into 4×4 sub regions. In each of these sub regions, horizontal and vertical Haar wavelet responses dx and dy are calculated at 5×5 regularly placed sample points as shown in Fig. 2. These responses are summed up in a particular interval to get Σdx, Σdy. Also the absolute values of these responses are summed up in a particular interval which gives $\Sigma|dx|$, $\Sigma|dy|$. Using these values, a 4 dimensional feature vector $V = \left(\Sigma dx, \Sigma dy, \Sigma|dx|, \Sigma|dy|\right)$ is constructed for each sub region. Thus, each extracted SURF point is associated with a $4\times(4\times4)$ descriptor, which is a 64 dimensional descriptor. This 64 dimensional descriptor is used for performing the matching operation.

2.2 Support Vector Machine (SVM)

SVM was introduced by Vapnik [15] with the aim of finding a highly generalized classifier. It is a very popular binary classifier and has strong mathematical backing behind its working principle. The working principle of classical SVM is for a linearly separable case where SVM tries to find that hyper-plane which separates the two classes with maximal margin. Maximal margin ensures lower VC dimension and hence higher generalization. For non-separable cases, C-SVM was introduced which allows misclassifications to take place, but with a penalty of cost C. The optimization function of C-SVM that has been used here as well is shown in Eqn. 5, where C is cost and ξ_i is the margin error of ith sample. SVM solves this optimization problem easily by solving its dual form instead. The dual form always accesses data in the form of dot products only. This comes to advantage as it allows one to use Kernel tricks [16] while solving non-linearly separable problem.

$$\min_{w,b} \frac{1}{2} w.w + C \sum_{i=1}^{l} \xi_i \tag{5}$$
$$s.t \quad y_i(w.x_i + b) \geq 1 - \xi_i, \quad i = 1, 2, \ldots, l$$
$$\xi_i \geq 0, \quad i = 1, 2, \ldots, l$$

3 Methodology

The entire algorithm can be divided into the steps given below. Each of these steps is explained elaborately in the preceding section.

3.1 Selection of Stable SURF Points

The algorithm begins with detection of SURF points in the prototype image. As shown in Fig. 4, these features are far too many in number. There is an inherent redundancy in these overcrowded features. Moreover, handling unnecessary features adversely affects the computational efficiency of the algorithm. Hence it becomes important to avoid the unnecessary features and select only the important ones. Stable SURF points are defined as those which have highest probability of being extracted, even when conditions are far from ideal. For identifying these stable points, prototype images of different sizes are generated i.e. 0.9, 0.8, ..., 0.1 times the original image and do SURF point detection on them. As can be observed in Fig. 3, as the resolution of the image reduces, the number of SURF points also reduce. For each SURF point, a counter is maintained for counting its presence in different resolutions of the prototype image. SURF points with higher count are chosen and others discarded. It has been empirically observed that 10 to 20 SURF point are sufficient for the algorithm to work.

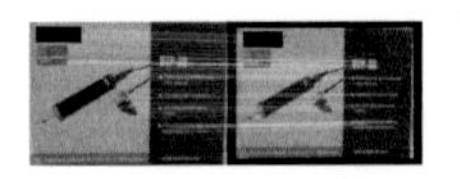

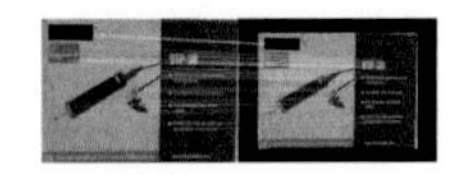

(a) Prototype at scale 0.9 (b) Prototype at scale 0.8

(c) Prototype at scale 0.7

Fig. 3 SURF points at different scales

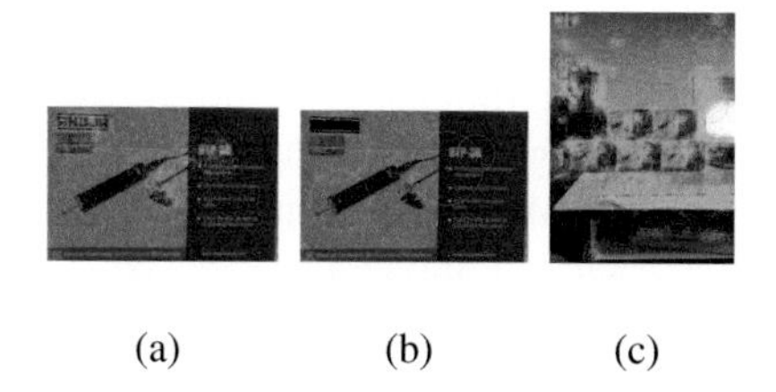

(a) (b) (c)

Fig. 4 (a) Prototype with SURF points (b) Prototype with stable SURF points (c) Scene with SURF points

3.2 Mapping of SURF Points

After selection of stable SURF points in the prototype image, all SURF points are detected in the scene image as shown in Fig. 4. As the purpose here is to identify multiple instances of prototype in the scene image, a one to many mapping of SURF points is done on the scene image. Matching between two SURF points is specified in terms of Normalized Squared Error (NSE) between the descriptors of the two points. Two SURF points are considered to be matched, if their NSE is above a threshold value of 0.5. All other SURF points that are not matched to any of the SURF point in the prototype are removed. Here again a lenient threshold is chosen as this helps in obtaining more matches and reduces the possibility of missing any SURF point that is present on a true instance of the prototype. This need not be a matter of concern because spatial filters and the SVM based false alarms detection mechanism takes care of incorrectly matched SURF points.

3.3 Application of Size Filter

To ensure that no more than one SURF points lie in a particular instance of the prototype in the scene image, filtering is applied on the matched SURF points in order. This can happen as two close SURF points can have very similar descriptors. For any particular SURF point in the prototype image, all the matching SURF points in the scene image are arranged in decreasing order of their extent of matching specified by NSE. Thereafter the SURF point with the highest extent of matching is chosen and any other SURF point in its vicinity is removed. The vicinity for any SURF point in the scene image is defined in terms of the expected size of the instance

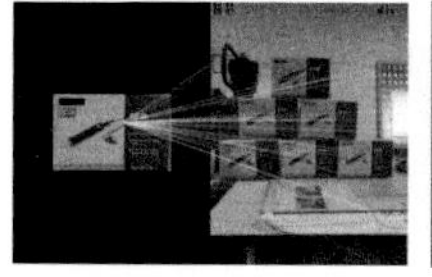

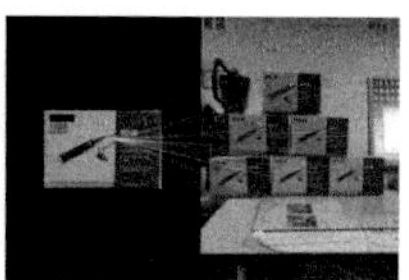

Before Size Filter After Size Filter

Fig. 5 Mapping of a particular SURF point from prototype scene

of prototype on which that SURF point lies. The expected size of this instance of the prototype is obtained by using the formula given in Eqn. 6, where S is the scale of SURF point in the prototype image and S' is the scale of the matched SURF point in the scene image. The vicinity of a SURF point in the scene is defined as a region of size 50 percent of the expected prototype instance size. The size is chosen as 50 percent of the expected size as it helps in not rejecting the points which undergo slight spatial rearrangement due to some rotation in the object instance. The same process is repeated for rest for the SURF points in the scene image. The effect of size filter for a particular SURF point in the prototype is shown in the Fig. 5.

$$\text{Expected Size} = \frac{\text{Size Of Prototype} \times S'}{S} \tag{6}$$

3.4 *Formation of Feature Grid Vectors (FGVs)*

Feature Grid Vectors (FGVs) are spatial vectors connecting two SURF points in an image. These vectors capture the information about spatial arrangement of the SURF points in an image. Feature Grid Vectors (FGVs) are formed from each stable SURF point in the prototype to every other stable SURF point on the prototype as shown in Fig. 6. Thus if we have n stable SURF points in the prototype then we will have $\binom{n}{2}$ FGVs in the prototype. Similar FGVs are defined for the matching SURF points that pass the size filter in the scene image. The FGV formed by connecting two SURF points in the prototype image is compared with the FGV formed by connecting the corresponding matched SURF points in the scene image and the matching factor between the two FGVs is determined in the way explained in the next paragraph. The FGVs on the scene image with matching factor in between 0.5 and 2 are considered as valid and the rest are discarded.

Let P_1,P_2 be the SURF points in prototype and S_1,S_2 be the SURF points in scene image. P_1 is matched with S_1 and P_2 is matched with S_2. X is the vector joining P_1 and P_2. Y is the vector joining S_1 and S_2. Y' is scaled vector of Y. Then the matching factor is given by Eqn. 7

$$\left.\begin{aligned} match(X,Y') &= \frac{|X-Y'|}{|X|},\; Y' = \frac{R_1+R_2}{2}Y \\ \text{where,}\quad R_1 &= \frac{scale(P_1)}{scale(S_1)},\; R_2 = \frac{scale(P_2)}{scale(S_2)} \end{aligned}\right\} \tag{7}$$

3.5 *Formation of Feature Grid Clusters (FGC)*

Once the set of valid FGVs in the scene image is obtained, the next task is the formation of Feature Grid Clusters (FGCs). A Feature Grid Cluster (FGC) is formed by grouping matched SURF points in the scene image. The FGCs are meant to represent instances of the prototype in the scene image. These are formed by clustering

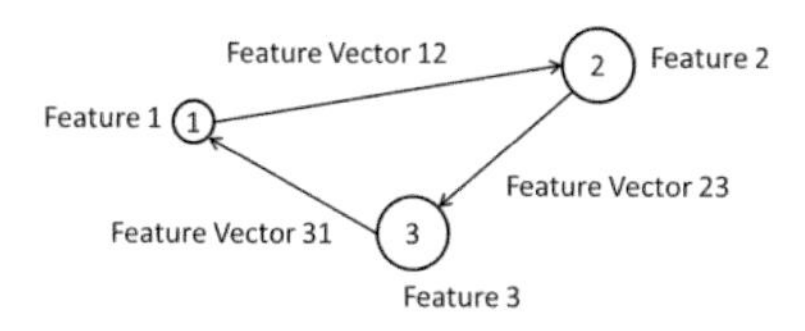

Fig. 6 Formation of Feature Grid Vectors

matched SURF points in the scene in a way such that each matched SURF Point in the cluster be connected to every other SURF point in the cluster by a valid FGV.

Its is clear that the number of SURF points in a FGC may vary from 2 to N, where N is the number of stable SURF points in the prototype image. Now for each FGC a feature vector of size $N + \binom{N}{2}$ is defined. The first N features of the feature vector represent the amount of matching of the SURF points in the FGC to the corresponding SURF point in the prototype. In case a SURF point is not present in some FGC the corresponding element in the feature vector is assigned a value of -1. The next $\binom{N}{2}$ entries store the match factor between FGVs of the prototype image and the FGC. In case a FGV of the prototype image is not present in the FGC it is assigned as -1.

3.6 Learning and Prediction

The objective in this step is to classify whether a FGC represents a true instance of the prototype or not. For this a model is trained with a Support Vector Machine based learning algorithm using the feature vector. The FGC belonging to an actual instance of the prototype are assigned a true value which is $+1$ and the FGC not belonging to an actual of the prototype are assigned a value -1. There can be the cases that two FGCs may represent the same object. In that case $+1$ is assigned to both the clusters. This process is done by manually inspecting the FGCs for multiple images to create training set. This training set is used to train the model.

Once training is complete, any scene image can be the algorithm which form FGC in that image and use the trained model to predict whether those FGC represent true instances of the prototype. For each positively classified FGC, a rectangle is constructed around the instance of the prototype it represents. Now in case two FGC belong to the same prototype instance, one of the them is eliminated since they would give almost the same rectangle. Once this is done detected objects are marked and there count is reported.

4 Experiments and Results

A Desktop application was developed to conveniently execute the whole algorithm. The setup includes a desktop PC with the application installed and a web-cam connected to the PC using an USB cable or some wireless means. The web-cam is positioned properly to capture the inventory scene. The user can operate the application using the Graphical User Interface (GUI) which is shown in the Fig. 7. Starting with the process first selecting the prototype image by cropping it from the scene image or choosing it from previously captured ones is done. By clicking compute button the application starts taking input video feed of the inventory scene from the web cam. After the computation is done the results are displayed on the GUI. There is also an auto-run mode in which a time can be specified and the process repeats until the time is elapsed. Results are displayed on the GUI after each computation is completed.

In all our experiments we have considered the object in Fig. 4 as our prototype. The results obtained for different test cases were tabulated in the Table 1 and Fig. 10. We have shown results for 3 different images in each test case. Test case 1 is where the objects are aligned similar to that of prototype. In test case 2 the objects are kept at different orientation angles. In test case 3 the objects are placed at different distances from camera resulting in different scales from that of prototype. The Before SVM model simply classifies all the FGC as true objects. It can be seen that the percentage error in most of the cases decreases after the SVM prediction which discards the falsely alamrs.

Results of the present method are compared with those of previous work [8] and are shown in the Table 2. Similar to Table 1 test case 1 has object aligned similar to prototype, teat case 2 has objects kept at different orientation w.r.t. the camera and test case 3 has objects kept at different distances from the camera. For each case a set 10 images were taken and tested by both the methods. Each test image has 5 to 10 objects. % Error is defined as the average percent error in all images, where the percent error in an image is defined as the percentage of |Actual Count - Predicted Count| to Actual Count. The errors obtained are very low for the present SURF method compared to previous Fuzzy histogram method for the test cases 2 and 3. Fig. 8 shows the comparison of results on the image containing objects placed at different distances resulting in different scales. Fig. 9 shows the results obtained for both the methods on an image containing object with different orientation. It is clearly evident the present SURF method outperforms the previous Fuzzy Histogram method.

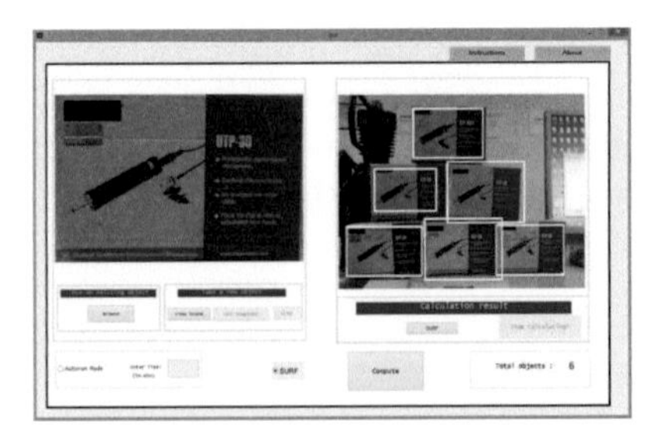

Fig. 7 Graphical User Interface

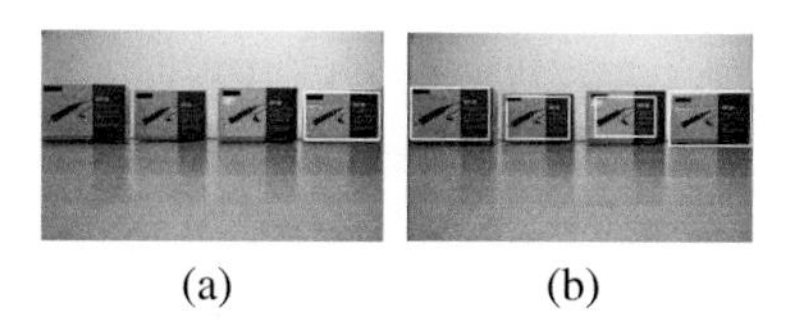

(a) (b)

Fig. 8 Counting of objects with variation in scale (a) using Fuzzy Histogram Method (b) using SURF based method

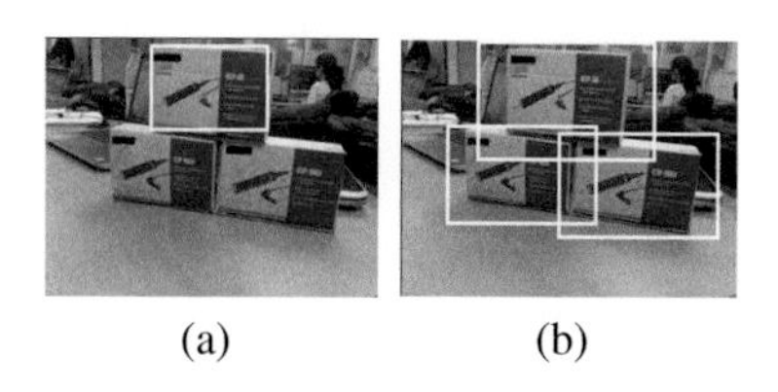

(a) (b)

Fig. 9 Counting of objects with variation in orientation (a) using Fuzzy Histogram Method (b) using SURF based method

Table 1 Object recognition accuracies in different test cases

Objects		Before SVM		After SVM	
Test Case	True Count	Count	% Error	Count	% Error
1	7	7	0	7	0
	7	8	12.5	7	0
	7	6	12.5	6	12.5
2	3	4	25	3	0
	3	3	25	3	0
	3	5	40	3	0
3	4	5	20	4	0
	6	8	25	6	0
	5	6	16.67	5	0

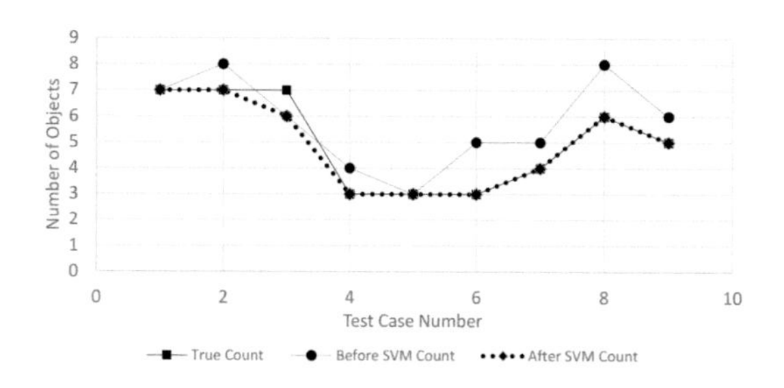

Fig. 10 Plot showing the number of actual object count and object count before and after SVM for different test cases

Table 2 Comparison between Fuzzy Histogram based method and SURF based method

Test Case	% Error	
	Fuzzy Histogram	SURF
1	2	0
2	32	1
3	29	0

5 Conclusion

Using SURF for matching makes the method much robust compare to the previous Fuzzy Histogram method. It is found that with very few training examples the algorithm is able to achieve very high accuracy. This shows how powerful our feature vector representation is. Also the choice of low thresholds while filtering SURF points and Feature Grid Vectors in the prototype enhances the overall robustness of the algorithm to rotation variations and partial occlusions. The issue of removing the false alarms is addressed by using the SVM learning model. Future work would be to improve performance on partially occluded objects and to make it work on low quality images as well.

Acknowledgments We would like to express our sincere thanks to The Boeing Company, USA for sponsoring this research work at IIT Kanpur.

References

1. Lempitsky, V., Zisserman, A.: Learning to count objects in images. In: Advances in Neural Information Processing Systems, pp. 1324–1332 (2010)
2. Barbedo, J.G.A.: Automatic object counting in neubauer chambers. In: Embrapa Informática Agropecuária-Artigo em anais de congresso (ALICE), SIMPÓSIO BRASILEIRO DE TELECOMUNICAÇÕES 2013, Fortaleza. Rio de Janeiro, September 2013 (2014)
3. Guo, X., Yu, F.: A method of automatic cell counting based on microscopic image. In: 2013 5th International Conference on Intelligent Human-Machine Systems and Cybernetics (IHMSC), pp. 293–296. IEEE (2013)
4. Fabic, J., Turla, I., Capacillo, J., David, L., Naval, P.: Fish population estimation and species classification from underwater video sequences using blob counting and shape analysis. In: 2013 IEEE International Conference Underwater Technology Symposium (UT), pp. 1–6. IEEE (2013)
5. Mezei, S., Darabant, A.S.: A computer vision approach to object tracking and counting. Studia Universitatis Babes-Bolyai, Informatica **55**(1) (2010)
6. Rabaud, V., Belongie, S.: Counting crowded moving objects. In: 2006 IEEE Computer Vision and Pattern Recognition, pp. 705–711. IEEE (2006)
7. Pornpanomchai, C., Stheitsthienchai, F., Rattanachuen, S.: Object detection and counting system. In: 2008 Congress on Image and Signal Processing (CISP 2008), vol. 2, pp. 61–65. IEEE (2008)

8. Verma, N.K., Goyal, A., Chaman, A., Sevakula, R.K., Salour, A.: Template matching for inventory management using fuzzy color histogram and spatial filters. In: 10th IEEE Conference on Industrial Electronics and Applications (2015) (accepted)
9. Lowe, D.G.: Distinctive image features from scale-invariant keypoints. International Journal of Computer Vision **60**(2), 91–110 (2004)
10. Bay, H., Tuytelaars, T., Van Gool, L.: Surf: speeded up robust features. In: Computer Vision, ECCV 2006, pp. 404–417. Springer (2006)
11. Huijuan, Z., Qiong, H.: Fast image matching based-on improved surf algorithm. In: 2011 International Conference on Electronics, Communications and Control (ICECC), pp. 1460–1463. IEEE (2011)
12. Juan, S., Qingsong, X., Jinghua, Z.: A scene matching algorithm based on surf feature. In: 2010 International Conference on Image Analysis and Signal Processing (IASP), pp. 434–437. IEEE (2010)
13. Lindeberg, T.: Discrete scale-space theory and the scale-space primal sketch. PhD thesis, Royal Institute of Technology (1991)
14. Brown, M., Lowe, D.G.: Invariant features from interest point groups. In: BMVC, Number s 1 (2002)
15. Vapnik, V.N.: An overview of statistical learning theory. IEEE Transactions on Neural Networks **10**(5), 988–999 (1999)
16. Burges, C.J.: A tutorial on support vector machines for pattern recognition. Data Mining and Knowledge Discovery **2**(2), 121–167 (1998)

Author Index